Green Chemistry and Sustainable Technology

Aims and Scope

The series Green Chemistry and Sustainable Technology aims to present cutting-edge research and important advances in green chemistry, green chemical engineering and sustainable industrial technology. The scope of coverage includes (but is not limited to):

- Environmentally benign chemical synthesis and processes (green catalysis, green solvents and reagents, atom-economy synthetic methods etc.)
- Green chemicals and energy produced from renewable resources (biomass, carbon dioxide etc.)
- Novel materials and technologies for energy production and storage (bio-fuels and bioenergies, hydrogen, fuel cells, solar cells, lithium-ion batteries etc.)
- Green chemical engineering processes (process integration, materials diversity, energy saving, waste minimization, efficient separation processes etc.)
- Green technologies for environmental sustainability (carbon dioxide capture, waste and harmful chemicals treatment, pollution prevention, environmental redemption etc.)

The series Green Chemistry and Sustainable Technology is intended to provide an accessible reference resource for postgraduate students, academic researchers and industrial professionals who are interested in green chemistry and technologies for sustainable development.

Guoliang Zhang · Annemie Bogaerts · Jingyun Ye ·
Chang-jun Liu
Editors

Advances in CO_2 Utilization

From Fundamentals to Applications

Springer

Editors
Guoliang Zhang
Center for Membrane and Water Science and Technology
Zhejiang University of Technology
Hangzhou, China

Jingyun Ye
Department of Chemistry and Biochemistry
Duquesne University
Pittsburgh, PA, USA

Annemie Bogaerts
PLASMANT, Department of Chemistry
University of Antwerp
Antwerpen, Belgium

Chang-jun Liu
School of Chemical Engineering and Technology
Tianjin University
Tianjin, China

ISSN 2196-6982 ISSN 2196-6990 (electronic)
Green Chemistry and Sustainable Technology
ISBN 978-981-99-8821-1 ISBN 978-981-99-8822-8 (eBook)
https://doi.org/10.1007/978-981-99-8822-8

This Springer imprint is published by the registered company Springer Nature Singapore Pte Ltd.
The registered company address is: 152 Beach Road, #21-01/04 Gateway East, Singapore 189721, Singapore

Paper in this product is recyclable.

Preface

Carbon dioxide is the primary anthropogenic greenhouse gas. With the rapid progress in renewable energy technologies, the prospect of harnessing CO_2-free energy has gained significant attention. This book has been curated to foster the research and development of technologies aimed at utilizing carbon dioxide.

This book covers all the key topics in the activation and conversion of CO_2 for the synthesis of fuels, chemicals and polymers. The book contains twelve chapters. The first chapter focuses on CO_2 conversion over metal organic framework (MOF)-based catalysts. As MOFs have been proven to be very promising materials for CO_2 capture, MOF-based catalysts will have more practical applications in CO_2 conversion in the near future.

The second chapter covers photo- and photo-thermal catalytic conversion of CO_2. Coupling heat and light in catalytic CO_2 conversion efficiently enhances the activity, manipulates the selectivity and/or improves the stability. This chapter explains why the combination is needed and how it works.

Renewable energy-derived 'green hydrogen' promises affordability and perpetual availability. Therefore, the following three chapters deal with thermo-catalytic CO_2 hydrogenation, which has the potential for CO_2 utilization at large scale. Chapter 3 is about CO_2 hydrogenation to liquid fuels. Chapter 4 reviews CO_2 hydrogenation to methanol and carbon monoxide, and Chap. 5 summarizes CO_2 hydrogenation to methane (i.e., CO_2 methanation).

A continuous effort in carbon dioxide reforming of methane (also called 'dry reforming of methane', DRM) and other low alkanes is being made. Chapter 6 describes CO_2 reforming of alkanes.

Chapter 7 is about electrocatalytic conversion of carbon dioxide, which recently attracts increasing attention. In this chapter, outlooks, pitfalls and scientific gaps of the electrocatalytic CO_2 conversion are addressed.

Chapter 8 presents the current status of research and development in the bio-conversion of carbon dioxide, and the challenges and future developments are also addressed.

Chapter 9 reviews the synthesis, properties and applications of CO_2-sourced polymers.

Plasma can activate CO_2 at room temperature and provide a high conversion. It can be readily turned on and off. No external heating is needed. Chapter 10 presents the progresses in the research and development of plasma-based CO_2 conversion.

In Chap. 11, a review on homogeneous CO_2 conversion is presented, which has been applied for the syntheses of many fine chemicals.

Finally, in Chap. 12, the approaches for CO_2 recycling at large scale are reviewed.

All chapters were written by researchers who are worldwide experts in their field. We thank all the authors of the chapters for their great contributions. We thank the peer reviewers for their hard work in the evaluation and improvement of the chapters. We also appreciate the great efforts of the editors of Springer Nature, who helped to improve the quality of the book.

We hope the reader will enjoy the book and find it informative and helpful in handling the important CO_2 utilization issues.

Hangzhou, China — Guoliang Zhang
Antwerpen, Belgium — Annemie Bogaerts
Pittsburgh, USA — Jingyun Ye
Tianjin, China — Chang-jun Liu

Contents

Contributors

Michele Aresta Innovative Catalysis for Carbon Recycling-IC2R, Bari, Italy

Annemie Bogaerts Research Group PLASMANT, Department of Chemistry, University of Antwerp, Antwerp, Belgium

Gabriele Centi Department of ChiBioFarAm, University of Messina, Messina, Italy;
European Research Institute of Catalysis (ERIC) aisbl, Brussels, Belgium

Huanhao Chen State Key Laboratory of Materials-Oriented Chemical Engineering, College of Chemical Engineering, Nanjing Tech University, Nanjing, China

Haiyang Cheng State Key Laboratory of Electroanalytical Chemistry, Changchun Institute of Applied Chemistry, Chinese Academy of Sciences, Changchun, China;
Jilin Province Key Laboratory of Green Chemistry and Process, Changchun Institute of Applied Chemistry, Chinese Academy of Sciences, Changchun, China

Sovik Das Department of Civil Engineering, Indian Institute of Technology Delhi, New Delhi, India

Huahua Fan State Key Laboratory of Fine Chemicals, Frontier Science Center for Smart Materials, PSU-DUT Joint Center for Energy Research, School of Chemical Engineering, Dalian University of Technology, Dalian, China

Xiaolei Fan Department of Chemical Engineering, School of Engineering, The University of Manchester, Manchester, UK;
Nottingham Ningbo China Beacons of Excellence Research and Innovation Institute, University of Nottingham Ningbo China, Ningbo, China

Xinwen Guo State Key Laboratory of Fine Chemicals, Frontier Science Center for Smart Materials, PSU-DUT Joint Center for Energy Research, School of Chemical Engineering, Dalian University of Technology, Dalian, China

Liang-Nian He State Key Laboratory and Institute of Elemento-Organic Chemistry, College of Chemistry, Nankai University, Tianjin, People's Republic of China

Evgenii V. Kondratenko Leibniz-Institut für Katalyse e. V., Rostock, Germany

Hui Li State Key Laboratory of Electroanalytical Chemistry, Changchun Institute of Applied Chemistry, Chinese Academy of Sciences, Changchun, China;
School of Applied Chemistry and Engineering, University of Science and Technology of China, Hefei, China;
Jilin Province Key Laboratory of Green Chemistry and Process, Changchun Institute of Applied Chemistry, Chinese Academy of Sciences, Changchun, China

Chang-jun Liu School of Chemical Engineering and Technology, Tianjin University, Tianjin, China

Xiaowa Nie State Key Laboratory of Fine Chemicals, Frontier Science Center for Smart Materials, PSU-DUT Joint Center for Energy Research, School of Chemical Engineering, Dalian University of Technology, Dalian, China

Tomohiro Nozaki Department of Mechanical Engineering, Tokyo Institute of Technology, O-Okayama, Tokyo, Japan

Deepak Pant Separation and Conversion Technology, Flemish Institute for Technological Research (VITO), Mol, Belgium

Siglinda Perathoner Department of ChiBioFarAm, University of Messina, Messina, Italy;
European Research Institute of Catalysis (ERIC) aisbl, Brussels, Belgium

Neha Sharma Department of Civil Engineering, Indian Institute of Technology Delhi, New Delhi, India

Chenyang Shen School of Chemical Engineering and Technology, Tianjin University, Tianjin, China

Chunshan Song Department of Chemistry, Faculty of Science, The Chinese University of Hong Kong, Shatin, Hong Kong, NT, China

Hui Song TJU-NIMS International Collaboration Laboratory, School of Material Science and Engineering, Tianjin University, Tianjin, P. R. China

Xin Tu Department of Electrical Engineering and Electronics, University of Liverpool, Liverpool, UK

Hao Wang State Key Laboratory of Fine Chemicals, Frontier Science Center for Smart Materials, PSU-DUT Joint Center for Energy Research, School of Chemical Engineering, Dalian University of Technology, Dalian, China

Mingrui Wang State Key Laboratory of Fine Chemicals, Frontier Science Center for Smart Materials, PSU-DUT Joint Center for Energy Research, School of Chemical Engineering, Dalian University of Technology, Dalian, China

Shuang Wang State Key Laboratory of Fine Chemicals, Frontier Science Center for Smart Materials, PSU-DUT Joint Center for Energy Research, School of Chemical Engineering, Dalian University of Technology, Dalian, China

Xiaorui Wang State Key Laboratory of Chemical Resource Engineering, Beijing Key Laboratory of Energy Environmental Catalysis, Beijing University of Chemical Technology, Beijing, P. R. China

Xinrui Wang Department of Chemical Engineering, School of Engineering, The University of Manchester, Manchester, UK

Zhiqun Wang State Key Laboratory of Fine Chemicals, Frontier Science Center for Smart Materials, PSU-DUT Joint Center for Energy Research, School of Chemical Engineering, Dalian University of Technology, Dalian, China

Zhou-jun Wang State Key Laboratory of Chemical Resource Engineering, Beijing Key Laboratory of Energy Environmental Catalysis, Beijing University of Chemical Technology, Beijing, P. R. China

An-Guo Wu State Key Laboratory and Institute of Elemento-Organic Chemistry, College of Chemistry, Nankai University, Tianjin, People's Republic of China

Guiming Xie State Key Laboratory of Chemical Resource Engineering, Beijing Key Laboratory of Energy Environmental Catalysis, Beijing University of Chemical Technology, Beijing, P. R. China

Zehai Xu Institute of Oceanic and Environmental Chemical Engineering, Center for Membrane and Water Science and Technology, Zhejiang University of Technology, Hangzhou, China

Qingxin Yang Leibniz-Institut für Katalyse e. V., Rostock, Germany

Jinhua Ye International Center for Materials Nanoarchitectonics (WPI-MANA), National Institute for Materials Science (NIMS), Tsukuba, Ibaraki, Japan

Shihang Yu Nottingham Ningbo China Beacons of Excellence Research and Innovation Institute, University of Nottingham Ningbo China, Ningbo, China

Guanghui Zhang State Key Laboratory of Fine Chemicals, Frontier Science Center for Smart Materials, PSU-DUT Joint Center for Energy Research, School of Chemical Engineering, Dalian University of Technology, Dalian, China

Guoliang Zhang Institute of Oceanic and Environmental Chemical Engineering, Center for Membrane and Water Science and Technology, Zhejiang University of Technology, Hangzhou, China

Miao Zhang State Key Laboratory of Fine Chemicals, Frontier Science Center for Smart Materials, PSU-DUT Joint Center for Energy Research, School of Chemical Engineering, Dalian University of Technology, Dalian, China

Fengyu Zhao State Key Laboratory of Electroanalytical Chemistry, Changchun Institute of Applied Chemistry, Chinese Academy of Sciences, Changchun, China; School of Applied Chemistry and Engineering, University of Science and Technology of China, Hefei, China;
Jilin Province Key Laboratory of Green Chemistry and Process, Changchun Institute of Applied Chemistry, Chinese Academy of Sciences, Changchun, China

Man Zhao State Key Laboratory and Institute of Elemento-Organic Chemistry, College of Chemistry, Nankai University, Tianjin, People's Republic of China

Chapter 1
CO_2 Conversion via MOF-Based Catalysts

Guoliang Zhang and Zehai Xu

Abstract Metal–organic framework (MOF)-based catalysts have attracted significant attentions for their distinguished characteristics in CO_2 conversion, such as very high surface area, recyclability, permanent porosity, and multiple active sites. Although numerous publications have emerged in recent years focusing on MOFs and their catalytic properties in CO_2 conversion, this chapter specifically emphasizes the enhancement of MOF-based catalysts through micro/macrostructure and multicomponent design. The challenges encountered in MOFs-based CO_2 conversion and the essential requirements for practical catalytic applications are addressed. A brief summary of strategies is provided for designing catalysts with desired microstructure, macrostructure, and multicomponent characteristics by using MOFs as precursors. The latest progresses in novel MOF and MOF-derived catalysts with complex structures, specific compositions, and improved MOF stability for CO_2 conversion are also reviewed. The broad application prospects of MOF-based catalysts in CO_2 conversion are introduced, including CO_2 organic transformation, hydrogenation, photocatalysis, electrocatalysis, and photoelectroreduction. Moreover, the challenges faced in the practical application of MOF-based catalysts in CO_2 conversion are presented, and the promising future directions are outlined.

1.1 Introduction

Anthropogenic CO_2 emissions have been a major contributor to greenhouse gases in the atmosphere, leading to severe consequences such as global warming, ocean acidification, and species extinction [1, 2]. Numerous methods have been implemented to CO_2 capture, storage, and transformation so as to cut down its atmospheric concentration and improve the existing energy supply [3]. However, the presence of the C=O bond in carbon dioxide with a bond enthalpy of 805 kJ/mol offers additional

G. Zhang (✉) · Z. Xu
Institute of Oceanic and Environmental Chemical Engineering, Center for Membrane and Water Science and Technology, Zhejiang University of Technology, Hangzhou 310014, China
e-mail: guoliangz@zjut.edu.cn

G. Zhang et al. (eds.), *Advances in CO_2 Utilization*, Green Chemistry and Sustainable Technology, https://doi.org/10.1007/978-981-99-8822-8_1

stability and requires significant amount of energy to break the bond [4]. As emissions continue to rise, CO_2 utilization has emerged as a promising solution. Among various CO_2 utilization pathways, converting CO_2 to fuel or chemical products is particularly appealing since it allows for CO_2 storage and the production of valuable commodities. Hydrogenation, thermocatalysis, electrocatalysis, photocatalysis, photoelectrocatalysis, and chemical fixation are some useful technologies employed in recent years for CO_2 utilization [5–8]. However, most catalysts have obvious defects such as low reactivity, high activation barriers, and structural instability. Hence, development of efficient, stable, and economical catalysts is crucial for CO_2 conversion.

The pursuit of desirable nanoporous materials has prompted the investigation of a variety of materials for CO_2 conversion, including porous carbon, zeolites, ionic liquids, and metal–organic frameworks (MOFs) [9]. Among them, MOFs have gained tremendous attention since their definition in 1995 by Yaghi, owing to their superior characteristics including high surface area, recyclability, permanent porosity, and multiple active sites [10]. However, as MOFs act as heterogeneous catalysts, the confined channel size may impede adsorbate–adsorbent interaction, leading to low CO_2 selectivity [11]. Additionally, their poor water and thermal stability usually limit their practical applications in the chemical industry [12]. The original MOFs also have limited electrical conductivity, which hinders the direct applications in catalysis [13]. Nevertheless, pristine MOFs have been used as substrates for the incorporation of metal oxide nanoparticles, metal complexes, and semiconductors within their pores by using post-synthetic modification approaches [14]. The formation of MOF composites provides favorable conditions for CO_2 adsorption in MOFs. One potential approach to catalytic CO_2 reduction is the use of MOF composites, which combine the pore characteristics of MOFs with the exceptional durability of newly added constituents. Furthermore, MOFs can also be transformed into various inorganic functional materials, such as metal compounds and carbon composites, through annealing or wet-chemical methods [15, 16]. The MOF-derived strategy offers unparalleled possibilities for designing and tailoring the structure and composition of materials. In recent studies, composites derived from MOFs have been investigated as catalysts for various catalytic reactions and have displayed promising performance in converting CO_2 [17, 18]. Understanding the relationship between structure, composition, and catalytic performance of MOF-derived catalysts is essential for designing and fabricating catalysts that meet the practical requirements of CO_2 conversion.

Despite the fact that numerous publications focusing on MOFs and their catalytic properties for CO_2 conversion have emerged in recent years, we focus on the enhancement of MOF-based catalysts for CO_2 conversion, specifically in terms of microstructure, macrostructure, and multicomponent design in this chapter (Fig. 1.1). By addressing the challenges faced by MOFs-based CO_2 conversion and the crucial requirements for catalysts in practical applications, we summarize the strategies for designing catalysts with desired microstructure, macrostructure, and multicomponent features by using MOFs as precursors. We also review the latest advances in

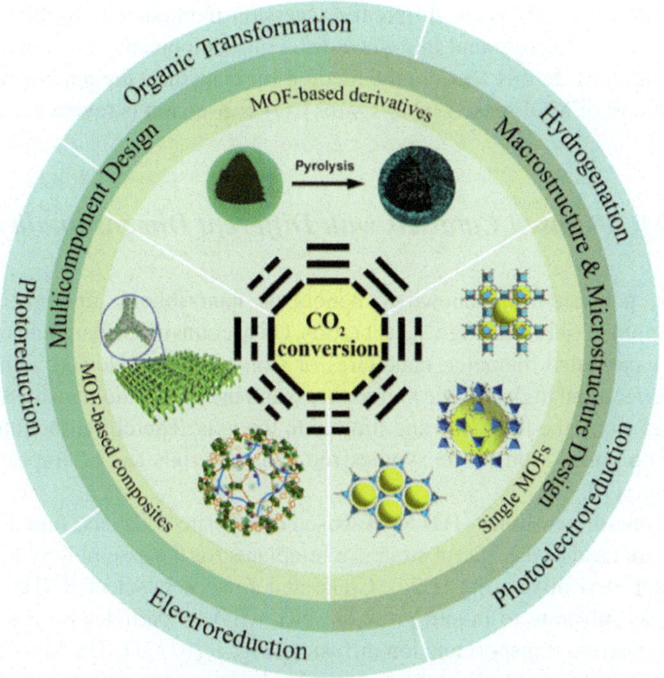

Fig. 1.1 Composition, structural engineering, and application of MOF-based catalysts for CO_2 conversion

MOF and MOF-derived catalysts with complex structures and specific compositions for CO_2 conversion, as well as the improvements in the stability of MOFs. The application of MOF-based catalysts for CO_2 conversion is briefly introduced, including CO_2 organic transformation, hydrogenation, photocatalysis, electrocatalysis, and photoelectroreduction. At last, we aim to provide some guidance for the development of high-performance MOF-based catalysts that meet the urgent demand for achieving large-scale production and controllable derivation processes in industrial settings. We hope that the perspective may facilitate the precise design of catalysts with optimal performance, ultimately promoting the practical application of CO_2 conversion.

1.2 Macrostructure and Microstructure Design of MOF-Based Catalysts

Porous and tunable MOFs have emerged as promising precursors for the controlled synthesis of catalyst materials with tailored structures and compositions Specifically, MOF composites with adjustable morphology, pore structure, and particle size are first obtained. Then, different conversion methods are applied to control the micro/macrostructure and introduce different components and functions to the catalyst materials. In this section, the method for designing the macrostructure and microstructure of MOF-based catalysts for CO_2 conversion is reviewed.

1.2.1 MOF-Based Catalysts with Different Dimensionality

Nanorods, nanoneedles, nanowires, nanobelts, nanoribbons, and nanotubes are typical nanomaterials with high aspect ratios. On account of the length in one dimension and nanoscaled diameter, nanowires or nanorods are endowed with desirable active interface and high specific surface area, and the electronic transmission can be maintained along the long axis and limited to the axis. Therefore, the ion diffusion and electron conduction of the synthesized nanomaterials can be improved due to the structure [19, 20].

As for one-dimensional (1D) MOF-based composite catalysts, functional nanomaterials are used as the 1D substrates or templates for the assembly of MOFs crystals or MOF derivatives (Fig. 1.2a). Crake and coworkers selected 1D TiO_2 phase nanofiber as substrate to in-situ grow the NH_2-UiO-66 particles on the surface to maximize electron transport and ion diffusion (Fig. 1.2b) [21]. The TiO_2/NH_2-UiO-66 composite was synthesized through a microwave-antenna strategy based on the TiO_2 nanofibers for both CO_2 capture and photocatalytic conversion. These NH_2-UiO-66 particles directly formed on the fiber owing to this rapid technique with low energy requirements (Fig. 1.2c), resulting in the generation of strong interactions between TiO_2 and NH_2-UiO-66 particles, thus improving charge separation as electrons transfer from TiO_2 to the NH_2-UiO-66. In particular, the prepared catalyst with the highest concentration of NH_2-UiO-66 exhibited prominent catalytic activity, which was 5 times higher than that of titanate.

Two-dimensional (2D) nanostructures refer to nanosheets, nanoplates, membranes, and nanofilms with a large lateral size and an atomic thickness in the third

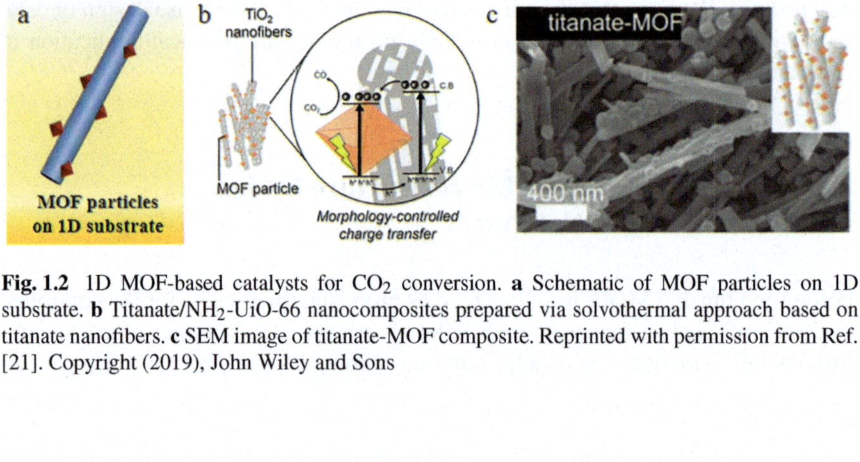

Fig. 1.2 1D MOF-based catalysts for CO_2 conversion. **a** Schematic of MOF particles on 1D substrate. **b** Titanate/NH_2-UiO-66 nanocomposites prepared via solvothermal approach based on titanate nanofibers. **c** SEM image of titanate-MOF composite. Reprinted with permission from Ref. [21]. Copyright (2019), John Wiley and Sons

dimension [22, 23]. The distinctive two-dimensional structure of material restricts electrons to two dimensions in the ultra-thin area. This structure offers a well-defined surface-to-volume ratio, exceptional mechanical strength and flexibility, and a large number of surface sites that can be altered through surface functionalization [24]. As a result, materials with 2D nanostructures typically demonstrate better electrochemical activities than their bulk counterparts.

MOF composites with 2D architectures are becoming increasingly popular, for example, MOFs loading or coating on 2D substrates (Fig. 1.3a). Li et al. developed a straightforward hydrothermal method to prepare zirconium-based NH_2-UiO-66 particles on 2D polyethylene terephthalate (PET) fiber textiles (Fig. 1.3b) [25]. The observation of tightly-bound NH_2-UiO-66 particles on the surface of the PET fiber provided compelling evidence for the successful coating of MOF particles onto the textile fiber (Fig. 1.3c), leading to the formation of a PET@NH_2-UiO-66 film. The PET@NH_2-UiO-66 film showed comparable catalytic activity to the NH_2-UiO-66 powder but with better catalytic repeatability. The film remained stable under 60 h of irradiation due to the close bonding between the MOF particles and PET fiber.

2D nanomaterials are well known for their strong in-plane bonding and their ability to assemble via interactions between their layers. When two different types of 2D layered nanomaterials are synthesized separately and then mixed and filtered, the interactions among the components can lead to the generation of a self-assembled hybrid material (Fig. 1.4a). Unlike a simple mixture, the components are tightly integrated via in-plane interactions with each other. This sort of approach has been widely used in the fabrication of self-standing membranes for CO_2 conversion, using one or two 2D layered components. For instance, zinc porphyrin-based MOF (Zn-MOF) and $BiVO_4$ nanosheets were combined to form an S-scheme 2D/2D heterojunction (Zn-MOF/$BiVO_4$) using a hydroxyl-induced strategy (Fig. 1.4b) [26]. The Zn-MOF was linked to $BiVO_4$ via strong interfacial interactions between the Zn in the ligand and hydroxyl groups in $BiVO_4$. Thus, a heterojunction of dimensionally compatible Zn-MOF/$BiVO_4$ was developed to effectively separate charges and provide ample catalytic sites, leading to exceptional performance in converting CO_2.

To create composite materials, how to maintain the precise dimensional scale of 0–100 nm remains a challenge when dealing with 0D, 1D, or 2D nanostructures after

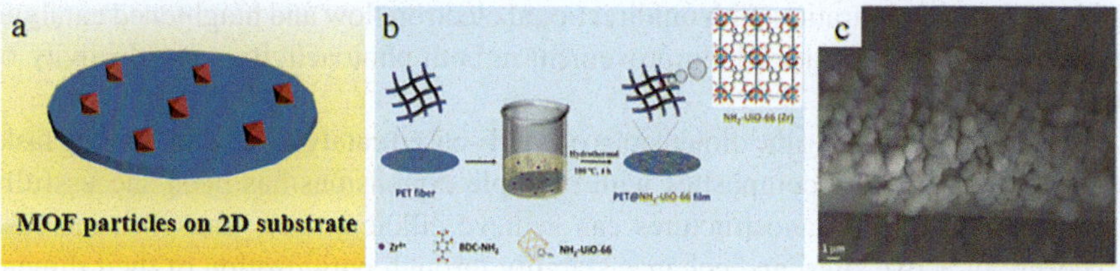

Fig. 1.3 2D MOF-based catalysts for CO_2 conversion. **a** Schematic of MOF particles on 1D substrate. **b** MOF particles on the 2D PET films via a hydrothermal approach. **c** SEM image of the MOF particles on the 2D PET films. Reprinted with permission from Ref. [25]. Copyright (2021), American Chemical Society

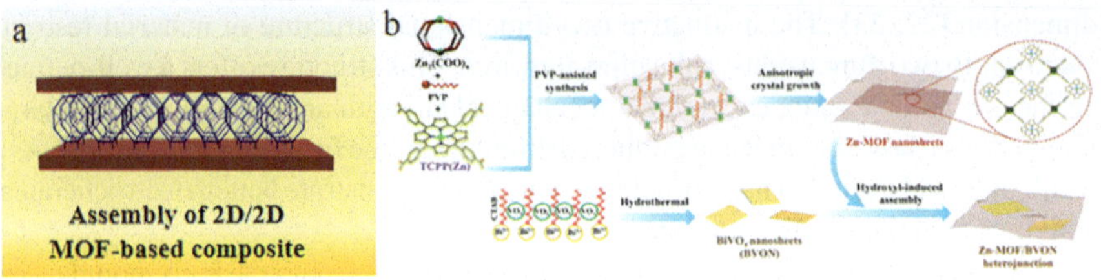

Fig. 1.4 **a** Schematic illustration of assembly of composite constructed by 2D MOF nanosheets and functional nanomaterials. **b** Schematic of Zn-MOF composites via assembly of 2D Zn-MOF and $BiVO_4$ nanosheets. Reprinted with permission from Ref. [26]. Copyright (2020), American Chemical Society [26]

coating, incorporation, assembly, or hybridization. While many MOF composites share the morphologies of spheres, rods, and sheets, they typically fall within the 3D structural scope. In the case of MOF composites with three-dimensional structures, the fabrication process involves assembling nanomaterials of various dimensions (0D, 1D, and 2D) at multiple length scales. Similar results happened in the creation of hierarchically porous structures such as hollow shapes and coralloid superstructures [27, 28]. The incorporation of hierarchical structures in MOF composites served to prevent the formation of aggregates and stacks, which made reduced porosity, poor permeability of electrolytes/reactants, and restricted access to inner active sites.

The coupling of hollow semiconductors with MOFs presents an excellent opportunity for the promotion of exceptionally efficient catalysts designed for the reduction of CO_2. Mo et al. demonstrated that when ZIF-8 was deposited onto H-CdS, it caused band bending that facilitated the separation of electrons and holes (Fig. 1.5) [29]. Despite this, the energy band mismatch between H-CdS and ZIF-8 created a hindrance for interfacial charge transfer, leading to a negligible increase in photocatalytic activity and CO_2 reduction selectivity. This challenge can be addressed by incorporating plasmonic Au into the semiconductor/MOF heterointerface to create the heterostructure. The embedded Au facilitated interfacial electron transfer along a specific pathway, generating energetic hot electrons that were shifted to ZIF-8. Additionally, the incorporation of Au promoted the creation of defects in ZIF-8, which served as exceptionally active and selective sites for converting CO_2 into CO. The stacked configuration with unidirectional electron flow and heightened catalytic reactivity led to a remarkable improvement in both photoactivity and selectivity of CO production.

Although controlling the dimensions of MOF-based catalysts is a challenging task, construction of MOF composites with multiple dimensions has been successfully achieved. Advanced nanostructures can achieve efficient mass transport, increase exposure of active sites, and enhance stability through optimization of their dimensional parameters. In comparison with bulk and isotropic counterparts, MOF-based catalysts with ideal dimensionalities have shown enhanced performance in areas such as catalysis and rechargeable batteries. Thus, there is a pressing need for the rational

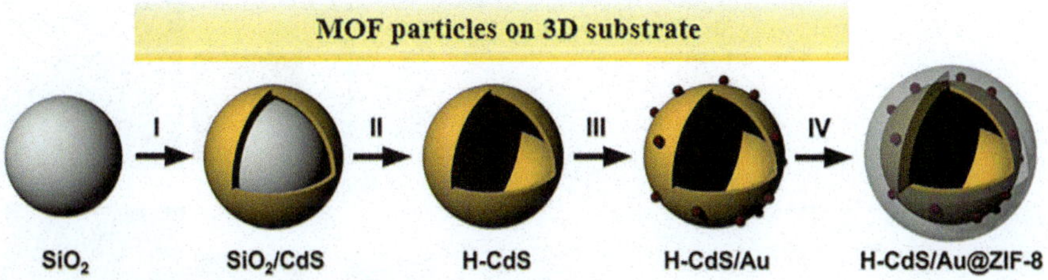

Fig. 1.5. 3D MOF-based catalysts for CO_2 conversion. H-CdS/Au@ZIF-8 composites prepared by ZIF-8 coverage on 3D hollow H-CdS/Au substrate. Reprinted with permission from Ref. [29]. Copyright (2021), Chemical Industry Press

design of MOF-based catalysts with specific morphologies and dimensionalities in order to achieve enhanced performance.

1.2.2 MOFs with Complex Structure

The pursuit of greater complexity in nanostructures, with respect to their structural subunits, architecture and composition, has become an area of intense interest for both practical applications and fundamental studies [30, 31]. By carefully and rationally designing complex nanostructures, materials can acquire the desired characteristics for a wide range of current and emerging applications [32]. In the field of electrochemical applications, advanced architectures have displayed better performance when compared to simple nanoparticle assemblies. For example, intricate porous structures composed of secondary subunits, which provide high surface area with exposed active sites, are anticipated to optimize catalytic activity.

Core–shell composites based on MOF not only can preserve the exceptional characteristics of the core and shell but also effectively address the limitations of individual materials, offering distinctive benefits in various applications. Utilizing MOF as the shell material is critical in safeguarding the core, preventing leaching and independent agglomeration during the reaction process. Moreover, MOF creates a relatively stable and consistent environment for the reaction, which remains largely unaffected by external factors. As an example, heterogeneous MOF-801-AgNC nanocrystals were fabricated through a seed-mediated synthetic method (Fig. 1.6) [33]. Initially, polyvinyl pyrrolidone-coated AgNCs were synthesized at the water–air interface and subsequently transferred onto a polystyrene matrix. The AgNCs acted as templates for MOF formation and stabilized them during the process, resulting in partially embedding of MOF in the polystyrene matrix. The polystyrene matrix was employed as a heterogeneous reaction interface for the formation of MOF-801 in the aqueous solution. In the synthesis of MOF-801 on AgNCs' surface, organic linkers in the form of free acid functional groups were necessary. They were introduced

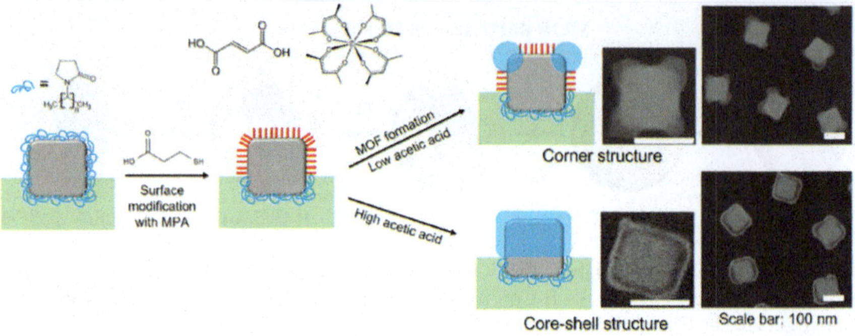

Fig. 1.6 MOFs with complex structure for CO_2 conversion. Heterogeneous corner MOF-AgNC and core–shell MOF-AgNC synthesized through a seed-mediated synthetic method. Reprinted with permission from Ref. [33]. Copyright (2022), Elsevier

by modifying the surface of AgNCs with mercaptopropionic acid. The CO_2 reduction reactivity of the MOF-AgNC nanocomposites obtained showed a remarkable sensitivity to the wavelength and intensity of the irradiation, which was due to the fluctuation of the number and position of hot electrons in AgNCs with the irradiation intensity. Importantly, it was observed that the corner MOF-AgNC nanocomposite had a faster CO_2 adsorption rate than the core–shell MOF-AgNC nanocomposite, primarily attributed to the higher surface area-to-volume ratio of the MOF in the former.

1.2.3 MOF-Based Derivatives with Complex Structure

MOF-derived materials with complex structures, such as hollow or hierarchical architectures, is striking in catalytic applications due to their specific properties and efficient mass transfer characteristics. These complex MOF derivatives have demonstrated excellent catalytic activity and selectivity toward CO_2 conversion. Additionally, their tunable pore sizes and surface functionalization offer great flexibility in designing catalysts with tailored properties for specific catalytic applications. The use of MOF derivatives with complex structure in catalysis represents a promising approach for creating sustainable and high-performance catalytic systems.

By employing novel synthesis techniques and precise regulation of the shell formation process, it is possible to produce hollow particles with diverse shell architectures derived from MOFs (Fig. 1.7a) [34]. For instance, a successful synthesis of multi-shelled hollow Co_3O_4 dodecahedron was achieved through the sequential templating approach by using MOFs as templates (Fig. 1.7b–g). The specific arrangement of cobalt atoms within MOFs allowed for precise orientation control of Co_3O_4 nanocrystals in the resulting hollow multi-shelled structures. This resulted in a distinct shell structure with a high degree of exposure of (111) facets. Notably,

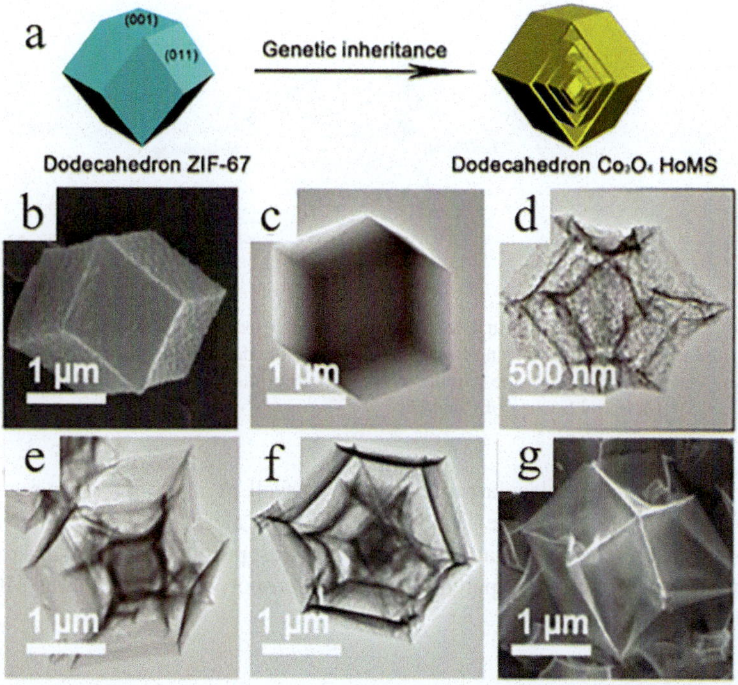

Fig. 1.7 MOF derivatives with complex structure for CO_2 conversion. **a** Co_3O_4 dodecahedron with hollow multi-shelled structures prepared via sequential templating approach. TEM and SEM images of ZIF-67 (**b**, **c**). TEM images of double-shelled (**d**) and triple-shelled (**e**) Co_3O_4 hollow dodecahedron. TEM and SEM images of quadruple-shelled Co_3O_4 hollow dodecahedron (**f**, **g**). Reprinted with permission from Ref. [34]. Copyright (2019), American Chemical Society

the catalytic activity of the quadruple-shelled Co_3O_4 hollow sample was approximately three and five times higher than that of Co_3O_4 hollow multi-shelled structures without facet control and Co_3O_4 nanoparticles, respectively.

1.2.4 Pore Engineering

To achieve a high chemical activity on a larger scale, it is important to consider mass and electron transport, which are closely dependent on the pore structures of the catalysts. Recent studies have illustrated that the hierarchical porous structure is ideal for catalysts to achieve high catalytic activity [35, 36]. As an example, Babu and coworkers developed a new method for fixing CO_2 at room temperature by using a dual-porous MOF, specifically the micro-mesoporous MOF, UMCM-1-NH_2 [36]. A solvent-free method was employed to synthesize a range of five-membered cyclic carbonates. UMCM-1-NH_2 with combination of mesopores for

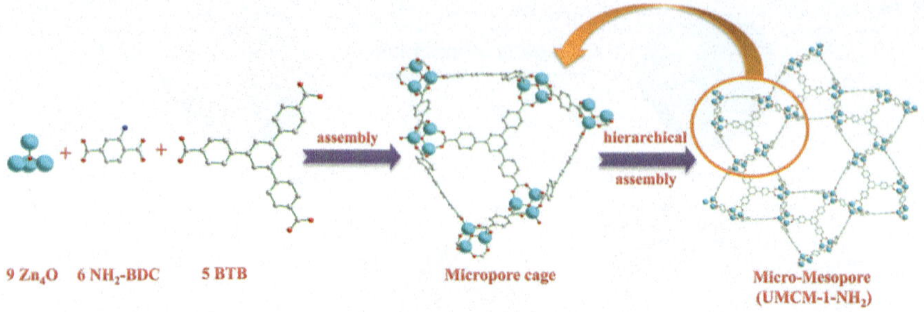

Fig. 1.8 Hierarchical assembly in UMCM-1-NH_2 is represented, in which the micropores are constructed using Zn_4O clusters and linkers, followed by the development of 1D mesopores from the hexagonal arrangements of the microporous cages. Reprinted with permission from Ref. [36]. Copyright (2016), Royal Society of Chemistry

guest diffusion and molecular accessibility, as well as micropores for regulating catalytic interactions at active sites, was found to be a crucial factor in the success of this approach. The combination of micro and mesoporous characteristics resulted in superior performance compared to pristine microporous or mesoporous MOFs in cycloaddition reactions (Fig. 1.8).

1.3 Multicomponent Design of MOF-Based Catalysts

1.3.1 Mixed-Metal MOFs

When a single MOFs is used as catalyst, the coordination environment of the metal at the node often changes, resulting in a break between the metal and the organic ligands, thus leading to a sharp decline in catalytic activity [37]. It is a good idea to fabricate mixed-metal MOFs, in which the high-valent metal facilitates the structural stability and the low-valent metal serves as the catalytic active center [38, 39]. Mixed-metal MOFs can exhibit synergistic interactions among various metal ions to activate the reactive species and lower the reaction energy barrier, so as to achieve high reactivity, selectivity, and stability in the process of CO_2 conversion [40].

The most common strategy for the construction of mixed-metal MOFs is to use heterometallic clusters as metal sources, such as trinuclear Fe_2M, bimetallic AB_2OX_3 clusters, and Zr_6-oxo-M clusters, etc. Lan and coworkers chose stable $Fe_2M(\mu_3\text{-}O)(OAc)_6(H_2O)_3$ (M = Co, Ni, Zn) clusters as the metal source to synthesize bimetallic MOFs, $[Fe_2M(\mu_3\text{-}O)(TCA)_2(H_2O)_3]$ (NNU-31) by one-pot synthesis (Fig. 1.9a) [41]. The obtained samples achieved high performance for CO_2 reduction without extra photosensitizers. Among them, the HCOOH yield of NNU-31-Zn was the highest (26.3 μmol g^{-1} h^{-1}) coupled with high selectivity (about 100%) (Fig. 1.9b). Typically, high-valent Fe(III) are deemed to serve the characters of oxidation as well as low-valent M(II) ions for reduction active sites (Fig. 1.9c). A number of isostructural mixed-metal MOF photocatalysts based on Fe_2M clusters were directly synthesized via a one-pot reaction for CO_2 conversion [42]. Zhang et al. constructed Ni/Ti mixed bimetallic MOFs [43], which had tunable physicochemical properties and improved catalytic behavior. The introduction of Ni enhanced the Ti/Ni metal-to-metal charge transfer and formed a more pronounced channel for carrier recombination. Ezugwu et al. synthesized a family of stable bimetallic MOFs using Zr(IV)-O-M clusters as metal nodes [44]. The existence of Cu-oxo clusters caused a significant reduction of the band gap from 2.95 eV (Zr-ATA) to 1.93 eV (Zr/Cu-ATA) and displayed outstanding performance for CO_2 conversion.

In addition to using heterometallic clusters, another direct way is to apply metalloligands as bridge ligands to introduce additional fixed metal atoms. Common metalloligands such as metalloporphyrin units have outstanding effects on CO_2 conversion, especially the photocatalytic conversion [45]. Lan and coworkers precisely designed a series of MOF materials based on metalloporphyrin organic ligands (Fig. 1.9d) [46]. The combination of reductive polyoxometalates (POMs, like Zn-ε-Keggin cluster) with metalloporphyrins gave these mixed-metal MOFs a big advantage in terms of electron harvesting, electron transport, and catalytic activity during CO_2 reduction reactions. Unsurprisingly, these PMOFs, especially for Co-PMOF, achieved a 99% Faraday efficiency in selective electrocatalytic reduction of CO_2–CO. Other mixed-metal MOF catalysts have also been constructed for CO_2 conversion, like Rh-PMOF-1(Zr) [47], PCN-224 (Cu) [48], PCN-222 (Fe) [49], $[Co(Ni\text{-}H_7TPPP)_2]\cdot 8H_2O$ [50],

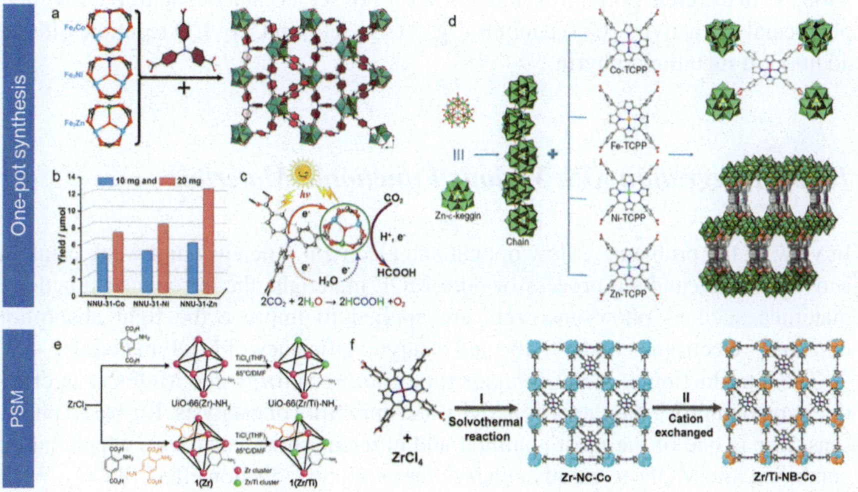

Fig. 1.9 **a** Mixed-metal NNU-31-M assemble by the Fe_2M cluster and TCA. Reprinted with permission from Ref. [41]. Copyright (2020), John Wiley and Sons. **b** Performance of NNU-31-M converting CO_2 into HCOOH. **c** The mechanism of NNU-31-M in CO_2 conversion. **d** M-PMOFs (M = Co, Fe, Ni, Zn) constructed by zigzag POM chains and 4-connected TCPP ligand. Reprinted with permission from Ref. [46]. Copyright (2018), Springer Nature

InTCP-Co, and FeTCPCo [51]. Besides, metal phthalocyanine units were also investigated to construct a laminated bimetallic MOFs and showed good performance in converting CO_2–CO [52]. Mixed-metal MOF catalysts synthesized by one-pot method for CO_2 conversion are summarized in Table 1.1.

1.3.2 Mixed-Ligand MOFs

Due to the diversity of ligand functionalities, designable pore environments, and potential internal synergistic catalytic sites, mixed-ligand MOFs exhibit superior properties over other conventional porous materials [53]. In terms of CO_2 application, mixed-ligand MOFs are considered to be an efficient route to enhance light absorption and electron transport and play a key role in decreasing the recombination rate of photo-generated carriers and increasing the current density [54].

Owing to their strong interaction with CO_2, excellent light absorption, electron conduction efficiency, and good catalytic activity, porphyrins and metalloporphyrins have become important functional ligands in the construction of MOF catalysts [55, 56]. The addition of porphyrin-based organic ligands to the MOF structure can avoid the use of rare precious metal-based photosensitizer. Kong and coworkers used the linear porphyrin ligand $DCPP^{2-}$ to replace $NDIDB^{2-}$ ligand in the synthesis of BUT-110 based on the size- and geometry-matched proprieties (Fig. 1.10) [57]. Compared with BUT-109(Zr), the chemical stability enhancement of BUT-110 series varied with different porphyrin contents. BUT110 series catalysts achieved excellent photocatalytic activity (64.0 $\mu mol\ h^{-1}\ g^{-1}$ CO yield) in CO_2–CO reaction with the addition of metalloporphyrin.

1.3.3 Integration with Various Functional Materials

In view of the problems of low optical and electrical efficiency and weak catalytic activity in the catalytic process of pure MOF materials, the introduced functional materials such as photosensitizers are applied to improve the light absorption efficiency, electronic conductivity, and catalytic efficiency of MOF materials.

The introduction of metal complex type photosensitizers into MOFs is an effective way to improve the catalytic activity and durability of catalysts. Ru-based photosensitizer is one of the most common and efficient photosensitizers, which can be combined into MOFs to greatly enhance the catalytic reduction effect of CO_2. Wang and colleagues introduced $Ru(bpy)_3]Cl_2{\cdot}6H_2O$ into a range of MOF materials, in which the two of them are used as photosensitizers, catalysts and cavities, respectively [58]. Benefiting from the introduction of photosensitizers, CO_2 conversion efficiency of Co-ZIF-9 was obviously improved. Purposefully embedding photosensitizers into the MOF pores increased photocatalytic CO_2 reduction activity. Through the guest efficacy of MOFs, Fisher et al. captured photosensitivities in a stable

Table 1.1 Various mixed-metal MOFs catalysts synthesized by one-pot method for CO_2 reduction reported in recent years

No.	Mixed-metal MOFs	Metal	Organic ligand	One-pot method	Ref
1	NNU-31-M (M = Co, Ni, Zn)	$Fe_2M(\mu_3\text{-}O)(OAc)_6(H_2O)_3$ (Fe_2M) clusters	TCA	NMF, acetic acid, 423 K, 32 h	[41]
2	PCN-250-Fe_2M (M = Mn, Zn, Ni, Co)	Fe_2M clusters	H_4abtc	DMF, acetic acid, 413 K, 2 h	[42]
3	Ni/Ti bimetallic MOFs	$Ti(OC_4H_9)_4$, $Ni(NO_3)_2{\cdot}6H_2O$	ATA	DMF, MeOH, 423 K, 20 h	[43]
4	NH_2-UiO-66(Zr/M) (M = Fe, Co, or Cu)	Zr(IV)-O-M clusters	ATA	DMF, benzoic acid, 393 K, 48 h	[44]
5	M-PMOF (M = Co, Fe, Ni, Zn)	Zn-ε-Keggin cluster ($\{\varepsilon\text{-}PMo_8{}^{V}Mo_4{}^{VI}O_{40}Zn_4\}$)	M-TCPP	H_2O, 453 K, 72 h	[46]
6	Rh-PMOF-1(Zr)	$ZrCl_4$	Rh(TCPP)Cl	DMF, benzoic acid, acetic acid 393 K, 48 h	[47]
7	PCN-224 (Cu)	$ZrCl_4$	CuTCPP	DMF, benzoic acid, 393 K, 48 h	[48]
8	PCN-222(Fe)	$ZrCl_4$	$TPPCO_2H$-Fe(III)Cl	DMF, benzoic acid, 413.15 K, 72 h	[49]
9	$[Co(Ni\text{-}H_7TPPP)_2]{\cdot}8H_2O$	$Co(CH_3COO)_2{\cdot}4H_2O$	Ni-H_8TPPP	H_2O, 433 K, 3 days	[50]
10	FeTCPCo	$FeCl_2$	3,4-TDHPP-Co	DMF/H_2O, 413 K, 72 h	[51]
11	PcCu–O_8–Zn	ZnO_4	CuN_4	DMF/H_2O, 393 K, 40 h	[52]

Ligands are abbreviated as

TCA = 4,4′,4″-tricarboxytriphenylamine; H_4abtc = 3,3′,5,5′-azobenzene tetra-carboxylic acid; ATA = 2-aminoterephthalic acid; M-TCPP = tetrakis[4-carboxyphenyl]-porphyrin-M; $TPPCO_2H$-Fe(III)Cl = [5,10,15,20-tetrakis(4-carboxyphenyl)porphyrinato]-Fe^{III} chloride; TPPP = 5,10,15,20-tetrakis-[p-phenylphosphonic acid]porphyrin; 3,4-TDHPP-Co = 5,10,15,20-tetrakis(3,4-dihydroxyphenyl)porphyrin-Co; CuN_4 = copper-phthalocyanine

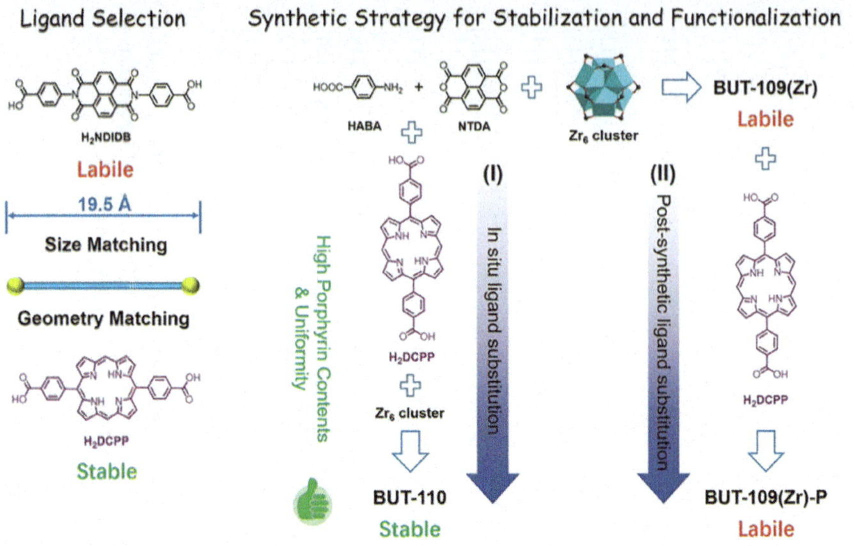

Fig. 1.10 Ligand selection and synthetic strategies of BUT-110 through (I) in-situ ligand substitution and (II) post-synthetic ligand substitution. Reprinted with permission from Ref. [57]. Copyright (2021), John Wiley and Sons [57]

(MIL-101)-NH_2(Al) cage [59], which further improved the stability of the MOF catalyst in addition to enhancing the catalytic activity. They also tried to immobilize both the CO_2 conversion catalyst (ReBr$(CO)_3$(4,4′-dcbpy)) and the photosensitizer (Ru-$(bpy)_2$(5,5′-dcbpy))Cl_2 onto isoreticular series of UiO-66, UiO-67, and UiO-68 (Fig. 1.11) [60]. Once the photosensitizer and catalyst were immobilized to UiO-66 on the surface, rapid CO_2-to-CO reduction was recorded under visible light. But the catalyst was deactivated after 1.5 h. Longer organic ligand in the UiO-68 structure led to a larger cavity space. When the photosensitizer and catalyst were fixed to the UiO-68, the MOF composite catalyst showed a longer activity (~ 9 h) but correspondingly slower reaction rate. The progresses demonstrated that the MOF pore size determined the electronic communication between the reaction center and the photosensitizer, providing fundamental insights into the design of composites with reticular structures.

1.3.4 Hybrid Carbon Materials Derived from MOFs

Because of the inexpensive cost, high surface area, tunable conductivity, and electrochemical activity, metal-free porous carbon has recently gained a lot of attention as electrocatalysts for CO_2RR. These materials are among the most promising alternatives to the pricey noble metal catalysts. Moreover, the N-doped carbon materials

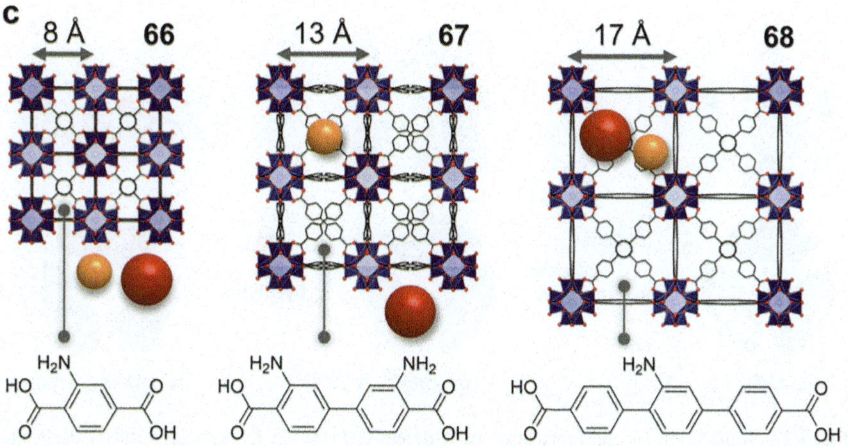

Fig. 1.11 MOF containing different lengths of linker as host anchoring catalyst [ReBr(CO)$_3$(4,4′-dcbpy)] (yellow ball) and photosensitizer [Ru(bpy)$_2$(5,5′-dcbpy)]Cl$_2$ (red ball). Reprinted with permission from Ref. [60]. Copyright (2021), John Wiley and Sons [60]

are significantly more desirable for selective CO_2 conversion due to their ability to suppress the opposing H_2 evolution reaction.

Huang et al. reported the fabrication of N-doped porous carbon (NPC) for the ECO_2RR by using Zn-MOF-74 as the template [61]. The obtained NPC had abundant active N atoms, and the type of N atoms can be adjusted with the adjustment of pyrolysis conditions, especially the highly active pyridine type N atoms and graphite type N atoms. The NPC catalysts with the best composition had a FE_{CO} of 98.4% at − 0.55V (vs. RHE). Additionally, Zheng et al. studied the catalytic mechanism of N-doped carbon (NC) derived from ZIF-8 and the influence of pyrolysis temperature [62]. Results showed that increasing the pyrolysis temperature can improve the performance of CO_2RR, and the enhancement in catalytic activity was contributed to the formation of pyridinic-N, which offered more attractive reactive sites.

So far, studies on metal-free catalysis of ECO_2RR have focused on heteroatoms as active sites, while ignoring the inherent carbon defect sites. Actually, the insertion of carbon defects is comparable to the heteroatom functionalization technique, which can cause charge redistribution and disrupt the symmetry in the molecular plane, increasing the possibility to activate CO_2 during the catalytic process. Exploring the specific structure–activity connections of intrinsic carbon defects throughout the catalytic process is still difficult since the production of these defects frequently results in heteroatom-doping effects. Under the circumstances, Jiang et al. proposed a K^+ assisted synthetic technique [63]. By restraining the pyrolysis process of MOF with K^+, porous carbon with abundant inherent carbon defects and complete removal of heteroatoms (K-defect-C) was obtained (Fig. 1.12). The obtained K-defect-C had a significant amount of 12-vacancy-type carbon defects. In the ECO_2RR process, the K-defect-C achieved a high CO FE up to 99% at − 0.45 V (vs. RHE) and outperformed MOF-derived porous carbon without K^+ etching. DFT calculations further revealed

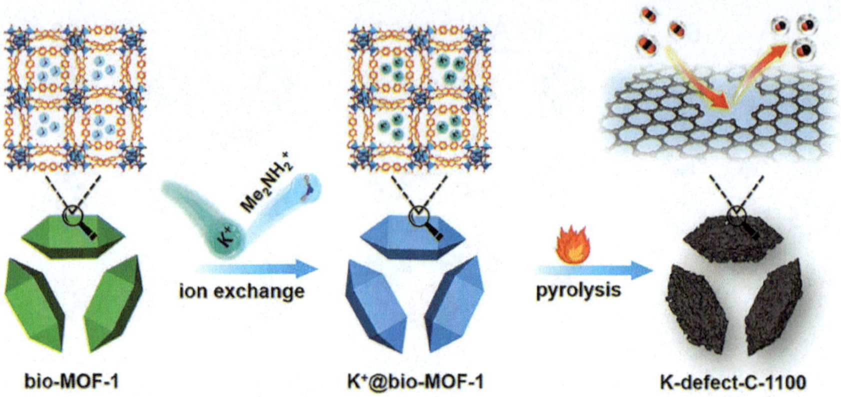

Fig. 1.12 Illustration for the fabrication of K-defect-C-1100 via K^+-assisted strategy. Reprinted with permission from Ref. [63]. Copyright (2022), John Wiley and Sons [63]

that the carbon defect in K-defect-C was conducive to the adsorption and activation of CO_2, thus speeding up the generation of *COOH during ECO_2RR process.

1.4 Improvement of Stability of MOF in CO_2 Conversion

Due to the high thermodynamic and kinetic stability of CO_2, harsh or critical reaction conditions such as high temperature and high pressure are often faced in CO_2 conversion. This requires higher chemical and thermal stability of catalysts. But unfortunately, MOF usually has poor stability in air or moisture environment. The stability of MOF has become an important factor hindering its application in CO_2 conversion. There are two main approaches to enhance the stability of MOF. Firstly, spatial arrangement is critical, such as interpenetrated structures and encapsulated structures. Secondly, MOF stability can be improved by enhancing its hydrophobicity.

1.4.1 Interpenetrated Structure

During the synthesis of MOF, particularly those containing long organic chain ligands, numerous void spaces are formed which are typically occupied by solvent molecules, counter ions, or additional networks. These molecules may reassemble to form the MOF network, namely interpenetration. Interpenetration has a significant effect on the size, shape, and surface area of the pores in the MOF [64]. The interpenetrated networks can reduce the MOF pore size and increase the wall thickness, and its higher strength non-covalent bonds can effectively lock the structure [65]. The aforementioned factors collectively contribute to the enhanced stability

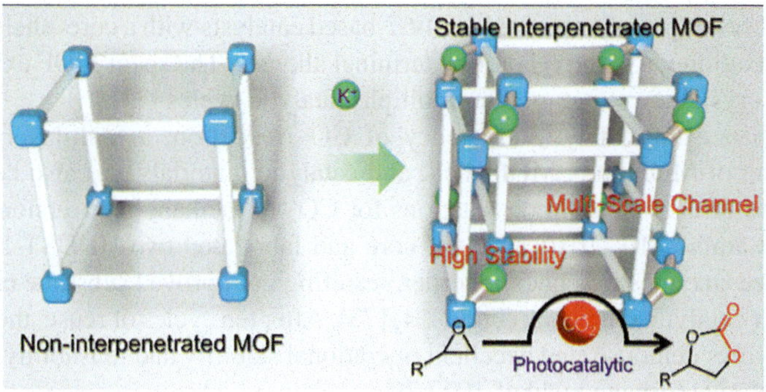

Fig. 1.13 K^+ induced framework interpenetration. Reprinted with permission from Ref. [71]. Copyright (2020), American Chemical Society

of MOF, consequently enabling them to manifest superior catalytic longevity and reusability in CO_2 reduction. In addition, although interpenetrated structure reduces the aperture and volume of the framework, the higher density allows it to provide more catalytic sites per unit volume compared to non-interpenetrated MOF [66]. Based on the aforementioned reasons, constructing an interpenetrated structure is an effective method for enhancing the thermal stability, water stability, and catalytic performance of MOF materials in CO_2 conversion. [67, 68]

To demonstrate the stability of the interpenetrated MOF structure, Zhang et al. prepared Ru-based MOFs with and without interpenetrating structures. In the photocatalytic reaction of CO_2, the exchange of solvent molecules and guest molecules in MOF without interpenetrating structures weakened the supramolecular interaction between guest molecules and frameworks, leading to structural collapse. In contrast, the strong π-stacking interactions between adjacent networks of MOF with interpenetrated structures effectively stabilized the structure of MOF [69]. Another study also proved that interpenetrated MOF possessed excellent crystal stability even after immersion in boiling water for 21 days during CO_2 cycloaddition reaction [70]. However, the interaction between networks may still not be able to completely prevent surface deformation or sliding due to the removal of guest molecules. Further stabilization of the interpenetrated MOF can be achieved through connecting the metal ions in a suitable manner. For example, MOF with good thermal stability and hydrolysis stability in the photocatalytic conversion of CO_2 were synthesized by connecting two adjacent uranyl ions with K^+ due to the significant shortening of the distance between uranyl ions in the interpenetrated structure (Fig. 1.13) [71].

1.4.2 Encapsulation

The encapsulation of MOF nanoparticles into stable material shells with physical and chemical stability is an effective way to obtain MOF with improved stability [72].

Gong and colleagues synthesized UiO-67-based catalysts with a core–shell structure for CO_2 fixation via carboxylation of terminal alkynes. The spent MOF exhibited an excellent crystalline structure after multiple catalytic cycles [73].

To enhance the catalytic efficiency of CO_2 reduction, a multilayer structure comprising of MOF core, MOF shell, and catalytic materials was also be utilized. In the case of enzyme cascade reactions for CO_2 to methanol conversion, Li et al. employed amine-MIL-101(Cr) as the core and fabricated two HKUST-1 layers to immobilize enzymes in sequential order, resulting in improved substrate channeling and higher catalytic efficiency (Fig. 1.14) [73]. After ten cycles of reuse, the immobilized enzymes demonstrated excellent operational stability and reusability, resulting in a cumulative formate yield of 1077.7%.

Although this structure can significantly increase MOF stability, it may affect catalytic activity because the active site is coated. Due to the presence of the shell, incomplete diffusion of reactants and products may also affect the reaction rate. The core–shell structure can be effectively improved by introducing cavities or active sites, such as yolk–shell structure [74]. Nano-shells with mesoporous pores provide a large number of paths for the transport of reactants and products. The yolk–shell structure protects the catalytic activity sites of MOF and slows down the decomposition of MOF in the catalytic reaction.

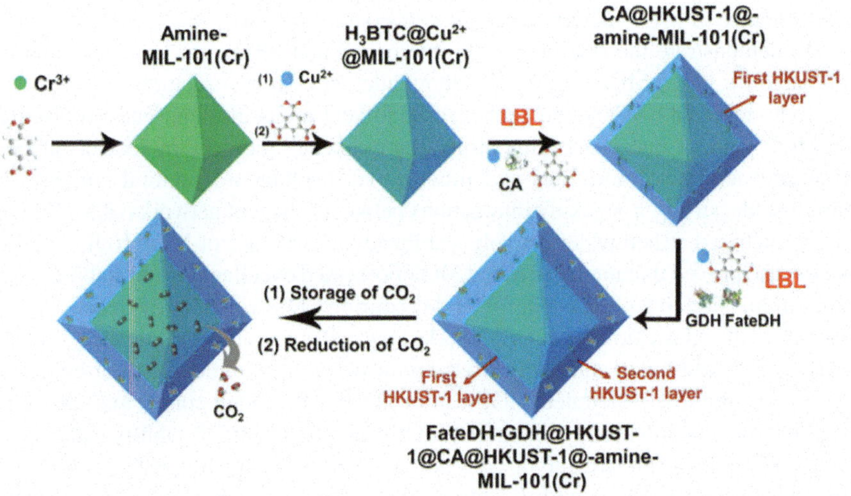

Fig. 1.14 Schematic illustration of the preparation of HKUST-1@amine-MIL-101(Cr)-based multienzymes for the reduction of adsorbed CO_2. Reprinted with permission from Ref. [73]. Copyright (2019), Frontiers

1.4.3 Design of Hydrophobic Properties

Due to the hydrophilic nature of the organic ligands in MOF, they are susceptible to attack and degradation by water molecules, which may lead to deactivation and decomposition of the MOF. On the other hand, it is possible that the presence of water molecules decreases the ability of MOF to capture CO_2. This is because water molecules can compete with CO_2 molecules for the active sites on the MOF. The competition for active sites results in reduced uptake of CO_2, which can negatively impact their efficiency as catalyze materials [75]. In addition, hydrophobic MOF exhibits higher current densities over a wide range of potential range compared to hydrophilic MOF, indicating that the preparation of hydrophobic MOF has a positive impact on improving the CO_2RR efficiency [76]. In summary, developing MOF with hydrophobic properties is of great significance for enhancing the stability of MOF and their CO_2 catalytic activity.

The instability of MOF in humid environments is primarily attributed to the attack of water molecules on metal nodes, resulting in framework degradation and functional loss. To mitigate this issue, post-modification has also been explored as a potential solution. For instance, the introduction of fluorination, methyl, and phenyl groups can effectively enhance the water stability of MOF by reducing the adsorption and penetration of water molecules onto the framework [77]. In the study by Yang et al., perfluoroalkyl chains were added to an iron-based MOF to enhance its hydrophobicity during electrocatalytic reduction of CO_2 [76]. A strategy of pre-activating MOF followed by post-modification seems to be a promising approach for incorporating hydrophobic compounds. For example, appending portal diamines to the open metal sites of the MOF platform allows for the ring-opening reaction of an epoxide, resulting in enhanced hydrophobicity of the MOF (Fig. 1.15) [78].

Overall, the stability of MOF is a critical factor in their application in CO_2 conversion. Through the design of encapsulated and interpenetrated structures, ligands and metals, and hydrophobic modification, the stability of MOF can significantly be improved. The design of encapsulated and interpenetrated structures focuses mainly on physical strength. Interpenetrated structures reduce the porosity of MOF, and the interlocking of multiple MOF frameworks effectively improves the stability of MOF. Encapsulated structures involve wrapping an unstable material with a stable material to prevent external attacks by molecules. The design of ligands and metals is the most commonly discussed topic in building stable MOF. Strengthening the coordinating bond strength can significantly increase the strength of MOF in essence. Furthermore, the multifunctional design of ligands has a significant improvement effect on CO_2 catalysis. Hydrophobic modification is similar to encapsulated structures. By promoting the hydrophobicity of MOF, it can avoid attacks by water molecules and greatly improve the water stability of MOF. In summary, the stability of MOF can be enhanced from various perspectives. Meanwhile, improving catalytic efficiency from the functional groups is a more promising research direction in the future.

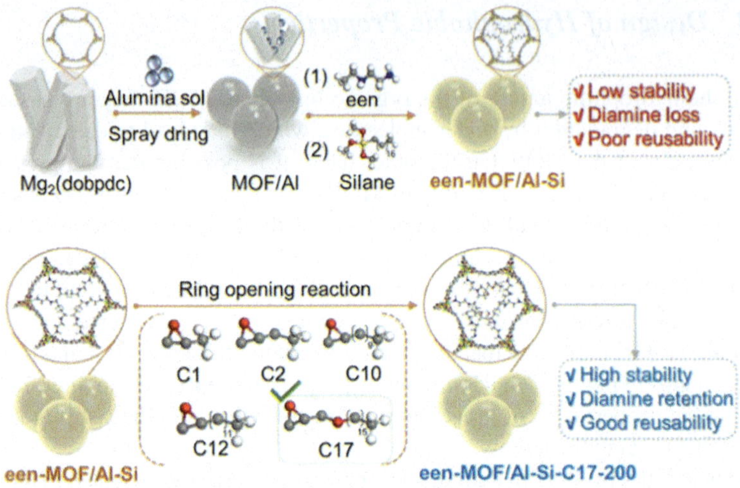

Fig. 1.15 Overall synthetic procedure representing (1) Diamine functionalization and (2) Silane coating. Reprinted with permission from Ref. [78]. Copyright (2022), American Chemical Society

1.5 Applications in CO_2 Conversion via MOF-Based Catalysts

1.5.1 CO_2 Organic Transformation

The cyclization of CO_2 with epoxides to cyclic organic carbonates (COCs) is a prospective method for CO_2 organic transformation to develop C1 chemical feedstock from CO_2. Cyclic organic carbonates are widely used as intermediates in the production of polycarbonate monomers, secondary battery electrolyte solvents, and engineering plastics. Homogeneous and heterogeneous catalysts such as ionic liquids, metal oxides, titanosilicates, and zeolite are currently used for CO_2 cycloaddition to produce COCs. Nevertheless, most of these catalysts require high temperature/high pressure, resulting in significant energy consumption and cost. Therefore, it is imperative to produce catalysts that are both highly effective and recyclable for the chemical transformation of CO_2 and epoxides into cyclic organic carbonates using mild conditions.

Owing to their functionalized porous structures and wide range of physical and chemical properties, MOFs are increasingly being considered as promising contenders for catalytic applications and has been widely explored in the synthesis of cyclic organic carbonates from CO_2 over the past years. According to the different catalytic sites of MOF, MOF materials are further divided into three subcategories. MOFs with surface or structural defects can generate Lewis acidic or basic sites; however, identifying the active sites in these defective MOFs may be challenging [79]. Active metal nodes are ligand unsaturated metal nodes or SBUs acting as Lewis acid sites. Typically, a Lewis base co-catalyst such as tetrabutylammonium bromide (TBAB) is usually required to assist the system to obtain the desired results [80]. MOF catalysts are constructed using functional linkers that contain Lewis acid, base, or ionic sites. These sites are incorporated into the linker to enhance the catalytic efficiency [81]. For instance, Liang et al. developed a MOFs-TBAB system and proposed an initial mechanism for synthesizing cyclic organic carbonates through the cyclization of CO_2 with epoxide (Fig. 1.16) [79]. This work has contributed to the development of MOFs catalysts for the chemical immobilization of CO_2 with epoxides.

Another promising strategy for the CO_2 organic transformation is the synthesis of carboxylic acids by activating terminal alkyne molecules, which is called carboxylation reactions [82]. The products obtained from this reaction, including alkyl carboxylic acids or esters, have potential applications in the pharmaceutical and chemical sectors [83, 84]. Liu and his colleagues used zeolite-based MOF (MIL-101)-loaded Ag NPs for the first time to catalyze the reaction of carbon dioxide with terminal alkynes to form propionic acid [85]. The load capacity of Ag was 1.66, 2.58, 4.16, and 6.97 wt%, respectively. They discovered that under the mild reaction conditions (50 °C and 1 atm CO_2), the yield of 3-phenylpropionic acid was 96.5% which proved that the synthesis of carboxylic acids by using NPs@MOF catalysts was a promising approach. In the past few decades, MOFs have been broadly investigated for the organic conversion of CO_2, including the cycloaddition reaction of

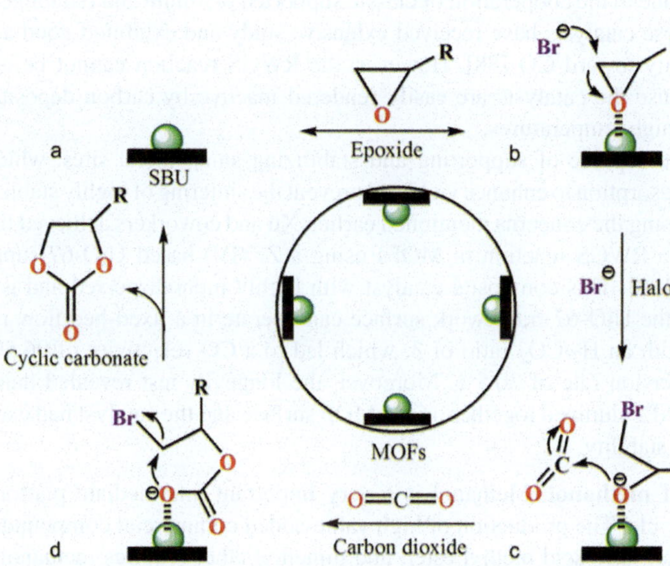

Fig. 1.16 Proposed mechanism for the catalytic cycloaddition of CO_2 with epoxide to produce cyclic carbonates using a typical MOF-TBAB system. Reprinted with permission from Ref. [79]. Copyright (2015), Royal Society of Chemistry [79]

CO_2 and the immobilization of CO_2 with terminal alkynes to carboxylic acids, and this area is rapidly becoming a prominent field in the MOF community.

1.5.2 Hydrogenation of CO_2

To put forward efficient means to lower CO_2 concentrations, selective hydrogenation of CO_2 to high value-added compounds should be a good choice which offers the precursors needed to boost industrial chemical and fuel production. The well-defined reticular frameworks and versatile modifiability of MOFs make them sort of prime materials for creating CO_2 hydrogenation-enabled catalysts. In this part, we try to summarize the most recent developments in MOF-based catalysts for CO_2 hydrogenation toward a variety of products.

Reverse water–gas shift to CO: The reverse water–gas shift (RWGS) reaction is a highly promising approach for converting CO_2, and the resulting CO can be a vital component of syngas used in the Fischer–Tropsch process to produce methanol, light olefins, and liquid fuels. It can also serve as a significant raw material for DMF production [86, 87]. The RWGS reaction, on the other hand, is a classic endothermic reaction, favoring itself thermodynamically and kinematically at high temperatures. To put it another way, the RWGS reaction requires a lot of energy to operate. On the other hand, due to the cooperation of classic supported platinum and reducible oxide supports, these catalysts have received extensive study and exhibited good activity and selectivity toward CO [88]. However, the RWGS reaction cannot be applied widely because the catalysts are easily rendered inactive by carbon deposition or sintering at high temperatures.

MOFs are capable of supporting and stabilizing small metal sites, which can reduce CO adsorption to enhance yield and prevent the sintering of highly stable metal sites, addressing the concerns mentioned earlier. Xu and coworkers achieved the first example of a RWGS reaction in MOFs using a Zr (IV)-based UiO-67 supported by Au NPs [89]. This composite catalyst with highly monodispersed and isolated Au NPs on the UiO-67 framework surface can operate in a fixed-bed flow reactor at 405 °C with an H_2/CO_2 ratio of 3, which led to a CO selectivity of 96.5% and a CO_2 conversion rate of 30.5%. Moreover, the longevity test revealed that there were no AuNPs clumped together on the MOF surface and the catalyst had excellent temperature stability.

Synthesis of methanol: Methanol is a very important intermediate platform for the carbon cycle. The production of high value-added commercial compounds such as acetic acid, fatty acid methyl ester, and dimethyl ether requires methanol as an industrial raw material [90, 91]. Syngas is usually used to produce methanol in industry. However, the use of CO_2 hydrogenation to produce methanol is not only an environmentally friendly synthetic route, but also an important step to achieve carbon cycle [92]. Some traditional heterogeneous catalysts for CO_2 hydrogenation, such as $Cu/ZnO/Al_2O_3$, are widely used in the production of industrial methanol because of their large reserves on the earth and low prices. However, these catalysts have low methanol conversion (~ 30%) and selectivity (30–70%) even at 200–300 °C and high pressure (5–10 MPa) [93, 94]. In recent years, the research on MOFs-based materials as CO_2 hydrogenation catalysts have gradually increased. The main reason is that when MOFs-based catalysts are used for CO_2 hydrogenation reaction, even if the reaction conditions are relatively mild, the selectivity and conversion of methanol are high, and the stability is good. Moreover, because MOFs have a well-defined structure, it is helpful to explain the nature of the active site and the structure-performance relationship.

In order to synthesize specific MOF-based catalysts with stable surface and interface structures, Wang and coworkers prepared ultra-small Cu/ZnOx NPs. They used UiO-bpy (bpy = 2,20-bipyridine) MOFs containing Zr_6 SBUs $\{Zr_6\ (\mu_3\text{-}O)_4\ (\mu_3\text{-}OH)_4\}$ as carriers and pre-assembled Cu (II) ions and Zr6 clusters on the bpy sites in the MOF to modify Zn (II) ions. At 250 °C and 4 MPa, the metallized UiO-bpy was synthesized by in-situ reduction using a reducing agent with a H_2/CO_2 ratio of 3 [95]. The anchoring of UiO-bpy MOF facilitated the formation of ultra-small Cu/ ZnO_x nanoparticles, which prevented Cu NP agglomeration. This is mainly due to the effective separation of Cu NPs and ZnOx and their confinement within the MOF cavity. When Cu/ZnO_x @ MOF catalyst was applied to the hydrogenation of carbon dioxide, extremely high catalytic activity was achieved, in which its space–time yield (2.59 $g_{MeOH}\ kg_{Cu}{}^{-1}\ h^{-1}$) was 2 times higher than that of commercial catalyst (Cu/ ZnO/Al_2O_3) (Fig. 1.17). In addition, the obtained catalyst not only achieved 100% selectivity to methanol, but also remained stable after continuous electrolysis for 100 h.

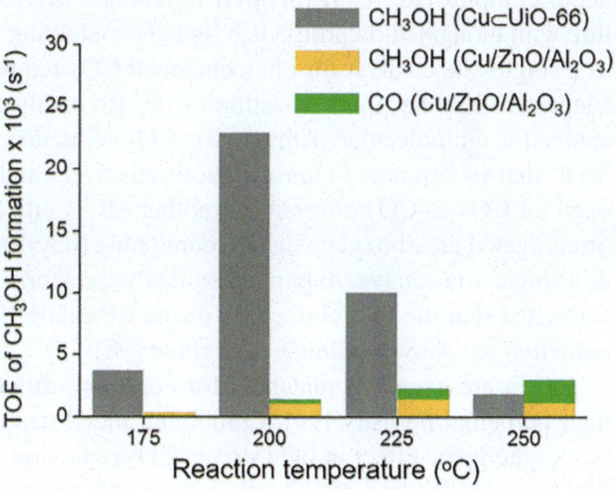

Fig. 1.17 TOFs for product formation were measured at different reaction temperatures over Cu/ZnO/ Al_2O_3 and Cu ⊂ UiO-66 catalyst. Reprinted with permission from Ref. [95]. Copyright (2017), Royal Society of Chemistry [95]

1.5.3 Photocatalytic CO_2 Reduction

MOFs have been widely explored to be utilized for photocatalysis. This section focuses on photocatalytic products, such as (i) the conversion of CO_2 into CO and (ii) the conversion of CO_2 into organic compounds (HCOOH, CH_3OH, and CH_4). The manipulation of the components in MOFs is a key factor in determining photocatalytic efficiency. To construct MOF composites for photocatalysis, those MOFs with little or no photoactivity can be combined with additional elements. Moreover, MOFs can be transformed into related derivatives to provide active sites with improved activity. In addition to partially inheriting the advantages of MOFs, MOF derivatives can also offer some extra benefits compared with parent MOFs.

Reduction of CO_2–CO: Numerous studies have indicated that modifying the organic ligands and metal clusters of MOFs can decrease the HOMO–LUMO energy gaps, leading to increased light absorption capabilities. In the realm of CO_2-to-CO photocatalytic reduction, various approaches are employed, such as incorporating linkers with photosensitizer functionalities, using electron-rich conjugated linkers and utilizing metalloligands [96–98].

Functionalized organic linkers can have a significant impact in increasing the photocatalytic properties of MOF materials. In a very preliminary study, the UiO-67 type MOF utilizing photosensitizer-functionalized linker [ReI(CO)$_3$(dcbpy)Cl] was investigated in the light-driven conversion of CO_2–CO [96]. In recent studies, it was found that a specific MOF exhibited an impressive turnover number (TON) of 10.9, making it a highly active catalyst for photocatalytic CO_2 reduction to CO. Remarkably, this value is more than three times greater than that of the photoactive homogeneous complex [ReI(CO)$_3$(dcbpy)Cl]. Another MOF, featuring a comparable structure with elongated dicarboxylate ligands containing (bpy)Re(CO)$_3$Cl, was created as a single-site catalyst for photochemical CO_2 reduction to CO [97]. Both reports suggested that the immobilization of the Re catalyst within the MOF framework enabled a unimolecular pathway for CO_2 reduction. Moreover, a multifunctional MOF that incorporated cuprous photosensitizers and Re catalysts was also developed for CO_2-to-CO conversion. Another MOF with a similar structure constructed on elongated dicarboxylate ligands containing (bpy)Re(CO)$_3$Cl was further designed as a single-site catalyst to photochemically transform CO_2 to CO. These progresses suggested that the immobilization of the Re catalyst in the MOF enabled the CO_2 reduction to follow a unimolecular route [98].

MOFs are excellent materials for confining different guest species because of their perpetual porosity [99]. Combining metal nanoparticles (MNPs) with MOFs has a synergistic effect in light-driven CO_2 reduction. Yaghi et al. reported Re-UiO-67 for photoreduction of CO_2. To enhance the photoreduction activity under visible light and maintain long-term stability, Re-UiO-67 was coated onto Ag nanocubes to form Ag-Re-UiO-67 (Fig. 1.18) [100], which demonstrated a sevenfold improvement in activity compared to Re-UiO-67. The stability was maintained for up to 48 h. Another innovative photocatalyst Ag@Co-ZIF-9 was also developed for CO_2-to-CO reduction under visible light. By utilizing a photosensitizer to aid in the process,

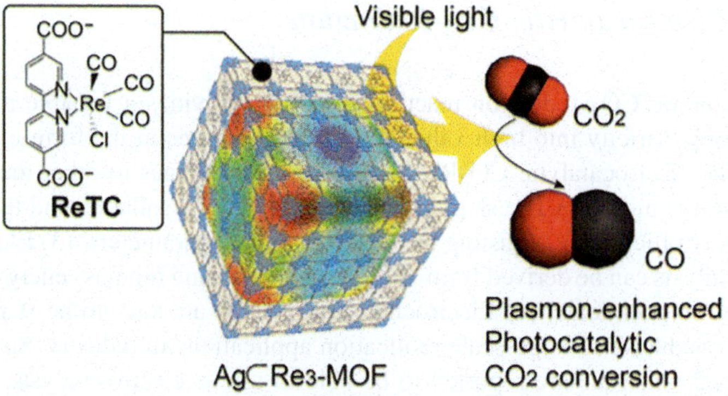

Fig. 1.18 Ren-MOF and Ag Ren-MOF structures have been designed and synthesized for the purpose of plasmon-enhanced photocatalytic CO_2 reduction. These innovative structures represent a significant advancement in the field of photocatalysis. Reprinted with permission from Ref. [100]. Copyright (2017), American Chemical Society

Ag NPs doping may offer potential method for enhancing the effectiveness and selectivity of MOF materials in CO_2 photoreduction [101].

Reduction of CO_2 to organic chemicals: Apart from reducing CO_2–CO, MOF-based photocatalysts can also efficiently reduce CO_2 to various organic compounds, primarily HCOOH, CH_3OH, and CH_4.While the liquid chemicals (HCOOH and CH_3OH) are major feedstocks for certain chemical processes, the generated gas product (CH_4) can directly be used as fuel. Tuning MOF-based photocatalysts is crucial for controlling the product selectivity and yield in order to satisfy practical requirements.

MOFs have significant benefits in photocatalytic CO_2 reduction due to their clearly defined structures and structural controllability. The most popular techniques involve functionalizing organic linkers and modifying metal clusters. Inspired by the arrangement of porphyrin pigments as light-harvesting antennae in natural photosynthesis, Jiang et al. employed a porphyrin-based Zr-MOF (PCN-222) to capture and convert CO_2 to HCOOH by using solar energy [102]. The lowest unoccupied molecular orbital (LUMO) potential of PCN-222 was more negative than the reduction of CO_2 to the formate anion, and after 10 h of visible light irradiation produced 30 μM of $HCOO^-$, which is remarkably higher than that of a homogeneous porphyrin ligand under similar conditions (2.4 μM of $HCOO^-$). The enhanced photocatalytic activity was found to originate from an ultra-long-lived electron trap state in PCN-222, as determined by ultrafast transient absorption and photoluminescence spectroscopy measurements.

1.5.4 Electrocatalytic CO_2 Reduction

Electrocatalytic CO_2 reduction reaction (CO_2RR) provides a feasible method for converting electricity into high value-added hydrocarbons in the form of chemical bonds. The electrocatalytic CO_2RR has superior advantages to other methods: (1) the reaction is mild; (2) The steps in the reaction are controllable, and the reaction can be more efficient by adjusting the electrochemical parameters. (3) Electricity in electrocatalysis can be derived from wind, water, solar, and biomass energy in nature and stored in chemicals; (4) Electrochemical reactors are easy to be standardized, which is conducive to large-scale replication applications in industry. So far, many MOF-based materials are reported to be able to use as electrocatalysts. Table 1.2 summarizes the electrocatalytic performance of some MOF-based catalysts for CO_2 reduction.

Among many products of electrocatalytic CO_2, CO_2 is reduced to CO, which has high selectivity and conversion efficiency, and CO is easily separated. It is one of the most economical and promising conversion routes. Silver (Ag) can be used as a catalyst for electrocatalytic CO_2RR. However, on the silver catalyst, the Faraday effect of CO_2 reduction to CO is low. At present, there are many studies on the use of silver synthetic composites as electrocatalysts. For example, Li et al. prepared Ag_2O/ZIF-7 as a CO_2RR electrocatalyst [103]. The catalyst performance test showed that the current density of CO on the composite was 26.2 mA/cm [2], and the Faraday efficiency reached 80.5% under CO_2 saturated 0.25 M K_2SO_4 electrolyte solution.

In order to design effective MOF catalysts with high conductivity, electroactive metal complex blocks (such as metal phthalocyanines and metal porphyrins) and/or functionalized elements can be loaded into porous two-dimensional MOFs. Cao and coworkers synthesized a two-dimensional conductive MOF using nickel phthalocyanine molecule with macrocyclic conjugated structure as the basic building unit (Fig. 1.19), which has excellent conductivity and structural stability [104]. After CO_2 electroreduction performance was tested in the H-type electrolytic cell aqueous medium, it was found that the CO selectivity reached 98.4% and the current density was up to 34.5 mA cm^{-2}, exceeding most other MOFs materials reported under the same system.

Table 1.2 Summary of electrocatalytic performance of MOF-based materials for CO_2 reduction

MOF-based materials	Electrolyte	Potential (vs. RHE) (V)	Main carbon product	Faradaic efficiency (%)	Refs.
Ag_2O/layered ZIF	0.25 M K_2SO_4	– 1.2	CO	80.5	[103]
NiPc-NiO4	0.5M $KHCO_3$	– 0.85	CO	98.4	[104]
S-HKUST-1	0.1M $KHCO_3$	– 1.2	C_2H_4	57.2	[105]
PTF(Ni)/Cu	0.1M $KHCO_3$/ 0.1 M KCl	– 1.1	C_2H_4	57.3	[106]

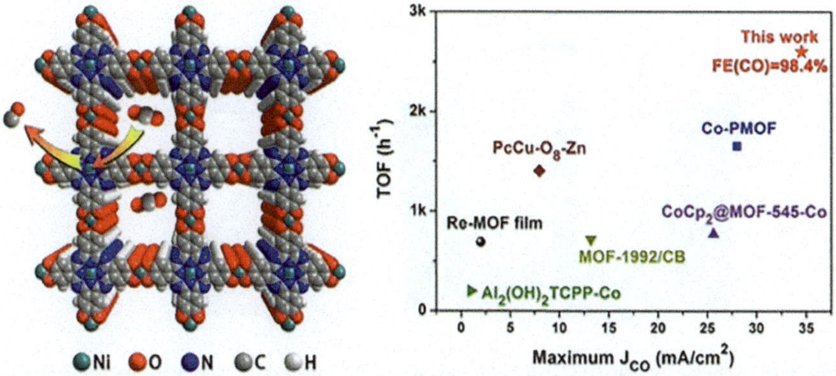

Fig. 1.19 Left image is the side view of the 2 × 2 square grid structure of NiPc-NiO_4 in AA stacking mode. The right figure is compared with the TOF and maximum CO partial pressure current density of reported MOF catalysts such as Co-PMOF, $CoCp_2$ @ MOF-545-Co, PcCu-O_8-Zn, MOF-1992/CB, Re-MOF film, and Al_2 (OH) ${}_2$TCPP-Co. Reprinted with permission from Ref. [104]. Copyright (2021), John Wiley and Sons

1.5.5 Photoelectroreduction

The field of CO_2 reduction is of particular interest in the utilization of photoelectrocatalysis (PEC) technology, which effectively merges the benefits of photocatalysis and electrocatalysis. Photoelectrocatalysis employ different kinds of semiconductors and bias potential to improve the reduction of CO_2 by effectively separating photogenerated electron and hole pairs under UV/visible light irradiation. In view of this, photoelectrocatalysis can be a favorable method in the conversion of CO_2 into useful chemicals. Generally, photoelectrocatalytic reactions occur in photoelectrochemical cells based on the interfacial behavior [107, 108]. The focus is the photoelectrocatalytic products, namely (i) the reduction of CO_2–CO and (ii) the conversion of CO_2 into organic compounds (CH_3OH, CH_3CH_2OH, and HCOOH).

Photoelectrocatalytic CO_2 reduction to CO: During a simulated AM 1.5 G solar irradiation, researchers observed that a Cu_2O cathode coated with HKUST-1 served the dual purpose of preventing photocorrosion which can provide active sites for CO_2 reduction (Fig. 1.20) [109]. To fabricate this photocathode, an indium tin oxide (ITO) glass surface was initially coated with a layer of Cu_2O, followed by a solution-phase process utilizing Cu_2O as a copper source to generate $Cu_3(BTC)_2$. The incorporation of $Cu_3(BTC)_2$ as a coating not only preserved the bandgap of Cu_2O but also significantly improved the separation and mobility of photogenerated charges within Cu_2O. In a three-electrode quartz PEC cell containing CO_2-saturated acetonitrile electrolyte with tetrabutylammonium hexafluorophosphate (TBAPF6), the reduction process exclusively yielded CO without any production of liquid hydrocarbons or other gaseous byproducts. Without light, the applied voltage enhanced the CO yield, although the yield of $Cu_3(BTC)_2/Cu_2O$ was two times greater than that of Cu_2O.

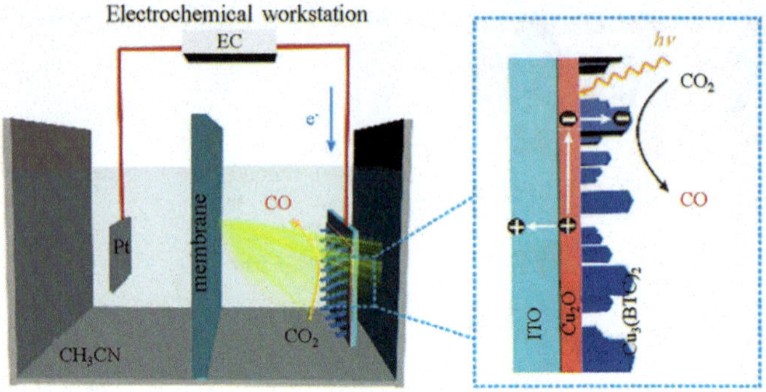

Fig. 1.20 Scheme of $Cu_3(BTC)_2/Cu_2O$ cathode in the process of photoelectrochemical CO_2 reduction. Reprinted with permission from Ref. [109]. Copyright (2019), American Chemical Society

The yield of $Cu_3(BTC)_2/Cu_2O$ was dramatically increased by roughly 4 times when the PEC cell was exposed to visible light.

Photoelectrocatalytic CO_2 reduction to organic chemicals: To investigate CO_2RR, Chen et al. synthesized a range of Ni^{2+}-doped NH_2-MIL-125-Ti materials with different Ni^{2+}/Ti^{4+} molar ratios [110]. The study exhibited the impact of Ni^{2+} dopants on the electronic structure of titanium oxo clusters within NH_2-MIL-125, which facilitated enhanced electron transfer and improved charge separation. In this approach, the performance of the photocatalyst can be altered. The CO_2 conversion rate with the highest value was 98.6% for NH_2-MIL-125-Ni1%/Ti. Costae Silva et al. utilized a Cu(BDC-NH_2) MOF coated on a Cu/Cu_2O electrode to investigate the selectivity of photoelectrocatalytic CO_2 conversion to methanol [111]. The enhanced selectivity was attributed to the MOF's large surface area and its strong capability for trapping CO_2. According to the researchers, the Cu/Cu_2O–Cu(BDC) electrode demonstrated a significant 42-fold improvement in methanol production compared to the control electrode of Cu/Cu_2O, when the Cu(BDC-NH_2) MOF was incorporated.

In PEC reduction of CO_2 into methanol, TiO_2 is frequently employed as a semiconductor in composite metal complex/semiconductor systems. Stülp et al. reported the utilization of TiO_2NT coated with Cu (II) aspirinate complex [112]. Electrochemical treatment or spin coating was used to decorate TiO_2NT with Cu (II) aspirinate. The electrodes of TiO_2NT treated with Cu (II) aspirinate by spin coating approach exhibited huge methanol production (0.8 mmol L^{-1}). Moreover, the formation of methanol decreased with the number of spin coating deposition cycles. The augmented surface coverage of nanotubes by subsequent layers of Cu(II) aspirinate led to a decline in the photogeneration of electron–hole pairs, thereby resulting in reduced production of the desired products. Cardoso et al. took photoelectrochemical (PEC) reduction of CO_2 into alcohols using TiO_2 nanotubes (TiO_2NT) decorated with ZIF-8 nanoparticles (NP) [113]. The presence of CO_2 in the experimental conditions resulted in a significant 97% increase in photocurrents for the TiO_2NT-ZIF-8 electrode. The interaction between CO_2 and ZIF-8 led to the formation of carbamate and facilitated improved charge separation. Compared with the Ti/TiO_2NT-ZIF-8 electrode, the conversion of CO_2 to methanol and ethanol exhibited a remarkable 20-fold and 430-fold increase, respectively. After 1 h of reaction at a potential of − 0.7 V. In contrast, the formation of methanol and ethanol using the Ti/TiO_2NT electrode was considerably lower. Notably, a higher production of methanol (0.7 mmol L^{-1}) and ethanol (10 mmol L^{-1}) was achieved at a potential of 0.1 V compared to − 0.7 V. The improved charge separation and more effective conduction band (CB) bending at 0.1 V enhanced the transfer of photogenerated electrons through the electrode, favoring the observed increase in alcohol production.

Current progresses have displayed that photoelectrocatalysis will have a wide range of possible applications. The benefits of these processes are combined in photoelectrocatalysis, which employs heterogeneous photocatalysis and electrocatalysis. The increasing interest in photoelectrocatalysis has led to significant interests toward the utilization of MOFs as photoelectrocatalysts. These MOFs possess desirable characteristics such as porosity, high surface area, and adjustable structures. Moreover, MOFs can form heterojunctions with other semiconductor materials and serve as effective co-catalysts in various photoelectrocatalytic systems. Furthermore, due to the porous features that are passed over from the original MOFs, derivatives of MOFs may also be utilized to build excellent photoelectrodes.

1.6 Summary and Perspectives

The practical application of MOF-based catalysts for CO_2 conversion is now facing several obstacles, including the issues related to poor stability, low mass transfer, and hard separation. For industrial-scale CO_2 conversion, precise control over the composition and structure of catalysts is necessary. In this regard, we provide a comprehensive understanding of the design approaches for MOF-based catalysts, covering diverse perspectives such as macrostructure, microstructure, and multicomponent control.. By utilizing the MOF-derived strategy, it is possible to achieve a controllable structure and composition, resulting in the creation of various derivatives with specific microstructure, macrostructures, and multicomponent characters that exhibit optimized performance from various angles in CO_2 conversion. The effective transport of reactants into the catalyst's interior and their adsorption onto active sites are characterized by multiple levels of pores and a significant surface area and facilitated by the presence of a porous microstructure,. Moreover, the catalytic oxidation efficiency for contaminant removal and overall stability can be greatly improved by employing intricate nanostructures and incorporating multiple components. This enhancement can be attributed to the synergistic effects generated by the interaction between different functional components and subunits. In addition, certain monolithic reactors derived from MOFs display the potential for practical CO_2 conversion applications, as they enable easy recovery of catalyst materials while maintaining mass transfer efficiency, high catalytic performance, and stability.

These progresses illustrate that the MOF-derived approach represents a promising avenue for the intentional design of catalysts tailored to meet specific CO_2 conversion application requirements.

Although the MOF-derived strategy for synthesizing and creating catalysts for CO_2 conversion has shown potential benefits, several challenges are still need to be mentioned for the practical implementation of MOF-based catalysts. The first challenge is related to preparation of MOF precursors in large scale. The predominant focus of research is on laboratory synthesis, which presents difficulties in ensuring a consistent and sufficient supply of precursors for downstream processes. In order to achieve large-scale production of MOF precursors, several factors must be taken into account, including the yield, solvent toxicity, reaction equipment, and cost. Therefore, it is essential to develop more efficient and sustainable synthetic methods that are cost-effective, easily scalable, environmentally friendly, and dependable. The second challenge is related to the precise control of the derivation process. To achieve the desired physicochemical properties and catalytic performance of MOF-derived composites, it is crucial to have precise control over the technological conditions during the derivation process. To this end, a thorough understanding of the mechanisms involved in wet-chemical reactions and pyrolysis is necessary. Moreover, it is crucial to examine how operating parameters affect the structure and composition characteristics of the resulting composites. This knowledge can then be used to optimize and control the production of high-quality catalyst materials.

Despite these challenges, the rapid development of MOFs, MOF composites, and MOF derivatives has demonstrated exited prospect in various applications for CO_2 conversion. Exploring the potential applications of materials with diverse dimensions in CO_2 conversion is currently a key focus and research trend. With ongoing efforts, it may be possible to realize the practical and industrial applications of MOF composites in diverse fields.

Acknowledgements This work was supported by National Key Research and Development Program of China (No. 2022YFB3804801), the National Natural Science Foundation of China (No. 21736009), Research and Development Program of Zhejiang Provincial Bureau of Science and Technology, China (Grants No. 2021C03169), SINOPEC Science and Technology Development Project from China Petrochemical Corporation (No. 33750000-20-ZC0607-0012 and 33750000-21-ZC0607-0005), and Tongjiang Scholarship from Fujian Quanzhou Government.

References

1. Xu Y, Ramanathan V, Victor DG (2018) Global warming will happen faster than we think. Nature 564:30–32
2. Zhang Z, Zheng Y, Qian L, Luo D, Dou H, Wen G, Yu A, Chen Z (2022) Emerging trends in sustainable CO_2 management materials. Adv Mater 34:2201547
3. Ding M, Flaig RW, Jiang H-L, Yaghi OM (2019) Carbon capture and conversion using metal–organic frameworks and MOF-based materials. Chem Soc Rev 48:2783–2828
4. Kidanemariam A, Lee J, Park J (2019) Recent innovation of metal-organic frameworks for carbon dioxide photocatalytic reduction. Polymers 11:2090

5. Schäffner B, Schäffner F, Verevkin SP, Börner A (2010) Organic carbonates as solvents in synthesis and catalysis. Chem Rev 110:4554–4581
6. Wang W, Wang S, Ma X, Gong J (2011) Recent advances in catalytic hydrogenation of carbon dioxide. Chem Soc Rev 40:3703–3727
7. Chen EX, Qiu M, Zhang YF, Zhu YS, Liu LY, Sun YY, Bu X, Zhang J, Lin Q (2018) Acid and base resistant zirconium polyphenolate-metalloporphyrin scaffolds for efficient CO_2 photoreduction. Adv Mater 30:1704388
8. Wu JX, Hou SZ, Zhang XD, Xu M, Yang HF, Cao PS, Gu Z (2019) Cathodized copper porphyrin metal–organic framework nanosheets for selective formate and acetate production from CO_2 electroreduction. Chem Sci 10:2199–2205
9. Jiang H-L, Xu Q (2011) Porous metal–organic frameworks as platforms for functional applications. Chem Commun 47:3351–3370
10. Yaghi OM, Li H (1995) Hydrothermal synthesis of a metal-organic framework containing large rectangular channels. J Am Chem Soc 117:10401–10402
11. Singh M, Solanki P, Patel P, Mondal A, Neogi S (2019) Highly active ultrasmall Ni nanoparticle embedded inside a robust metal–organic framework: remarkably improved adsorption, selectivity, and solvent-free efficient fixation of CO_2. Inorg Chem 58:8100–8110
12. Li WB, Su PC, Li ZJ, Xu ZH, Wang F, Ou HS, Zhang JH, Zhang G, Zeng E (2017) Ultrathin metal-organic framework membrane production by gel-vapour deposition. Nat Commun 8:406
13. Song X, Wang X, Li Y, Zheng C, Zhang B, Di CA, Li F, Jin C, Mi W, Chen L, Hu W (2020) 2D semiconducting metal–organic framework thin films for organic spin valves. Angew Chem Int Ed 59:1118–1123
14. Cohen SM (2012) Postsynthetic methods for the functionalization of metal-organic frameworks. Chem Rev 112:970–1000
15. Qin L, Xu ZH, Zheng YL, Li C, Mao JW, Zhang G (2020) Confined transformation of organometal-encapsulated MOFs into spinel $CoFe_2O_4$/C nanocubes for low-temperature catalytic oxidation. Adv Func Mater 30:1910257
16. Wang X, Chen Z, Zhao X, Yao T, Chen W, You R, Zhao C, Wu G, Wang J, Huang W, Yang J, Hong X, Wei S, Wu Y, Li Y (2018) Regulation of coordination number over single Co sites: triggering the efficient electroreduction of CO_2. Angew Chem Int Ed 130:1962–1966
17. Wang S, Guan BY, Lou XWD (2018) Construction of $ZnIn_2S_4$–In_2O_3 hierarchical tubular heterostructures for efficient CO_2 photoreduction. J Am Chem Soc 140:5037–5040
18. Wang S, Guan BY, Lu Y, Lou XW (2017) Formation of hierarchical In_2S_3–$CdIn_2S_4$ heterostructured nanotubes for efficient and stable visible light CO_2 reduction. J Am Chem Soc 139:17305–17308
19. Zhang G, Xiao X, Li B, Gu P, Xue H, Pang H (2017) Transition metal oxides with one-dimensional/one-dimensional-analogue nanostructures for advanced supercapacitors. J Mater Chem A 5:8155–8186
20. Bai Y, Liu C, Shan Y, Chen T, Zhao Y, Yu C, Pang H (2022) Metal-organic frameworks nanocomposites with different dimensionalities for energy conversion and storage. Adv Energy Mater 12:2100346
21. Crake A, Christoforidis KC, Gregg A, Moss B, Kafizas A, Petit C (2019) The effect of materials architecture in TiO_2/MOF composites on CO_2 photoreduction and charge transfer. Small 15:1805473
22. Zhao M, Huang Y, Peng Y, Huang Z, Ma Q, Zhang H (2018) Two-dimensional metal–organic framework nanosheets: synthesis and applications. Chem Soc Rev 47:6267–6295
23. Mao JW, Meng Q, Xu ZH, Xu LS, Fan Z, Zhang G (2022) MOF-on-MOF heterojunction-derived Co_3O_4–$CuCo_2O_4$ microflowers for low-temperature catalytic oxidation. Chem Commun 58:13600–13603
24. Sonia FJ, Jangid MK, Ananthoju B, Aslam M, Johari P, Mukhopadhyay A (2017) Understanding the Li-storage in few layers graphene with respect to bulk graphite: experimental, analytical and computational study. J Mater Chem A 5:8662–8679

25. Li PX, Yan XY, Song XM, Li JJ, Ren BH, Gao SY, Cao R (2021) Zirconium-based metal–organic framework particle films for visible-light-driven efficient photoreduction of CO_2. ACS Sustain Chem Eng 9:2319–2325
26. Qiu T, Liang Z, Guo W, Tabassum H, Gao S, Zou R (2020) Metal–organic framework-based materials for energy conversion and storage. ACS Energy Lett 5:520–532
27. Qin L, Ru R, Mao JW, Meng Q, Fan Z, Li X, Zhang G (2020) Assembly of MOFs/polymer hydrogel derived Fe_3O_4-CuO@hollow carbon spheres for photochemical oxidation: freezing replacement for structural adjustment. Appl Catal B Environ 269:118754
28. Mo W, Fan Z, Zhong S, Chen W, Hu L, Zhou H, Zhao W, Lin H, Ge J, Chen J (2023) Embedding plasmonic metal into heterointerface of MOFs-encapsulated semiconductor hollow architecture for boosting CO_2 photoreduction. Small 19:2207705
29. Liu BY, Chen MJ, Yang L, Zhao B, Xia T, Chang GG (2021) Hollow MOF capsule encapsulated amino-functionalized ionic liquid for excellent CO_2 catalytic conversion. Chin J Chem Eng 40:124–130
30. Liu H, Nosheen F, Wang X (2015) Noble metal alloy complex nanostructures: controllable synthesis and their electrochemical property. Chem Soc Rev 44:3056–3078
31. Knez M, Nielsch K, Niinistö L (2007) Synthesis and surface engineering of complex nanostructures by atomic layer deposition. Adv Mater 19:3425–3438
32. Hu H, Guan B, Xia B, Lou XW (2015) Designed formation of $Co_3O_4/NiCo_2O_4$ double-shelled nanocages with enhanced pseudocapacitive and electrocatalytic properties. J Am Chem Soc 137:5590–5595
33. Qin Z, Li H, Yang X, Chen L, Li Y, Shen K (2022) Heterogenizing homogeneous cocatalysts by well-designed hollow MOF-based nanoreactors for efficient and size-selective CO_2 fixation. Appl Catal B Environ 307:121163
34. Wang L, Wan J, Zhao Y, Yang N, Wang D (2019) Hollow multi-shelled structures of Co_3O_4 dodecahedron with unique crystal orientation for enhanced photocatalytic CO_2 reduction. J Am Chem Soc 141:2238–2241
35. Qin L, Liu MM, Wu YJ, Xu ZH, Guo XW, Zhang G (2016) Bioinspired hollow and hierarchically porous MO_X(M = Ti, Si)/carbon microellipsoids supported with Fe_2O_3 for heterogenous photochemical oxidation. Appl Catal B Environ 194:50–60
36. Babu R, Kathalikkattil AC, Roshan R, Tharun J, Kim D-W, Park D-W (2016) Dual-porous metal organic framework for room temperature CO_2 fixation via cyclic carbonate synthesis. Green Chem 18:232–242
37. Liu X, Wang X, Kapteijn F (2020) Water and metal–organic frameworks: from interaction toward utilization. Chem Rev 120:8303–8377
38. Abednatanzi S, Derakhshandeh PG, Depauw H, Coudert FX, Vrielinck H, Van Der Voort P, Leus K (2019) Mixed-metal metal–organic frameworks. Chem Soc Rev 48:2535–2565
39. Chen L, Wang HF, Li C, Xu Q (2020) Bimetallic metal–organic frameworks and their derivatives. Chem Sci 11:5369–5403
40. Mensah-Darkwa K, Ampong DN, Dzikunu P, Souza FM, Kumar A, Gupta RK (2023) Multimetallic organic framework-derived materials for electrocatalytic CO_2 reduction reaction. Fuel 335:127056
41. Dong LZ, Zhang L, Liu J, Huang Q, Lu M, Ji WX, Lan YQ (2020) Stable heterometallic cluster-based organic framework catalysts for artificial photosynthesis. Angew Chem Int Ed 59:2659–2663
42. Dong H, Zhang X, Lu Y, Yang Y, Zhang YP, Tang HL, Zhang F-M, Yang ZD, Sun X, Feng Y (2020) Regulation of metal ions in smart metal-cluster nodes of metal-organic frameworks with open metal sites for improved photocatalytic CO_2 reduction reaction. Appl Catal B: Environ 276:119173
43. Chen S, Xu X, Gao H, Wang J, Li A, Zhang X (2021) Fine-tuning the metal oxo cluster composition and phase structure of Ni/Ti bimetallic MOFs for efficient CO_2 reduction. J Phys Chem C 125:9200–9209
44. Ezugwu CI, Ghosh S, Bera S, Faraldos M, Mosquera ME, Rosal R (2023) Bimetallic metal-organic frameworks for efficient visible-light-driven photocatalytic CO_2 reduction and H_2 generation. Sep Purif Technol 308:122868

45. Liang Z, Wang HY, Zheng H, Zhang W, Cao R (2021) Porphyrin-based frameworks for oxygen electrocatalysis and catalytic reduction of carbon dioxide. Chem Soc Rev 50:2540–2581
46. Wang YR, Huang Q, He CT, Chen Y, Liu J, Shen FC, Lan YQ (2018) Oriented electron transmission in polyoxometalate-metalloporphyrin organic framework for highly selective electroreduction of CO_2. Nat Commun 9:4466
47. Liu J, Fan YZ, Li X, Wei Z, Xu YW, Zhang L, Su CY (2018) A porous rhodium. (III)-porphyrin metal-organic framework as an efficient and selective photocatalyst for CO_2 reduction. Appl Catal B Environ 231:173–181
48. Wang L, Jin P, Huang J, She H, Wang Q (2019) Integration of copper (II)-porphyrin zirconium metal–organic framework and titanium dioxide to construct Z-scheme system for highly improved photocatalytic CO_2 reduction. ACS Sustain Chem Eng 7:15660–15670
49. Smith MR, Martin CB, Arumuganainar S, Gilman A, Koel BE, Sarazen ML (2023) Mechanistic elucidations of highly dispersed metalloporphyrin metal-organic framework catalysts for CO_2 electroreduction. Angew Chem Int Ed 62:e202218208
50. Qin MF, Mu QQ, Bao SS, Liang X, Peng Y, Zheng LM (2021) Chemically exfoliated semiconducting bimetallic porphyrinylphosphonate metal–organic layers for photocatalytic CO_2 reduction under visible light. ACS Appl Energy Mater 4:4319–4326
51. Chen EX, Qiu M, Zhang YF, He L, Sun YY, Zheng HL, Wu X, Zhang J, Lin Q (2022) Energy band alignment and redox-active sites in metalloporphyrin-spaced metal-catechol frameworks for enhanced CO_2 photoreduction. Angew Chem Int Ed 61:202111622
52. Zhong H, Ghorbani-Asl M, Ly KH, Zhang J, Ge J, Wang M, Liao ZQ, Makarov D, Zschech E, Brunner E, Weidinger IM, Zhang J, Krasheninnikov AV, Kaskel S, Dong RH, Feng XL (2020) Synergistic electroreduction of carbon dioxide to carbon monoxide on bimetallic layered conjugated metal-organic frameworks. Nat Commun 11:1409
53. Viciano-Chumillas M, Liu X, Leyva-Pérez A, Armentano D, Ferrando-Soria J, Pardo E (2022) Mixed component metal-organic frameworks: heterogeneity and complexity at the service of application performances. Coord Chem Rev 451:214273
54. Dhakshinamoorthy A, Asiri MA, Garcia H (2016) Mixed-metal or mixed-linker metal organic frameworks as heterogeneous catalysts. Catal Sci Technol 6:5238–5261
55. Lin S, Diercks CS, Zhang YB, Kornienko N, Nichols EM, Zhao Y, Paris AR, Kim D, Yang P, Yaghi OM, Chang CJ (2015) Covalent organic frameworks comprising cobalt porphyrins for catalytic CO_2 reduction in water. Science 349:1208–1213
56. Hod I, Sampson MD, Deria P, Kubiak CP, Farha OK, Hupp JT (2015) Fe-porphyrin-based metal–organic framework films as high-surface concentration, heterogeneous catalysts for electrochemical reduction of CO_2. ACS Catal 5:6302–6309
57. Kong XJ, He T, Zhou J, Zhao C, Li TC, Wu XQ, Wang K, Li JR (2021) In situ porphyrin substitution in a Zr(IV)-MOF for stability enhancement and photocatalytic CO_2 reduction. Small 17:2005357
58. Xu G, Zhang H, Wei J, Zhang HX, Wu X, Li Y, Li C, Zhang J, Ye J (2018) Integrating the g-C_3N_4 nanosheet with b–h bonding decorated metal–organic framework for CO_2 activation and photoreduction. ACS Nano 12:5333–5340
59. Stanley PM, Thomas C, Thyrhaug E, Urstoeger A, Schuster M, Hauer J, Rieger B, Warnan J, Fischer RA (2021) Entrapped molecular photocatalyst and photosensitizer in metal–organic framework nanoreactors for enhanced solar CO_2 reduction. ACS Catal 11:871–882
60. Stanley PM, Haimerl J, Thomas C, Urstoeger A, Schuster M, Shustova NB, Casini A, Rieger B, Warnan J, Fischer RA (2021) Host–guest interactions in a metal–organic framework isoreticular series for molecular photocatalytic CO_2 reduction. Angew Chem Int Ed 60:17854–17860
61. Ye L, Ying Y, Sun D, Zhang Z, Fei L, Wen Z, Qiao J, Huang H (2020) Highly efficient porous carbon electrocatalyst with controllable N-species content for selective CO_2 reduction. Angew Chem Int Ed 59:3244–3251
62. Zheng Y, Cheng P, Xu J, Han J, Wang D, Hao C, Rezvani Alanagh H, Long C, Shi X, Tang Z (2019) MOF-derived nitrogen-doped nanoporous carbon for electroreduction of CO_2 to CO the calcining temperature effect and the mechanism. Nanoscale 11:4911–4917

63. Ling LL, Jiao L, Liu X, Dong Y, Yang W, Zhang H, Ye B, Chen J, Jiang HL (2022) Potassium-assisted fabrication of intrinsic defects in porous carbons for electrocatalytic CO_2 reduction. Adv Mater 34:2205933
64. Li WB, Zhang YF, Zhang CY, Meng Q, Xu ZH, Su PC, Li QB, Shen C, Fan Z, Qin L, Zhang G (2016) Transformation of metal-organic frameworks for molecular sieving membranes. Nat Commun 7:11315
65. Jiang Y, Yu Y, Zhang X, Weinert M, Song X, Ai J, Han L, Fei H (2021) N-heterocyclic carbene-stabilized ultrasmall gold nanoclusters in a metal-organic framework for photocatalytic CO_2 reduction. Angew Chem Int Ed 60:17388–17393
66. Kim H, Moon HS, Sohail M, Yoon YN, Shah SFA, Yim K, Moon JH, Park YC (2019) Synthesis of cyclic carbonate by CO_2 fixation to epoxides using interpenetrated MOF-5/*n*-Bu_4NBr. J Mater Sci 54:11796–11803
67. Rani P, Husain A, Bhasin KK, Kumar G (2022) Metal-organic framework-based selective molecular recognition of organic amines and fixation of CO_2 into cyclic carbonates. Inorg Chem 61:6977–6994
68. Zhou HF, Liu B, Hou L, Zhang WY, Wang YY (2018) Rational construction of a stable Zn_4O-based MOF for highly efficient CO_2 capture and conversion. Chem Commun 54:456–459
69. Zhang S, Li L, Zhao S, Sun Z, Luo J (2015) Construction of interpenetrated ruthenium metal–organic frameworks as stable photocatalysts for CO_2 reduction. Inorg Chem 54:8375–8379
70. Ma FX, Mi FQ, Sun MJ, Huang T, Wang ZA, Zhang T, Cao R (2022) A highly stable Zn_9-pyrazolate metal-organic framework with metallosalen ligands as a carbon dioxide cycloaddition catalyst. Inorg Chem Front 9:1812–1818
71. Huang ZW, Hu KQ, Mei L, Wang CZ, Chen YM, Wu WS, Chai ZF, Shi WQ (2020) Potassium ions induced framework interpenetration for enhancing the stability of uranium-based porphyrin MOF with visible-light-driven photocatalytic activity. Inorg Chem 60:651–659
72. Wang J, Dai Z, Wang L, Zhang D, Wang Y, Li J, Zhou F, Huang J (2023) Z-Scheme heterojunction of porphyrin-based core-shell Zr-MOF@ Pro-COF-Br hybrid materials for efficient visible-light-driven CO_2 reduction. J Mater Chem A 11:2023–2030
73. Li Y, Wen L, Tan T, Lv Y (2019) Sequential co-immobilization of enzymes in metal-organic frameworks for efficient biocatalytic conversion of adsorbed CO_2 to formate. Front Bioeng Biotechnol 7:394
74. Bao S, Li J, Guan B, Jia M, Terasaki O, Yu J (2020) A green selective water-etching approach to MOF@ mesoporous SiO_2 yolk-shell nanoreactors with enhanced catalytic stabilities. Matter 3:498–508
75. Geng C, Sun Y, Zhang Z, Qiao Z, Zhong C (2023) Mixed matrix metal-organic framework membranes for efficient CO_2/N_2 separation under humid conditions. AIChE J 69:e18025
76. Yang X, Li QX, Chi SY, Li HF, Huang YB, Cao R (2022) Hydrophobic perfluoroalkane modified metal-organic frameworks for the enhanced electrocatalytic reduction of CO_2. SmartMat 3:163–172
77. Hu P, Wang H, Xiong C, Liu H, Han J, Zhou J, Zhao Z, Wang Y, Ji H (2021) Probing the node chemistry of a metal-organic framework to achieve ultrahigh hydrophobicity and highly efficient CO_2/CH_4 separation. ACS Sustain Chem Eng 9:15897–15907
78. Choe JH, Kim H, Kang M, Yun H, Kim SY, Lee SM, Hong CS (2022) Functionalization of diamine-appended MOF-based adsorbents by ring opening of epoxide: long-term stability and CO_2 recyclability under humid conditions. J Am Chem Soc 144:10309–10319
79. Chen RY, Tian D, Li YW, Lv YB, Sun HW, Chang Z, Bu XH (2015) A new anionic metal–organic framework showing tunable emission by lanthanide (III) doping and highly selective CO_2 adsorption properties. RSC Adv 5:24655–24660
80. Li YW, Li JR, Wang LF, Zhou BY, Chen Q, Bu XH (2013) Microporous metal–organic frameworks with open metal sites as sorbents for selective gas adsorption and fluorescence sensors for metal ions. J Mater Chem A 1:495–499
81. Liu XT, Jia YY, Zhang YH, Ren GJ, Feng R, Zhang SY, Zaworotko MJ, Bu XH (2016) A new Co(ii) metal–organic framework with enhanced CO_2 adsorption and separation performance. Inorg Chem Front 3:1510–1515

82. Liang J, Xie YQ, Wang XS, Wang Q, Liu TT, Huang YB, Cao R (2018) An imidazolium-functionalized mesoporous cationic metal–organic framework for cooperative CO_2 fixation into cyclic carbonate. Chem Commun 54:342–345
83. Moon J, Jeong M, Nam H, Ju J, Moon JH, Jung HM, Lee S (2008) One-pot synthesis of diarylalkynes using palladium-catalyzed Sonogashira reaction and decarboxylative coupling of sp carbon and sp^2 carbon. Org Lett 10:945–948
84. Correa A, Martin R (2009) Metal-catalyzed carboxylation of organometallic reagents with carbon dioxide. Angew Chem Int Ed 48:6201–6204
85. Liu XH, Ma JG, Niu Z, Yang GM, Cheng P (2015) An efficient nanoscale heterogeneous catalyst for the capture and conversion of carbon dioxide at ambient pressure. Angew Chem Int Ed 127:1002–1005
86. Kattel S, Liu P, Chen JG (2017) Tuning selectivity of CO_2 hydrogenation reactions at the metal/oxide interface. J Am Chem Soc 139:9739–9754
87. Yao B, Xiao T, Makgae OA, Jie X, Gonzalez-Cortes S, Guan S, Edwards PP (2020) Transforming carbon dioxide into jet fuel using an organic combustion-synthesized Fe–Mn–K catalyst. Nat Commun 11:6395
88. Kattel S, Yan B, Chen JGG, Liu P (2016) CO_2 hydrogenation on Pt, PtiSiO($_2$) and Pt/TiO_2: importance of synergy between Pt and oxide support. J Catal 343:115–126
89. Xu H, Li Y, Luo X, Xu Z, Ge J (2017) Monodispersed gold nanoparticles supported on a zirconium-based porous metal–organic framework and their high catalytic ability for the reverse water–gas shift reaction. Chem Commun 53:7953–7956
90. Navarro-Jaén S, Virginie M, Bonin J, Robert M, Wojcieszak R, Khodakov AY (2021) Highlights and challenges in the selective reduction of carbon dioxide to methanol. Nat Rev Chem 5:564–579
91. Olah GA (2005) Beyond oil and gas: the methanol economy. Angew Chem Int Ed 44:2636–2639
92. Zhong J, Yang X, Wu Z, Liang B, Huang Y, Zhang T (2020) State of the art and perspectives in heterogeneous catalysis of CO_2 hydrogenation to methanol. Chem Soc Rev 49:1385–1413
93. Kattel S, Ramírez PJ, Chen JG, Rodriguez JA, Liu P (2017) Active sites for CO_2 hydrogenation to methanol on Cu/ZnO catalysts. Science 355:1296–1299
94. Li H, Qiu C, Ren S, Dong Q, Zhang S, Zhou F, Liang X, Wang J, Li S, Yu M (2020) Na^+-gated water-conducting nanochannels for boosting CO_2 conversion to liquid fuels. Science 367:667–671
95. Yang H, Zhang C, Gao P, Wang H, Li X, Zhong L, Wei W, Sun Y (2017) A review of the catalytic hydrogenation of carbon dioxide into value-added hydrocarbons. Catal Sci Technol 7:4580–4598
96. Wang C, Xie Z, deKrafft KE, Lin W (2011) Doping metal–organic frameworks for water oxidation, carbon dioxide reduction, and organic photocatalysis. J Am Chem Soc 133:13445–13454
97. Huang R, Peng Y, Wang C, Shi Z, Lin W (2016) A rhenium-functionalized metal–organic framework as a single-site catalyst for photochemical reduction of carbon dioxide. Eur J Inorg Chem 2016:4358–4362
98. Feng X, Pi Y, Song Y, Brzezinski C, Xu Z, Li Z, Lin W (2020) Metal–organic frameworks significantly enhance photocatalytic hydrogen evolution and CO_2 reduction with earth-abundant copper photosensitizers. J Am Chem Soc 142:690–695
99. Xiao JD, Jiang HL (2019) Metal–organic frameworks for photocatalysis and photothermal catalysis. Acc Chem Res 52:356–366
100. Choi KM, Kim D, Rungtaweevoranit B, Trickett CA, Barmanbek JTD, Alshammari AS, Yang P, Yaghi OM (2017) Plasmon-enhanced photocatalytic CO_2 conversion within metal–organic frameworks under visible light. J Am Chem Soc 139:356–362
101. Chen M, Han L, Zhou J, Sun C, Hu C, Wang X, Su Z (2018) Photoreduction of carbon dioxide under visible light by ultra-small Ag nanoparticles doped into Co-ZIF-9. Nanotechnology 29:284003

102. Xu HQ, Hu J, Wang D, Li Z, Zhang Q, Luo Y, Yu SH, Jiang HL (2015) Visible-light photoreduction of CO_2 in a metal–organic framework: boosting electron–hole separation via electron trap states. J Am Chem Soc 137:13440–13443
103. Jiang X, Wu H, Chang S, Si R, Miao S, Huang W, Li Y, Wang G, Bao X (2017) Boosting CO_2 electroreduction over layered zeolitic imidazolate frameworks decorated with Ag_2O nanoparticles. J Mater Chem A 5:19371–19377
104. Yi JD, Si DH, Xie R, Yin Q, Zhang MD, Wu Q, Cao R (2021) Conductive two-dimensional phthalocyanine-based metal-organic framework nanosheets for efficient electroreduction of CO_2. Angew Chem Int Ed 60:17108–17114
105. Wen CF, Zhou M, Liu PF, Liu Y, Wu X, Mao F, Dai S, Xu B, Wang XL, Jiang Z, Hu P, Yang S, Wang HF, Yang HG (2022) Highly ethylene-selective electrocatalytic CO_2 reduction enabled by isolated cu−s motifs in metal–organic framework based precatalysts. Angew Chem Int Ed 61:e202111700
106. Meng DL, Zhang MD, Si DH, Mao MJ, Hou Y, Huang YB, Cao R (2021) Highly selective tandem electroreduction of CO_2 to ethylene over atomically isolated nickel-nitrogen site/copper nanoparticle catalysts. Angew Chem Int Ed 60:25485–25492
107. Wu N (2018) Plasmonic metal–semiconductor photocatalysts and photoelectrochemical cells: a review. Nanoscale 10:2679–2696
108. Iqbal MZ, Siddique S (2018) Recent progress in efficiency of hydrogen evolution process based photoelectrochemical cell. Int J Hydrogen Energy 43:21502–21523
109. Deng X, Li R, Wu S, Wang L, Hu J, Ma J, Jiang W, Zhang N, Zheng X, Gao C, Wang L, Zhang Q, Zhu J, Xiong Y (2019) Metal–organic framework coating enhances the performance of Cu_2O in photoelectrochemical CO_2 reduction. J Am Chem Soc 141:10924–10929
110. Chen S, Hai G, Gao H, Chen X, Li A, Zhang X, Dong W (2021) Modulation of the charge transfer behavior of Ni(II)-doped NH2-MIL-125(Ti): Regulation of Ni ions content and enhanced photocatalytic CO_2 reduction performance. Chem Eng J 406:126886
111. Silva BC, Irikura K, Flor JBS, dos Santos RMM, Lachgar A, Frem RCG, Zanoni MVB (2020) Electrochemical preparation of Cu/Cu_2O–Cu(BDC) metal-organic framework electrodes for photoelectrocatalytic reduction of CO_2. J CO_2 Util 42:101299
112. Stülp S, Cardoso JC, de Brito JF, Flor JBS, Frem RCG, Sayão FA, Zanoni MVB (2017) An artificial photosynthesis system based on Ti/TiO_2 coated with cu(ii) aspirinate complex for CO_2 reduction to methanol. Electrocatalysis 8:279–287
113. Cardoso JC, Stulp S, de Brito JF, Flor JBS, Frem RCG, Zanoni MVB (2018) MOFs based on ZIF-8 deposited on TiO_2 nanotubes increase the surface adsorption of CO_2 and its photoelectrocatalytic reduction to alcohols in aqueous media. Appl Catal B Environ 225:563–573

Chapter 2
Photo-Thermo Catalytic Conversion of CO_2: What, Why, How and Future Perspectives

Zhou-jun Wang, Hui Song, Guiming Xie, Xiaorui Wang, and Jinhua Ye

Abstract Photo-thermo catalysis serves as a novel strategy for CO_2 conversion, wherein enhanced activity, manipulated selectivity and/or improved stability can be achieved with synergisms between photocatalysis and thermocatalysis. To clarify fundamental issues and promote innovative development in this new field, comprehensive summarization and up-to-date perspective have been presented herein. Specifically, the conception, principles and operating mechanisms of photo-thermo catalysis have been systematically discussed, which has been followed by thorough analysis of representative examples in photo-thermo CO_2 conversion. Moreover, challenges and perspectives have been proposed as implications for future development. Overall, this book chapter is expected to clear up theoretical basis and set up practical guidelines for photo-thermo catalytic conversion of CO_2.

Contributions All authors contributed to this book chapter. Z. W. drafted the manuscript. H. S., G. X. and X. W. revised the content. Z. W. and J. Y. conceived the content and finalized the manuscript.

Z. Wang (✉) · G. Xie · X. Wang
State Key Laboratory of Chemical Resource Engineering, Beijing Key Laboratory of Energy Environmental Catalysis, Beijing University of Chemical Technology, Beijing 100029, P. R. China
e-mail: wangzj@mail.buct.edu.cn

H. Song
TJU-NIMS International Collaboration Laboratory, School of Material Science and Engineering, Tianjin University, Tianjin 300072, P. R. China

J. Ye (✉)
International Center for Materials Nanoarchitectonics (WPI-MANA), National Institute for Materials Science (NIMS), 1-1 Namiki, Tsukuba 305-0044, Ibaraki, Japan
e-mail: Jinhua.YE@nims.go.jp

G. Zhang et al. (eds.), *Advances in CO_2 Utilization*, Green Chemistry and Sustainable Technology, https://doi.org/10.1007/978-981-99-8822-8_2

2.1 Introduction

To achieve carbon neutrality by the middle of this century, feasible carbon-negative technologies should be developed [1, 2]. Notably, catalytic conversion of CO_2 into value-added products serves as a promising one which not only lowers CO_2 concentration but also produces chemicals/fuels [3, 4]. Due to the high bond energy of C=O (750 kJ mol^{-1}) in CO_2 molecule, adequate energy should be supplied to initiate CO_2 conversion [5]. In terms of energy input (heat, light and electricity), CO_2 conversion is implemented via thermo-, photo- and electro- catalytic reactions, respectively [6]. At present, thermocatalytic CO_2 conversion is typically obsessed by high energy consumption and low selectivity, whereas photocatalytic and electrocatalytic CO_2 conversion are still limited by low activity and small production scale. To overcome drawbacks of single energy input, the integration of two energy input has been proposed as a potential solution [1, 7]. Recent pioneering works from other scientists and our group have demonstrated that combining heat and light in catalytic conversion of CO_2 would efficiently enhance activity, manipulate selectivity and/or improve stability [1, 8–10]. In the meantime, several critical reviews have been published with different emphasis [1, 11, 12]. Although useful insights have been provided by these reviews, several fundamental questions still remain to be clarified, such as "What is photo-thermo catalysis?", "Why does photo-thermo catalysis work?" and "How does photo-thermo catalysis work?". Therefore, systematic summarization and up-to-date outlook are still demanding.

Herein, fundamentals and perspectives of photo-thermo catalytic conversion of CO_2 will be reviewed. Initially, the conception of photo-thermo catalysis will be defined, which presents the scope of this book chapter as well. Subsequently, principles of photo-thermo catalysis will be discussed, which demonstrates the superiority of this technology. After that, operating mechanisms of photo-thermo catalysis will be expounded, which shows how photo-thermo catalysis works in CO_2 conversion. Finally, challenges and opportunities in this field will be proposed, which delivers useful implications for future development.

2.2 What Is Photo-Thermo Catalysis?

As a newly-developed field with a distinct feature of interdisciplinarity, the conception of catalytic processes involving heat and light appears relatively confusing. This confusion can be clearly noticed on the diverse nomenclature in the literature, wherein most frequently used terms include photothermal, photo-thermal, photo-thermo and thermo-photo catalysis [1, 11–19]. In essential, the primary difference among these terms lies in the function of light. In photothermal or photo-thermal catalysis, the dominant role of light is to provide thermal energy via the light-to-heat effect (Fig. 2.1a) [14]. Since light merely affords elevated reaction temperature but not involves in catalytic reactions, the catalytic mechanism therein is identical to that

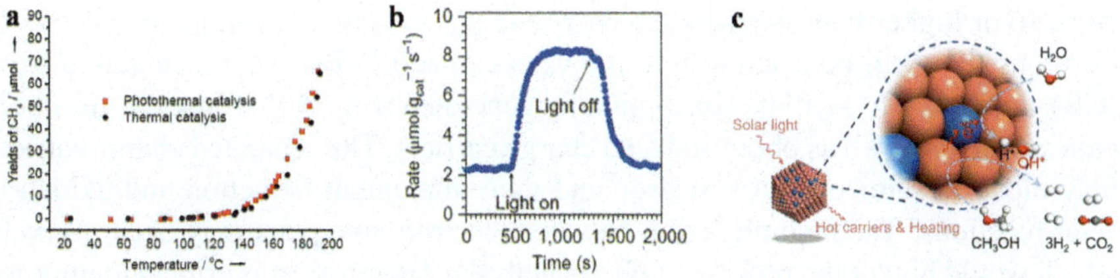

Fig. 2.1 **Three representative cases involving both light and heat**. **a** Photothermal catalysis. Adapted from Ref. [14]. **b** Photo-thermo catalysis. Adapted from Ref. [16]. **c** Photo-photothermal catalysis. Adapted from Ref. [18]

in traditional thermocatalytic reactions [14, 15]. In photo-thermo or thermo-photo catalysis, the light-to-heat effect is insignificant, and light participates in the reaction via mediating critical intermediates/active sites (Fig. 2.1b) [16]. The introduction of light as additional energy input leads to exceptional catalytic performances, which is not observed in individual photocatalysis or thermocatalysis [1, 16, 17]. Definitely, photo-thermo catalysis shows advantage over photothermal catalysis because effective coupling between solar energy and thermal energy provides brand-new opportunities for catalytic reactions. Besides, there are also some processes wherein light exhibits both thermal and non-thermal effects (Fig. 2.1c) [18]. Notably, no new function mechanisms for light occur, and the contribution from each effect is difficult to be disentangled [18, 19]; therefore, this kind of processes is not so attractive for fundamental research. To reflect the role of light in various circumstances, above-mentioned three cases are recommended to be termed as photothermal catalysis, photo-thermo catalysis and photo-photothermal catalysis, respectively. In this book chapter, only photo-thermo catalysis will be covered, wherein heat and light are synergized and the interference from the light-to-heat effect is excluded.

2.3 Why Does Photo-Thermo Catalysis Work?

To understand why photo-thermo catalysis works better than individual photocatalysis or thermocatalysis, principles of photocatalysis and thermocatalysis will be briefly revisited, and then synergies between photocatalysis and thermocatalysis will be thoroughly analyzed.

2.3.1 Principles of Photocatalysis and Thermocatalysis

Principles of photocatalysis and thermocatalysis are relatively clear, which have been exhaustively summarized in reviews and books [20–25]. Photocatalysis normally proceeds over semiconductors (Fig. 2.2a) [1, 20–22]. After photons with energy equal to or higher than the bandgap energy (E_{BG}) of semiconductors are absorbed, electrons (e^-) will be excited from the valence band (VB) to the conduction band (CB), and holes (h^+) will be concomitantly generated in the VB. These electron/hole pairs are also known as photo-induced charge carriers. The separated charge carriers may move to semiconductor surface and then implement reduction and oxidation semi-reactions. The recombination between electron/hole pairs may occur as well, which would hinder the process of photocatalysis. Therefore, it is critical to improve the separation and mobility capability of charge carriers.

Thermocatalysis consists of both physical and chemical reaction steps (Fig. 2.2b) [23–25]. Seven reaction steps exist in a typical gas reaction on a porous catalyst: film diffusion of reactants (step 1), pore diffusion of reactants (step 2), adsorption of reactants (step 3), surface reaction (step 4), desorption of products (step 5), pore diffusion of products (step 6) and film diffusion of products (step 7). Reaction steps 4–6 are closely related to chemical reactions, whereas the rest four steps are mainly associated with physical processes. The overall reaction rate is normally determined by the slowest reaction step, i.e., the rate-limiting step. Optimum strength in adsorption of

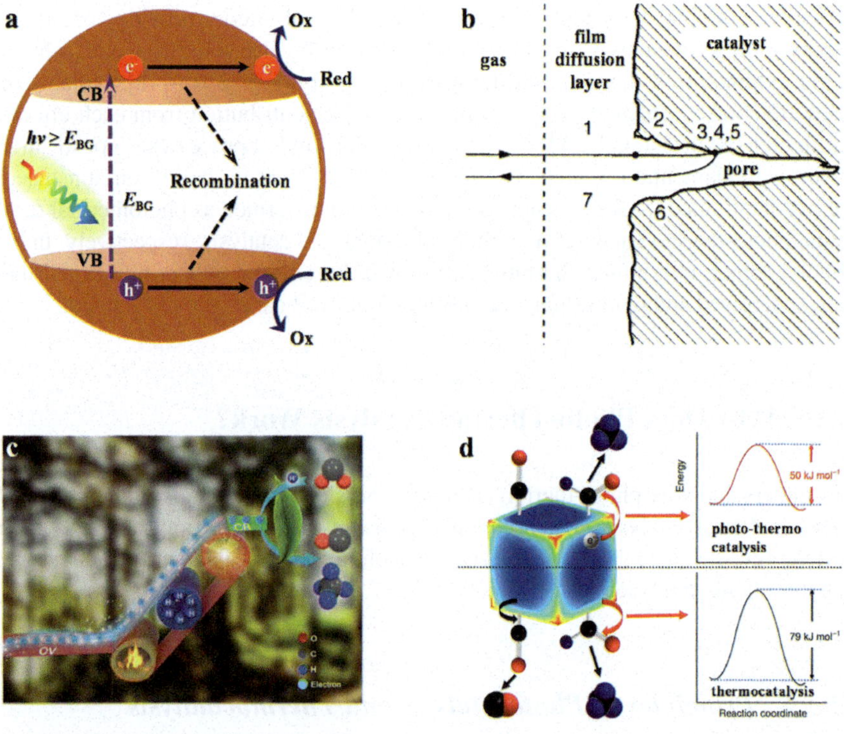

Fig. 2.2 Principles of photocatalysis and thermocatalysis as well as synergies between photocatalysis and thermocatalysis. **a** Principles of photocatalysis. **b** Principles of thermocatalysis. **c** Benefit from heat for photocatalysis. Adapted from Ref. [28]. **d** Benefit from light for thermocatalysis. Adapted from Ref. [9]

reactants and binding of intermediates/products is generally conducive to efficient thermocatalysis. Moreover, free valences or electron defects on catalyst surface are often identified as active sites.

2.3.2 Synergies Between Photocatalysis and Thermocatalysis

On the basis of principles of individual photocatalysis or thermocatalysis, synergies in photo-thermo catalysis will be expounded from two aspects, i.e., benefit from heat for photocatalysis and benefit from light for thermocatalysis.

Benefit from heat for photocatalysis is closely related to the excitation, separation and migration of photo-induced charge carriers [26–28]. For example, Sá's group observed more efficient hot electron harvesting in an Au/TiO_2 hybrid system at higher operation temperature, which was explained by the enhancement in hot carrier generation due to phonon coupling [27]. Zhang's group reported a thermal-assisted electron transfer in defected TiO_2 (Fig. 2.2c), which contributed to exceptional activity in photo-thermo CO_2 reduction and acetaldehyde removal [28]. In addition, heat may promote semi-reactions on semiconductor surface by accelerating mass transfer and/or optimizing chemical adsorption [29]. Moreover, heat may facilitate semi-reactions by modifying redox potentials relative to the CB and VB of semiconductors [30].

Benefit from light for thermocatalysis mainly involves in chemical reactions on catalyst surface, wherein light exerts influence on microsctructures of catalysts or elemental steps of reactions [26, 31, 32]. For instance, the seminal work from Linic's group documented that visible light illumination would induce reduction of Cu oxides and thus increase selectivity in propylene epoxidation [31]. Everitt's and Liu's groups found selectivity change over Rh catalyzed CO_2 hydrogenation after introducing ultraviolet light, which was explained by selective activation of CHO intermediates by photo-excited hot electrons (Fig. 2.2d) [9]. Photo-induced charge carriers may participate in adsorption of reactants, surface reaction and desorption of products by tuning bond activation, changing reaction paths and accelerating species removal, respectively, which ultimately boosts catalytic performance of thermocatalysis [32–34].

2.4 How Does Photo-Thermo Catalysis Work in CO_2 Conversion?

To elaborate how photo-thermo catalysis works in CO_2 conversion, typical routes for photo-thermo catalysis will be first introduced, and then relevant reactions for CO_2 conversion will be covered. Finally, representative examples for photo-thermo CO_2 conversion will be analyzed.

2.4.1 Routes for Photo-Thermo Catalysis

In terms of operating mechanisms over various catalytic materials, routes for photo-thermo catalysis can be categorized as plasmonic and non-plasmonic ones [1, 17, 35]. In plasmonic route, synergisms rely on the localized surface plasmon resonance (LSPR) effect [1, 36]. LSPR is an optical phenomenon that a coherent oscillation of free electrons in plasmonic nanoparticles is induced by light with wavelength longer than the particle size [37]. The excitation of LSPR results in dephasing of free electrons and generation of energetic charge carriers (electrons and holes). The resultant energetic electrons (also called hot electrons) participate in photo-thermo catalysis via electron transfer processes, which can not only mediate adsorbates on metal surface but also modify semiconductors interfaced with the metal (Fig. 2.3) [38]. There are two pathways for electron transfer processes. One is the indirect pathway wherein hot electrons are generated on plasmonic metals before being transferred to adjacent adsorbates or semiconductors. The other is the direct pathway wherein hot electrons are straightway transferred to adsorbates or semiconductors through an additional damping, chemical interface damping (CID). In comparison, the direct pathway is recognized to be more capable than the indirect pathway in maneuvering catalytic reactions because energy decay of hot electrons on the metal surface is avoided in the former one [32]. Besides, thermalization, back-transfer and relaxation processes occur together with electron transfer processes; therefore, care should be taken in disentangling thermal and non-thermal effect in plasmonic route. Several experimental signatures have been established to confirm the dominant contribution of non-thermal effect, such as reduced activation energy, linear or superlinear dependence of reaction rate on light intensity, enlarged kinetic isotope effect (KIE) and modified selectivity patterns [12, 15]. LSPR excitation is sensitive to size, shape and composition of plasmonic nanostructures that include coinage metals (Au, Ag and Cu), alkali metals, aluminum and some semiconductors [39, 40]. Overall, abundant LSPR-excited hot electrons, efficient electron transfer processes and tunable optical properties offer vast opportunities for modulating photo-thermo catalysis in plasmonic route.

Compared with plasmonic route, non-plasmonic route is less studied in photo-thermo catalysis [1, 17, 41]. In general, non-plasmonic route is realized through photo-excited electron/hole pairs over semiconductor nanomaterials [1, 17, 42]. On one hand, heat may facilitate the excitation, separation and migration of electron/hole pairs in photocatalysts, promoting photocatalysis [28]. On the other hand, light may modify vital microstructures of thermocatalysts [e.g., surface defects and frustrated Lewis pair (FLP)], promoting thermocatalysis [43]. Altogether, semiconductors are normally involved, and photo-excited electron/hole pairs are responsible for synergisms in non-plasmonic route.

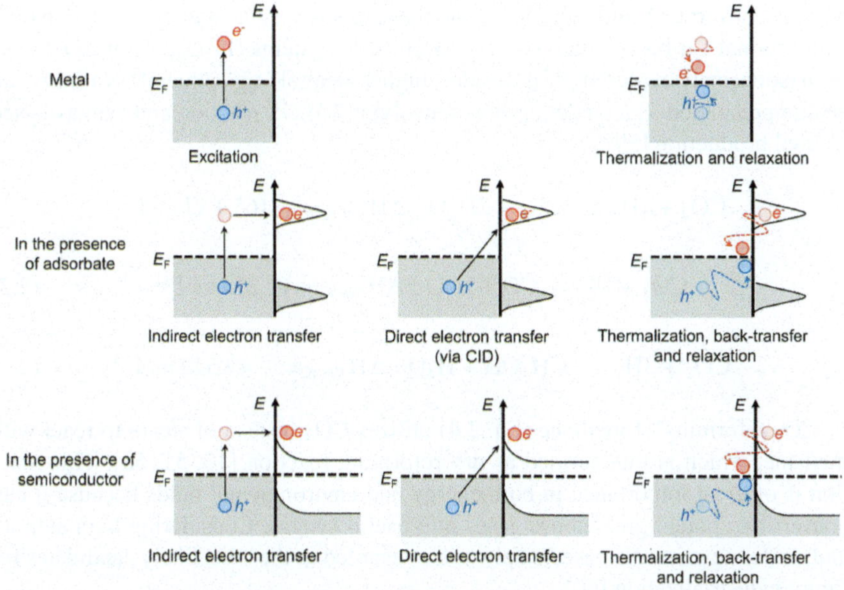

Fig. 2.3 LSPR-induced hot electron generation and electron transfer/back-transfer processes in metal, metal/adsorbate and metal/semiconductor systems. E_F denotes Fermi level of plasmonic metals. The right descriptor indicates the highest occupied molecular orbitals (HOMOs) and the lowest unoccupied molecular orbitals (LUMOs) for adsorbates, or the CB and VB for semiconductors. The gray parts represent electronic state population. Adapted from Ref. [38]

2.4.2 *Reactions for Catalytic Conversion of CO_2*

Depending on whether carbon keeps +4 oxidation state, CO_2 conversion can be categorized as non-reduction and reduction processes, corresponding to products of chemicals (e.g., urea or salicylic acid) and fuels (e.g., CH_4, CO and CH_3OH), respectively [44]. Due to the high reaction barrier, a catalyst is mandatory for reduction processes. Moreover, high energy substances should be employed to reduce the inert CO_2 [45]. Specifically, H_2, CH_4 and H_2O are primary reductants for photo-thermo CO_2 conversion, which proceed via reactions of hydrogenation, reforming and artificial photosynthesis [8–10, 46–54]. It is worth mentioning that all these reactions have not been commercialized yet.

Using "green" H_2 from renewable source, CO_2 hydrogenation reactions including CO_2 methanation, reverse water–gas shift (RWGS) and methanol/higher alcohol synthesis have been regarded as promising routes for CO_2 conversion [8, 9, 46–50, 55]. CO_2 methanation (Eq. 2.1) has potential applications in renewable energy storage based on power-to-gas (P2G) technology, wherein Ni- and Ru-based catalysts are most widely investigated [56]. RWGS reaction (Eq. 2.2) produces valuable CO for downstream syngas (CO/H_2) transformations, wherein Cu- and Fe-based catalysts are most competitive systems [57]. Methanol synthesis reaction (Eq. 2.3) can directly convert CO_2 into liquid fuel and thus attracts special attention, wherein both conventional Cu-based catalysts and novel oxide systems (In_2O_3, ZnO/ZrO_2) have been extensively explored [58]. Besides, higher alcohol (e.g., ethanol) synthesis can be comprehended as a tandem process involving RWGS reaction and syngas-based alcohol production [59].

$$CO_2 + 4H_2 \rightleftharpoons CH_4 + 2H_2O \quad \Delta H_{298\,K} = -165.1\ \text{kJ mol}^{-1} \tag{2.1}$$

$$CO_2 + H_2 \rightleftharpoons CO + H_2O \quad \Delta H_{298\,K} = 41.2\ \text{kJ mol}^{-1} \tag{2.2}$$

$$CO_2 + 3H_2 \rightleftharpoons CH_3OH + H_2O \quad \Delta H_{298\,K} = -49.5\ \text{kJ mol}^{-1} \tag{2.3}$$

CO_2 reforming of methane (Eq. 2.4) utilizes CO_2 instead of steam to react with methane, which is thus termed as dry reforming reaction [10, 51, 52]. This reaction is of great importance in both energy and environmental fields because it can convert two potent greenhouse gases into useful syngas. Considering high activity and moderate price, Ni-based catalysts are regarded as the most likely candidates for future industrialization [60].

$$CO_2 + CH_4 \rightleftharpoons 2CO + 2H_2 \quad \Delta H_{298\,K} = 247.3\ \text{kJ mol}^{-1} \tag{2.4}$$

Artificial photosynthesis (Eq. 2.5) refers to reaction of CO_2 with H_2O for the production of solar fuels (including CH_4, methanol and other hydrocarbons) [53, 54]. This reaction mimics the photosynthesis process in green plants that converts CO_2 and water into O_2 and organic compounds under light illumination. Oxides and perovskites have been reported as possible catalysts [61].

$$CO_2 + H_2O \rightleftharpoons CH_4 + CH_3OH + \cdots \tag{2.5}$$

2.4.3 *Representative Examples for Photo-Thermo Catalytic Conversion of CO_2*

In terms of catalytic outcome, the synergistic effect in representative examples for photo-thermo CO_2 conversion will be grouped as enhanced activity, manipulated selectivity and improved stability.

2.4.3.1 Enhanced Activity

Enhanced activity is the most prevailing benefit reported in photo-thermo catalytic conversion of CO_2. Many research groups have observed such a promotion via plasmonic route, among which our group has conducted pioneering work. Various plasmonic nanostructures including AuRh/SBA-15, PdAu/Al_2O_3, PtAu/SiO_2, Fe@C, Ni@SiO_2, Ni/Al_2O_3 and Cu/ZnO have been explored, and several critical reviews from different perspectives have been published by us in past decade [8, 10, 51, 62–65]. One of representative works is photo-thermo CO_2 hydrogenation to methanol over plasmonic Cu/ZnO catalysts [8]. Photo-thermo catalysis enhanced methanol production rate by 54% relative to thermocatalysis, and this enhancement was reversible upon switch of light (Fig. 2.4a). Several experimental signatures such as reduced apparent activation energy (Fig. 2.4b), linear dependence of methanol production rate on light intensity, rational correlation between light-contributed activity and LSPR-induced optical absorption, as well as enlarged KIE (Fig. 2.4c), consistently document the decisive role of plasmon-excited hot electrons in photo-thermo methanol synthesis. Electromagnetic field distribution simulations (Fig. 2.4d) together with transient absorption kinetic studies (Fig. 2.4e) reveal that abundant hot electrons would be excited on plasmonic Cu nanoparticles and then transferred to ZnO oxides through metal-oxide interfaces. In-situ diffuse reflectance infrared Fourier transform spectroscopy (DRIFTS) spectra manifest that hot electrons on Cu–ZnO interfaces would accelerate formate activation and thus enhance methanol synthesis (Fig. 2.4f).

In comparison, fewer research groups have reported enhanced activity via non-plasmonic route, wherein Ozin's group has carried out impressive studies. A series of works have been reported over hydroxylated indium oxide, $In_2O_{3-x}(OH)_y$, for photo-thermo CO_2 hydrogenation by this group [49, 66–68]. $In_2O_{3-x}(OH)_y$ nanostructures with abundant surface hydroxides and oxygen vacancies exhibited much higher CO production rate in photo-thermo RWGS reaction than in thermocatalytic process (Fig. 2.5a) [66]. Kinetic measurements, in-situ DRIFTS spectra together with density functional theory (DFT) calculations revealed a FLP mechanism in both photo-thermo and thermocatalytic RWGS reactions (Fig. 2.5b). A Lewis base hydroxide and a Lewis acid indium adjacent to an oxygen vacancy served as a FLP site, which would dissociate H_2 and then activate CO_2 to produce CO and H_2O. Time-dependent density functional theory (TDDFT) calculations combined with transient absorption spectroscopy and photoconductivity measurements further disclosed that FLP sites in the excited state (ES) would be more active than those in the ground state (GS) and thus led to reduced activation energy (Fig. 2.5c) [67]. In a following work, rod-like $In_2O_{3-x}(OH)_y$ nanocrystal superstructures were found to exhibit enhanced activity in photo-thermo methanol synthesis (Fig. 2.5d), wherein the enhancement was rationalized by the superior activity of FLP sites in the ES as well [68].

To summarize, enhanced activity in photo-thermo CO_2 conversion has been reported in both plasmonic and non-plasmonic routes. In plasmonic route, the enhancement is primarily related to hot electrons excited by the LSPR effect over plasmonic nanostructures, which would accelerate intermediate conversion and thus

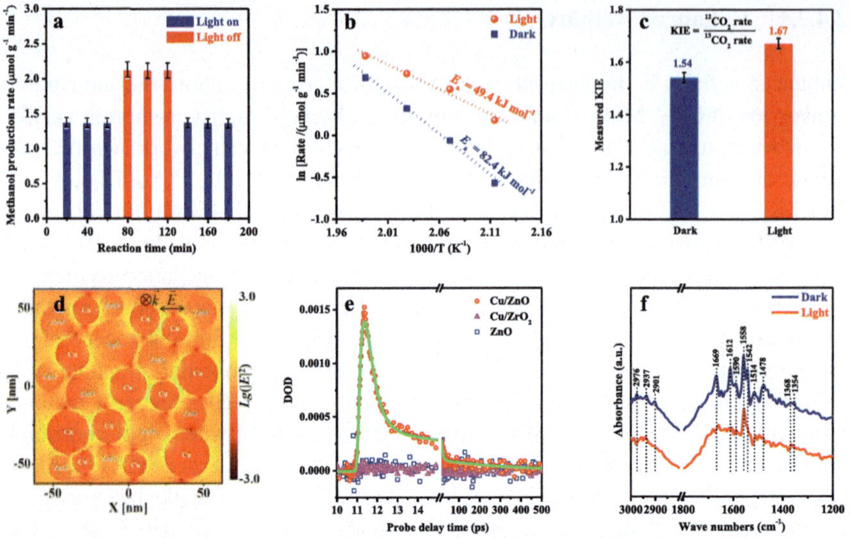

Fig. 2.4 Enhanced activity in photo-thermo CO_2 hydrogenation to methanol over Cu/ZnO catalysts via plasmonic route. **a** Enhanced activity. **b** Reduced activation energy. **c** Enlarged KIE value. **d** Electromagnetic field distribution simulations. **e** Transient absorption kinetic studies. **f** In-situ DRIFTS spectra. Adapted from Ref. [8]

boost catalytic activity. In non-plasmonic route, the enhancement is closely associated with photo-excited electron/hole pairs over semiconductor nanostructures, which would modify active sites and thus benefit catalytic reactions.

2.4.3.2 Manipulated Selectivity

Although selectivity control appears more attractive in photo-thermo catalytic conversion of CO_2, very limited work has been reported to date. Current studies mainly concentrate on hydrogenation reactions. For instance, methanation and RWGS reactions often compete over transition metals to produce CH_4 and CO, respectively. As mentioned earlier, Everitt's and Liu's groups accomplished selectivity control with plasmonic Rh nanocubes, wherein methanation pathway was selectively stimulated in photo-thermo CO_2 hydrogenation [9]. In contrast, Halas's group reported increased selectivity toward RWGS reactions over a plasmonic Al@Cu_2O catalyst after introducing a visible light [69]. The Al@Cu_2O catalyst was a typical antenna-reactor heterostructure consisting of a plasmonic antennas (Al) and a catalytically active component (Cu_2O). The unique selectivity therein was explained with plasmon-induced C–O bond dissociation and CO* desorption, wherein LSPR-excited hot electrons in Al deposited their energy into unoccupied states of adsorbed CO_2 on Cu_2O to facilitate direct C–O bond dissociation and subsequent CO* desorption. Consequently, RWGS reaction pathway was selectively

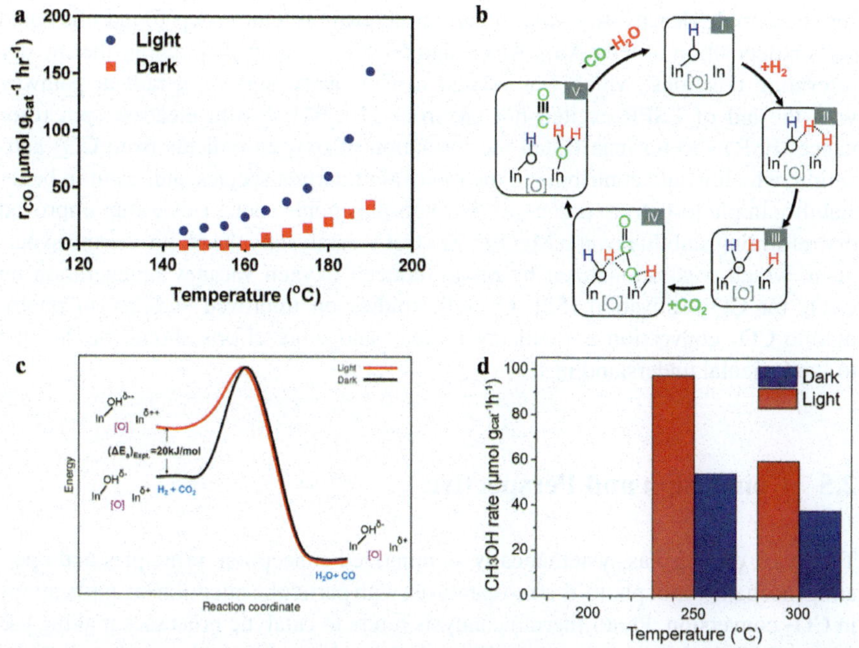

Fig. 2.5 Enhanced activity in photo-thermo CO_2 hydrogenation over $In_2O_{3-x}(OH)_y$ catalysts via non-plasmonic route. **a** Enhanced activity in photo-thermo RWGS. **b** The FLP mechanism in RWGS. **c** Photo-excited FLP sites with reduced activation energy in photo-thermo RWGS. **d** Enhanced activity in photo-thermo methanol synthesis. Parts a and b were adapted from Ref. [66]. Parts c and d were adapted from Refs. [67, 68], respectively

activated under light illumination. In another work, García's and Corma's groups reported higher selectivity of ethanol in photo-thermo CO_2 hydrogenation over a non-plasmonic Na-Co@C catalyst [50]. Mechanism studies on the basis of near-ambient pressure X-ray photoelectron spectroscopy (NAP-XPS) and in-situ Raman spectra revealed that photo-excited electrons activated CO_2–$CO_2^{\delta-}$ and promoted CO_2 dissociation into CO*. The resultant CO* was stabilized by carbon layers on catalyst surface and then transformed into ethanol in a CO* insertion manner. To sum up, the above examples document that selectivity patterns can be modulated via either plasmonic or non-plasmonic routes. Despite different photo-induced excitation exists in two routes, the reaction pathway is similarly changed by altering bond activation or intermediate conversion.

2.4.3.3 Improved Stability

Most of catalytic reactions for photo-thermo CO_2 conversion were operated for only several hours, wherein catalyst stability was severely overlooked. Among various reactions for photo-thermo CO_2 conversion, dry reforming reaction is notorious for susceptible deactivation due to carbon deposition. Our group found alleviated carbon deposition on $Pd_{90}Au_{10}/Al_2O_3$ and $Ni@SiO_2$ catalysts in photo-thermo dry reforming reactions, which was related to CO_2 activation via a radical pathway with the aid of LSPR-excited hot electrons [10, 64]. In-situ electron spin resonance (ESR) spectra manifested the formation of oxygen radicals from CO_2 activation, which would contribute to the removal of carbon species and result in better stability in photo-thermo processes. For non-plasmonic route, Li's group improved dry reforming stability over a $MgO/Pt/Zn–CeO_2$ catalyst by the photo-thermo synergism, which was rationalized by photo-induced oxygen vacancy regeneration on CeO_2 for CO_2 activation [52]. Overall, studies on improved stability in photo-thermo CO_2 conversion are still insufficient, and more efforts should be devoted to fundamental understandings.

2.5 Conclusion and Perspective

This book chapter has systematically summarized conception, principles and operating mechanisms of photo-thermo catalysis, with particular attention on applications in CO_2 conversion. Photo-thermo catalysis refers to catalytic processes making use of both light and heat without significant light-to-heat effect, wherein benefit from heat for photocatalysis and benefit from light for thermocatalysis work together to yield superior catalytic performance. Photo-thermo catalysis functions via either plasmonic or non-plasmonic route, both of which may afford enhanced activity, manipulated selectivity and/or improved stability in CO_2 conversion.

While applauding the achieved progress in this new field, the following challenges but also opportunities should be noted in future investigations.

Catalytic tests: (1) Mass transfer and heat management are rarely addressed in most cases. The interference from mass and heat diffusion may cause misleading performance evaluation. (2) The penetration depth of incident photons is typically below 100 nm [70]. Innovative reactors should be designed to maximize illuminated proportion of catalysts. (3) Potential temperature bias affects accurate comparisons between photo-thermo catalysis and thermocatalysis. Practical temperature detections with nanoscale resolutions should be developed [71]. (4) Activity data should be presented in a more unified manner. Since photo-thermo catalysis is generally compared with thermocatalysis, activity should be described with turnover frequency (TOF).

Mechanism studies: (1) In-situ characterizations should be developed to clearly "see" how light works with heat in driving photo-thermo catalytic reactions. Incorporating light illumination into commercial characterization facilities calls for delicate equipment modifications. (2) Theoretical methodologies suitable for ES electronic calculations should be established for photo-thermo catalysis. Current DFT calculations are tailored for GS electronic calculations, which should be remodeled to simulate photocatalytically involved processes. (3) Big data and machine learning should be paid sufficient attention for mechanism studies. Machine learning exhibits distinct advantage by analyzing big data from literature. (4) Scientists from chemistry, chemical engineering, physics and material science should work together to push forward the industrialization. With efforts from interdisciplinarity communities, bright future would be anticipated in photo-thermo catalytic conversion of CO_2.

Acknowledgements The financial support from National Natural Science Foundation of China (22378017, 22372118) and National Key Research and Development Project (2018YFB1501405, 2022YFC2105604) are acknowledged.

Competing Interests The authors declare no competing interests.

Additional Information

References

1. Wang Z-J, Song H, Liu H, Ye J (2020) Coupling of solar energy and thermal energy for carbon dioxide reduction: status and prospects. Angew Chem Int Ed 59:8016–8035
2. Guterres A (2050) Carbon neutrality by 2050: the world's most urgent mission. https://www.un.org/sg/en/content/sg/articles/2020-12-11/carbon-neutrality-2050-the-world%E2%80%99s-most-urgent-mission. Accessed 28 July 2023
3. Lin H, Luo S, Zhang H, Ye J (2022) Toward solar-driven carbon recycling. Joule 6:294–314
4. Gao W et al (2020) Industrial carbon dioxide capture and utilization: state of the art and future challenges. Chem Soc Rev 49:8584–8686
5. Li K, Peng B, Peng T (2016) Recent advances in heterogeneous photocatalytic CO_2 conversion to solar fuels. ACS Catal 6:7485–7527
6. Das S et al (2020) Core-shell structured catalysts for thermocatalytic, photocatalytic, and electrocatalytic conversion of CO_2. Chem Soc Rev 49:2937–3004
7. Tackett BM, Gomez E, Chen JG (2019) Net reduction of CO_2 via its thermocatalytic and electrocatalytic transformation reactions in standard and hybrid processes. Nat Catal 2:381–386
8. Wang Z-J et al (2019) Photo-assisted methanol synthesis via CO_2 reduction under ambient pressure over plasmonic Cu/ZnO catalysts. Appl Catal B Environ 250:10–16
9. Zhang X et al (2017) Product selectivity in plasmonic photocatalysis for carbon dioxide hydrogenation. Nat Commun 8:14542
10. Liu H et al (2016) Design of PdAu alloy plasmonic nanoparticles for improved catalytic performance in CO_2 reduction with visible light irradiation. Nano Energy 26:398–404
11. Ghoussoub M, Xia M, Duchesne PN, Segal D, Ozin G (2019) Principles of photothermal gas-phase heterogeneous CO_2 catalysis. Energy Environ Sci 12:1122–1142
12. Mateo D, Cerrillo JL, Durini S, Gascon J (2021) Fundamentals and applications of photothermal catalysis. Chem Soc Rev 50:2173–2210
13. Keller N, Ivanez J, Highfield J, Ruppert AM (2021) Photo-/thermal synergies in heterogeneous catalysis: towards low-temperature (solar-driven) processing for sustainable energy and chemicals. Appl Catal B Environ 296:120320

14. Meng X et al (2014) Photothermal conversion of CO_2 into CH_4 with H_2 over group VIII nanocatalysts: an alternative approach for solar fuel production. Angew Chem Int Ed 53:11478–11482
15. Luo S, Ren X, Lin H, Song H, Ye J (2021) Plasmonic photothermal catalysis for solar-to-fuel conversion: current status and prospects. Chem Sci 12:5701–5719
16. Christopher P, Xin H, Linic S (2011) Visible-light-enhanced catalytic oxidation reactions on plasmonic silver nanostructures. Nat Chem 3:467–472
17. Fang S, Hu YH (2022) Thermo-photo catalysis: a whole greater than the sum of its parts. Chem Soc Rev 51:3609–3647
18. Luo S et al (2021) Triggering water and methanol activation for solar-driven H_2 production: interplay of dual active sites over plasmonic ZnCu alloy. J Am Chem Soc 143:12145–12153
19. Chen Y et al (2021) Cooperative catalysis coupling photo-/photothermal effect to drive Sabatier reaction with unprecedented conversion and selectivity. Joule 5:3235–3251
20. Song H, Meng X, Wang Z-J, Liu H, Ye J (2019) Solar-energy-mediated methane conversion. Joule 3:1606–1636
21. Wang Q, Domen K (2020) Particulate photocatalysts for light-driven water splitting: mechanisms, challenges, and design strategies. Chem Rev 120:919–985
22. Schneider J, Bahnemann D, Ye J, Li Puma G, Dionysiou DD (eds) (2016) Photocatalysis: fundamentals and perspectives. Royal Society of Chemistry
23. Yang F, Deng D, Pan X, Fu Q, Bao X (2015) Understanding nano effects in catalysis. Nat Sci Rev 2:183–201
24. Liu L, Corma A (2018) Metal catalysts for heterogeneous catalysis: from single atoms to nanoclusters and nanoparticles. Chem Rev 118:4982–5079
25. Hagen J (eds) (2020) Industrial catalysis: a practical approach, 3rd edn. Wiley-VCH
26. Hong J et al (2022) Photothermal chemistry based on solar energy: from synergistic effects to practical applications. Adv Sci 9:2103926
27. Hattori Y et al (2021) Phonon-assisted hot carrier generation in plasmonic semiconductor systems. Nano Lett 21:1083–1089
28. Yu F et al (2020) Enhanced solar photothermal catalysis over solution plasma activated TiO_2. Adv Sci 7:2000204
29. Xu M, Hu X, Wang S, Yu J, Zhu D, Wang J (2019) Photothermal effect promoting CO_2 conversion over composite photocatalyst with high graphene content. J Catal 377:652–661
30. Han B, Wei W, Chang L, Cheng P, Hu YH (2016) Efficient visible light photocatalytic CO_2 reforming of CH_4. ACS Catal 6:494–497
31. Marimuthu A, Zhang J, Linic S (2013) Tuning selectivity in propylene epoxidation by plasmon mediated photo-switching of Cu oxidation state. Science 339:1590–1593
32. Aslam U, Rao VG, Chavez S, Linic S (2018) Catalytic conversion of solar to chemical energy on plasmonic metal nanostructures. Nat Catal 1:656–665
33. Dong Y, Hu C, Xiong H, Long R, Xiong Y (2023) Plasmonic catalysis: new opportunity for selective chemical bond evolution. ACS Catal 13:6730–6743
34. Barba-Nieto I, Gómez-Cerezo N, Kubacka A, Fernández-García M (2021) Oxide-based composites: applications in thermophotocatalysis. Catal Sci Technol 11:6904–6930
35. Song C, Wang Z, Yin Z, Xiao D, Ma D (2022) Principles and applications of photothermal catalysis. Chem Catal 2:52–83
36. Liu L, Zhang X, Yang L, Ren L, Wang D, Ye J (2017) Metal nanoparticles induced photocatalysis. Nat Sci Rev 4:761–780
37. Lee H, Park Y, Song K, Park JY (2022) Surface plasmon-induced hot carriers: generation, detection, and applications. Acc Chem Res 55:3727–3737
38. Zhang Y et al (2018) Surface-plasmon-driven hot electron photochemistry. Chem Rev 118:2927–2954
39. Gellé A, Jin T, de la Garza L, Price GD, Besteiro LV, Moores A (2020) Applications of plasmon-enhanced nanocatalysis to organic transformations. Chem Rev 120:986–1041
40. Linic S, Aslam U, Boerigter C, Morabito M (2015) Photochemical transformations on plasmonic metal nanoparticles. Nat Mater 14:567–576

41. Sarina S et al (2014) Viable photocatalysts under solar-spectrum irradiation: nonplasmonic metal nanoparticles. Angew Chem Int Ed 53:2935–2940
42. Mateo D, Albero J, García H (2017) Photoassisted methanation using Cu_2O nanoparticles supported on graphene as a photocatalyst. Energy Environ Sci 10:2392–2400
43. Dong Y et al (2020) Shining light on CO_2: from materials discovery to photocatalyst, photoreactor and process engineering. Chem Soc Rev 49:5648–5663
44. Martin AJ, Larrazábal GO, Pérez-Ramirez J (2015) Towards sustainable fuels and chemicals through the electrochemical reduction of CO_2: lessons from water electrolysis. Green Chem 17:5114–5130
45. Gao P, Zhong L, Han B, He M, Sun Y (2022) Green carbon science: keeping the pace in practice. Angew Chem Int Ed 61:e202210095
46. Wang K, Jiang R, Peng T, Chen X, Dai W, Fu X (2019) Modeling the effect of Cu doped TiO_2 with carbon dots on CO_2 methanation by H_2O in a photo-thermal system. Appl Catal B Environ 256:117780
47. Mateo D, Albero J, García H (2018) Graphene supported NiO/Ni nanoparticles as efficient photocatalyst for gas phase CO_2 reduction with hydrogen. Appl Catal B Environ 224:563–571
48. Deng B, Song H, Peng K, Li Q, Ye J (2021) Metal-organic framework-derived Ga-Cu/CeO_2 catalyst for highly efficient photothermal catalytic CO_2 reduction. Appl Catal B Environ 298:120519
49. Hoch LB et al (2014) The rational design of a single-component photocatalyst for gas-phase CO_2 reduction using both UV and visible light. Adv Sci 1:140013
50. Liu L (2018) Sunlight-assisted hydrogenation of CO_2 into ethanol and C2+ hydrocarbons by sodium-promoted Co@C nanocomposites. Appl Catal B Environ 235:186–196
51. Liu H et al (2015) Conversion of carbon dioxide by methane reforming under visible-light irradiation: surface-plasmon-mediated nonpolar molecule activation. Angew Chem Int Ed 54:11545–11549
52. Pan F, Xiang X, Du Z, Sarnello E, Li T, Li Y (2020) Integrating photocatalysis and thermo-catalysis to enable efficient CO_2 reforming of methane on Pt supported CeO_2 with Zn doping and atomic layer deposited MgO overcoating. Appl Catal B Environ 260:118189
53. Wang L et al (2016) Hydrogen-treated mesoporous WO_3 as a reducing agent of CO_2 to fuels (CH_4 and CH_3OH) with enhanced photothermal catalytic performance. J Mater Chem A 4:5314–5322
54. Ha MN, Lu G, Liu Z, Wang L, Zhao Z (2016) 3DOM-$LaSrCoFeO_{6-\delta}$ as a highly active catalyst for the thermal and photothermal reduction of CO_2 with H_2O to CH_4. J Mater Chem A 4:13155–13165
55. Lv C et al (2023) Nanostructured materials for photothermal carbon dioxide hydrogenation: regulating solar utilization and catalytic performance. ACS Nano 17:1725–1738
56. Wang X, Jin R, Yan W, Li H, Wang Z-J (2022) An Al_2O_3-supported NiFe bimetallic catalyst derived from hydrotalcite precursors for efficient CO_2 methanation. Catal Today 402:38–44
57. Pahija E et al (2022) Experimental and computational synergistic design of Cu and Fe catalysts for the reverse water−gas shift: a review. ACS Catal 12:6887–6905
58. Xie G, Jin R, Ren P, Fang Y, Zhang R, Wang Z-J (2023) Boosting CO_2 hydrogenation to methanol by adding trace amount of Au into Cu/ZnO catalysts. Appl Catal B Environ 324:122233
59. Wang W, Wang S, Ma X, Gong J (2011) Recent advances in catalytic hydrogenation of carbon dioxide. Chem Soc Rev 40:3703–3727
60. Qi R, An L, Guo Y, Zhang R, Wang Z-J (2022) In-situ fabrication of ultrasmall Ni nanoparticles from $Ni(OH)_2$ precursors for efficient CO_2 reforming of methane. Ind Eng Chem Res 61:198–206
61. Du X, Liu L, Ye J (2021) Plasmonic metal nanoparticles for artificial photosynthesis: advancements, mechanisms, and perspectives. Solar RRL 5:2100611
62. Song H et al (2018) Light-enhanced carbon dioxide activation and conversion by effective plasmonic coupling effect of Pt and Au nanoparticles. ACS Appl Mater Interfaces 10:408–416

63. Zhang H et al (2016) Surface-plasmon-enhanced photodriven CO_2 reduction catalyzed by metal-organic-framework-derived iron nanoparticles encapsulated by ultrathin carbon layers. Adv Mater 28:3703–3710
64. Liu H et al (2017) Light assisted CO_2 reduction with methane over SiO_2 encapsulated Ni nanocatalysts for boosted activity and stability. J Mater Chem A 5:10567–10573
65. Liu H, Dao TD, Liu L, Meng X, Nagao T, Ye J (2017) Light assisted CO_2 reduction with methane over group VIII metals: universality of metal localized surface plasmon resonance in reactant activation. Appl Catal B Environ 209:183–189
66. Ghuman KK, Wood TE, Hoch LB, Mims CA, Ozin GA, Singh CV (2015) Illuminating CO_2 reduction on frustrated Lewis pair surfaces: investigating the role of surface hydroxides and oxygen vacancies on nanocrystalline $In_2O_{3-x}(OH)_y$. Phys Chem Chem Phys 17:14623–14635
67. Ghuman KK et al (2016) Photoexcited surface frustrated Lewis pairs for heterogeneous photocatalytic CO_2 reduction. J Am Chem Soc 138:1206–1214
68. Wang L et al (2018) Photocatalytic hydrogenation of carbon dioxide with high selectivity to methanol at atmospheric pressure. Joule 2:1369–1381
69. Robatjazi H et al (2017) Plasmon-induced selective carbon dioxide conversion on earth-abundant aluminum-cuprous oxide antenna-reactor nanoparticles. Nat Commun 8:27
70. Lorber K, Djinovic P (2022) Accelerating photo-thermal CO_2 reduction to CO, CH_4 or methanol over metal/oxide semiconductor catalysts. iScience 25:104107
71. Bian X, Zhao Y, Zhou C, Zhang T (2023) Minimizing temperature bias through reliable temperature determination in gas-solid photothermal catalytic reactions. Angew Chem Int Ed 62:e2022193402023

Chapter 3
Thermocatalytic CO_2 Hydrogenation to Liquid Fuels

Mingrui Wang, Huahua Fan, Miao Zhang, Hao Wang, Zhiqun Wang, Shuang Wang, Xiaowa Nie, Guanghui Zhang, Chunshan Song, and Xinwen Guo

Abstract Thermocatalytic CO_2 hydrogenation to liquid fuels, including ethanol and liquid hydrocarbons, has drawn global attention recently as a potential strategy to decrease CO_2 emissions and reduce the consumption of and dependence on fossil fuels. The nature of catalyst has an important impact on the conversion and selectivity and clarifying the catalyst structure-performance relationship is essential to synthesize the desired liquid products. Compared with C_1 products, the generation of ethanol and liquid hydrocarbons with two or more carbon atoms is more difficult owing to the high energy barrier for C–C coupling. Current studies show that the interfacial catalysts are suitable for CO_2 hydrogenation to ethanol and the active interface is generally responsible for key CO^* insertion, while the metal/carbide catalysts and oxide-zeolite tandem catalysts play vital roles in that to liquid hydrocarbons. In this chapter, we discuss the latest advances in the representative noble metals, transition metals and their carbides, and modified Cu-based catalysts for the synthesis of ethanol, Fe–, Co-based catalysts as well as tandem catalysts for that of long-chain hydrocarbons. Fundamental understanding on active sites, structural evolution, CO_2 activation, and reaction mechanism are discussed based on computational and experimental results. On this basis, we discuss some concepts on catalyst design as well as the challenges and opportunities for its development and potential industrial applications.

M. Wang · H. Fan · M. Zhang · H. Wang · Z. Wang · S. Wang · X. Nie · G. Zhang · X. Guo (✉)
State Key Laboratory of Fine Chemicals, Frontier Science Center for Smart Materials, PSU-DUT Joint Center for Energy Research, School of Chemical Engineering, Dalian University of Technology, Dalian 116024, China
e-mail: guoxw@dlut.edu.cn

C. Song
Department of Chemistry, Faculty of Science, The Chinese University of Hong Kong, Shatin, Hong Kong, NT, China

G. Zhang et al. (eds.), *Advances in CO_2 Utilization*, Green Chemistry and Sustainable Technology, https://doi.org/10.1007/978-981-99-8822-8_3

3.1 CO_2 Hydrogenation to Ethanol

Catalytic CO_2 conversion with renewable H_2 to methanol and higher alcohols (C_{2+} alcohols) has drawn a longstanding attention owing to the important applications of alcoholic products in the chemical and pharmaceutical sectors [1, 2]. Among them, methanol and ethanol (C_2H_5OH) are important liquid fuels while the synthesis of methanol from CO_2 has been introduced in the previous section. Compared with methanol, fuel ethanol has greater applications. While in United States and Brazil, ethanol becomes competitive with traditional gasoline fuel in terms of cost and energy [3]. However, the direct synthesis of ethanol from CO_2 hydrogenation is still significantly challenging owing to the difficulty on controlled C=O/C–O bond activation and C–C coupling [4]. Clarifying the fundamental structure-performance relationship and optimizing the catalyst structure to tune the balance of surface intermediates and to improve the key insertion/coupling steps are of uppermost priority for designing highly effective catalysts using in CO_2 hydrogenation to ethanol.

$$2CO_2 + 6H_2 \Leftrightarrow C_2H_5OH + 3H_2O \; \Delta H_{298\,K} = -86.7 \text{ kJ mol}^{-1} \quad (3.1)$$

As shown in Eq. (3.1), CO_2 hydrogenation to ethanol is an exothermic reaction while relatively low temperatures and high pressures profit the forward reaction. Considering the reverse water gas shift (RWGS) as the side reaction, the thermodynamic calculations by Xu et al. [1] indicate that the selectivity of ethanol gradually decreases with the increasing temperature, while the CO_2 conversion varies following a volcanic curve (Fig. 3.1a). The yield of ethanol achieves the optimum values at 250–320 °C, which agrees with the selected temperature range in the actual reaction. As illustrated in Fig. 3.1b, the fixed-bed and tank reactors are generally employed for the reaction. In comparison with tank reactors, the fixed-bed reactors are more likely to be further applied in the practical production, whereas the ethanol selectivity and catalyst lifetime constitute the main problems. The ethanol formation generally undergoes the CO^* insertion pathway. For instance, on Na modified Rh-based catalyst (Fig. 3.1c), Zhang et al. [5] inferred that CO_2 is initially activated to form the carbonate (CO_3^{2-}), and then converts to CO^* on the Rh^+ sites. Some CO^* intermediates migrate to neighboring Rh^0 sites and are further hydrogenated to CH_x^*. The key insertion of CO^* into Rh^0–CH_x would occur at the interface while the generated CH_x–CO^* species further converts to C_2H_5OH. This mechanism is similar to that of homogeneous hydroformylation. However, there are still some other potential reaction pathways, which are strongly dependent on the surface structure of catalysts. The current catalysts for CO_2 hydrogenation to ethanol can be divided into three classes: (1) Noble metal catalysts (Ir, Rh, Pd, and Pt); (2) Transition metal and carbide catalysts (Fe–/Co-based and other composite carbide catalysts); (3) Modified Cu-based catalysts. Moreover, their combination could further optimize the reaction route and increase the ethanol yield.

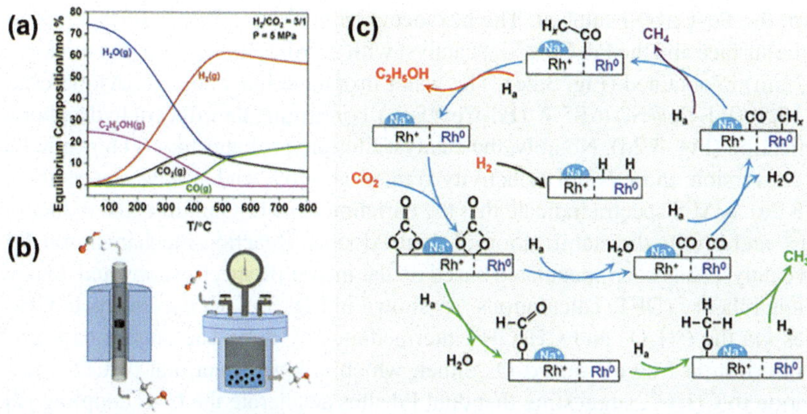

Fig. 3.1 Fundamental reaction properties and reactors for CO_2 hydrogenation to ethanol. **a** Thermodynamic calculations on the equilibrium composition of CO_2 hydrogenation to ethanol and CO with respect to reaction temperature (5 MPa, H_2/CO_2=3). Reproduced with permission from Ref. [1]. **b** Diagrammatic sketch of common reactors, including the fixed-bed (left) and tank reactors (right). Reproduced with permission from Ref [1]. **c** Classical insertion mechanism for ethanol formation as proposed on Na-modified Rh-based catalysts. Reproduced with permission from Ref. [5]

3.1.1 Noble Metal Catalysts

Noble metal materials exhibit preferable performance for CO_2 conversion to ethanol owing to its promoting effect on the adsorption and activation of CO^* intermediate. Accordingly, on noble metal catalysts, the ethanol formation commonly follows the CO^* insertion route. The noble metal is generally dispersed at atomic level to improve the atom availability. Caparrós et al. [6] reported a single atom Pd/Fe_3O_4 catalyst with the Pd loading of 0.1 wt.%. Pd atoms are anchored on the surface of Fe_3O_4 as discerned by the high angle annular dark field-scanning transmission electron microscopy (HAADF-STEM) images. Pd/Fe_3O_4 exhibits an excellent ethanol selectivity of 97.5% at 300 ºC, 0.1 MPa, but the CO_2 conversion is limited (conv. < 1%). Liu and co-workers developed a series of catalysts with Pd dimer using fixed-bed reactors. As displayed in Fig. 3.2a and b, they prepared a CeO_2 nanorod-supported Pd-based catalyst and according to the X-ray absorption fine structure (XAFS) spectra and catalytic performance, the Pd_2O_4 dimer has been evidenced as the active center for ethanol generation [7]. The initial ethanol selectivity achieves 99.2% with a CO_2 conversion of 9.2% at 240 ºC, 3 MPa. The space time yield (STY) of ethanol achieves 45.6 $g_{ethanol}$ gPd^{-1} h^{-1}, which is ~ 90 times as that on Pd nanoparticles/CeO_2 catalyst, highlighting the importance of dual Pd sites. However, Pd_2O_4/CeO_2 exhibits a poor catalyst stability, while the production selectivity shifts from ethanol to CO (~20%) in initial 5 h. The hydrophobic shell is further introduced to enrich the generated H_2O around the catalyst and to improve the structure stability of Pd_2O_4 dimer [8]. In brief, a hydrophilic SiO_2 layer is first coated at Pd_2Ce core to form the $Pd_2Ce@Si$ catalyst. The hexadecyl-trimethoxy-silane is further grafted onto the surface and the $Pd_2Ce@Si_{16}$ catalyst with a hydrophobic shell (the thickness of ~ 7 nm) is obtained (Fig. 3.2c). The water-droplet contact angle (CA) increases from 10.5° ($Pd_2Ce@Si$) to 85.7° ($Pd_2Ce@Si_{16}$), confirming the effect of hydrophobic modification (Fig. 3.2d). Notably, the catalyst lifetime prolongs to ~ 60 h, while the CO_2 conversion and ethanol selectivity remain at ~ 6% and ~ 95%, respectively (Fig. 3.2e). XAFS spectra indicate that the enrichment of H_2O inhibits the reduction of Pd^{2+} and profits the stabilization of dual Pd sites. Reaction mechanism on Pd-based catalysts has been speculated based on the in situ IR spectroscopy and density functional theory (DFT) calculations. As shown in Fig. 3.2f, CO^* converts to CH_x^* species via the CH_xO^* and CH_2OH^* intermediates by turn. The energy barrier of CO^* insertion is 1.09 eV on Pd_2O_4 dimer, which is lower than that of CO^* transformation to CH_xO^*, suggesting that dual Pd sites accelerate the C–C coupling [7]. By comparison, when the noble metals exist as the nanoparticles, the selectivity of ethanol significantly decreases while methanol is the dominant alcoholic product, suggesting that the key coupling potentially occurs at the active interfaces [9, 10].

Noble metal catalysts also exhibit outstanding selectivity to ethanol in tank reactors. In this system, the synergistic effect between metal atom and oxide support always plays an important role. Ye et al. [11] designed an Ir_1–In_2O_3 catalyst, integrating two active sites by anchoring the monatomic Ir onto the In_2O_3 support. Mechanistic investigations demonstrate that the Lewis acid–base pairs between single site $Ir^{\delta+}$ and adjacent oxygen vacancy (O_v) in partially In_2O_3 are responsible for ethanol formation. The CH_3O^*-O_v and $Ir^{\delta+}$–CO^* are jointly involved in C–C coupling, exhibiting an ethanol selectivity of 99.7% at an initial turnover frequency (TOF) of 481 h^{-1} at 200 °C, 3 MPa, and $H_2/CO_2 = 5$. Similarly, the synergistic effect is also observed between single atom Rh and Ti-doped CeO_2 support [12]. The addition

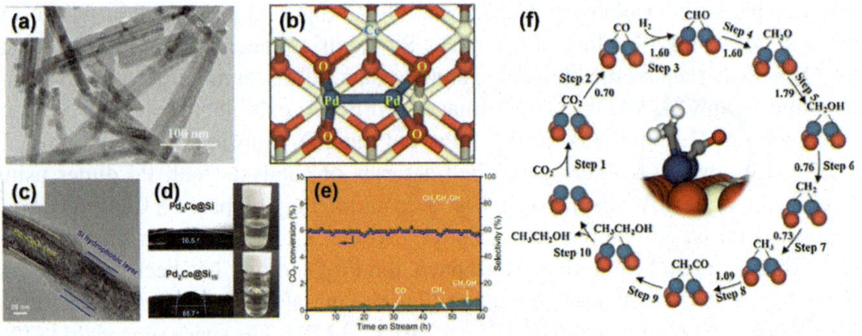

Fig. 3.2 CO_2 hydrogenation to ethanol on noble metal catalysts. **a** Morphology of Pd_2O_4/CeO_2 nanorod catalyst. Reproduced with permission from Ref. [7]. **b** Theoretical structure of Pd_2O_4 dimer. Reproduced with permission from Ref. [7]. **c** Core–shell microstructure of hydrophobic shell-modified $Pd_2Ce@Si_{16}$ catalyst. Reproduced with permission from Ref. [8]. **d** Water-droplet contact angle. Reproduced with permission from Ref. [8]. **e** Catalytic performance of CO_2 hydrogenation to ethanol on $Pd_2Ce@Si_{16}$ catalyst at 240 °C, 3 MPa. Reproduced with permission from Ref. [8]. **f** Potential reaction pathway on Pd_2O_4 dimer catalyst. Reproduced with permission from Ref. [7]

of Ti assists in the construction with CeO_2 to form the $CeTiO_x$ and enhances the metal-support interaction. As a result, the Rh-O bond is strengthened and the charge density of single atom Rh increases. The unique structure facilitates the cleavage of C–O bond in CH_xOH^* and $COOH^*$ species and the formation of CO^*/CH_x^* intermediates, leading to an excellent performance of which ethanol selectivity is 99.1% and its TOF achieves record 493.1 h^{-1} at 250 °C, 3 MPa.

Apart from single atom metal and support, bimetallic catalyst also shows an excellent performance for CO_2 to ethanol. For example, Bai et al. [13] introduced Cu into Pd-based catalyst and the Pd–Cu alloy shows an ethanol selectivity up to 92.0% and a TOF of 359.0 h^{-1} at 200 °C, 3.2 MPa. The incorporation of Cu enhances the reducibility of surface palladium oxides through the charge transfer, and threefold bridge-bonded CO^* species is mainly generated on bimetallic alloy. This adsorbed CO^* is facile to convert to HCO^*, resulting in a decreased CO^* coverage and avoiding the common poisoning phenomenon on Pd-based catalyst. The theoretical calculations further indicate that on Pd_2Cu (110) surface, the rate-determining step is the decomposition of CH_2OH^* to CH_2^* and OH^*, followed by the reaction of CO_2^* and H^* to $COOH^*$ and the decomposition of $COOH^*$ to CO^* and OH^* species [14].

As stated above, the strategy of constructing atom-precision sites is of great importance to promoting CO_2 hydrogenation to ethanol on noble metal catalysts. The interface of noble metal and oxide support is generally considered as the active center for C–C coupling, while with the rapid development of structural characterization, its intrinsic structure would be clearly clarified, facilitating the design and preparation of effective noble metal catalysts.

3.1.2 Transition Metal and Carbide Catalysts

Fe, Co and their carbides are traditionally employed to catalyze CO_x conversion to C_{2+} hydrocarbons due to their promoting effect on C–C coupling [15]. For ethanol formation, considering Fe–/Co-based catalysts generally lead to the rapid formation and accumulation of CH_x^* intermediate, the key issue here is how to control the balance of dissociation/non-dissociation of CO^* intermediate to ensure the balance of surface CO^* and CH_x^* species for following insertion process.

FeC_x-based catalysts have been proven to be well-responsible for promoting CO_2 conversion to ethanol and other C_{2+} alcohols, such as propanol and butanol. Guo and co-workers focus on using the K-modified MnO_x–FeC_x catalysts for the synthesis of ethanol in fixed-bed reactors. According to the in situ XRD investigations, they found that the Fe_5C_2 and Fe_3C coexist with MnO in the reaction [16]. The interface of FeC_x and MnO is responsible for the insertion of CO^* species, and an ethanol selectivity of ~ 7% with a CO_2 conversion of ~ 40% is obtained at 300 °C, 3 MPa, while some propanol and butanol are also detected, indicating the high ability of CO_2 activation and carbon chain growth on FeC_x. The introduction of Cu further enhances the hydrogenation process, and the assembled MnCuK–FeC_x and CuZnAlZr tandem catalyst (mass ratio = 1/1) increases the selectivity of ethanol and total higher alcohols to 9.3% and 15.5%, respectively [17].

Alkali metal promoters are crucial for the regulation of the structure and surface reactions on FeC_x-based catalysts. Xu et al. presented a series of alkali metal-modified Cu–Fe–Zn catalysts. For Cs-modified catalyst, the yield of ethanol shows a classical volcanic curve with respect to Cs dosage [18]. At the Cs loading of 3 wt.%, the optimized ethanol STY achieves ~ 16 mg g_{cat}^{-1} h^{-1} at 310 °C and 5 MPa, while excessive Cs addition (5%) causes the decline of CO_2 conversion and the increased CO selectivity. Similar phenomenon has also been observed on K-modified catalysts. The optimum ethanol yield on Cu–Mg–Fe–Zn quaternary catalyst reaches ~ 64 mg g_{cat}^{-1} h^{-1} at 4.6 wt.% K loading [19]. The addition of K promoter has been demonstrated to tune the H_2 activation and the reducibility of catalysts. As shown in Fig. 3.3a, the reduction peak of CuO shifts to high temperatures with the increasing K contents, suggesting that K promoter suppresses the H_2 activation and catalyst reduction. The investigations on the capacity of H_2 adsorption (Fig. 3.3b) further confirm that K addition reduces the amount of strongly adsorbed H_2 (> 300 °C), which prefers to assist in the reaction rather than weakly adsorbed H_2. Besides, K promoter also influences the adsorption and activation of CO_2 and CO^*. Combined with the temperature programmed desorption (TPD) and in situ diffuse reflectance infrared Fourier transform spectroscopy (DRIFTS) results, Xu et al. [19] confirmed that K promoter boosts the adsorption of CO_2 and its following conversion toward CO^* through the RWGS reaction, and it also enhances the non-dissociative activation of CO^* instead of its dissociation and hydrogenation to CH_x^*. Accordingly, as illustrated in Fig. 3.3c, the surface composition consisting of various intermediates can be tuned by the K content. Only at a moderate content (4.6 wt.% of K), the balance of surface CO^* and CH_x^* would lead to the effective formation of ethanol and other C_{2+} alcohols.

K promoter can also influence the catalyst structure via the electronic interaction. In brief, as shown in Fig. 3.3d, the strong interaction at K–ZrO_2 interface (K donates the electrons to ZrO_2) induces the formation of O vacancies and surface $Zr^{\delta+}$ sites [20]. The interface potentially weakens the dissociation of CO^* and more non-dissociative CO^* species exists on the surface. As a result, an outperformed ethanol yield is achieved up to ~ 90 mg g_{cat}^{-1} h^{-1} on K-modified K–Cu–FeC_x/ZrO_2 catalyst at 320 °C, 4 MPa. Moreover, the K-modified FeC_x-based catalyst can be further coupled with Cu–Zn–Al ternary catalyst (CZA) and the synergistic effect in composite catalyst has been shown in Fig. 3.3e [21]. CZA component accelerates the formation of CO^*, which migrates to K–CuMgZnFe (K-CMZF) and takes parts in the CO^* insertion process for ethanol formation. The proximity of two components is one of the most important factors. The optimum ethanol yield reaches ~ 110 mg g_{cat}^{-1} h^{-1} at 320 °C and 4 MPa while the two components are packaged in the form of powder mixing. Wang et al. also reported a tandem Na–Fe@C and CZA catalyst, while the similar migration of CO^* species from CZA to FeC_x surface has been examined [22]. Mechanistic analysis, including the in situ DRIFTS and DFT calculations, suggest the aldehyde species is an important intermediate which leads to the high CO_2 conversion (39.2%) and ethanol selectivity (35.0%). Besides,

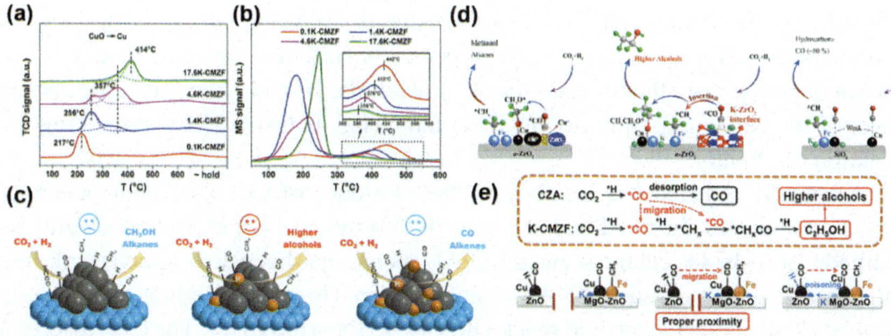

Fig. 3.3 CO_2 hydrogenation to ethanol on K-modified FeC_x-based catalysts. **a** H_2-temperature programmed reduction (TPR) profiles for clarifying the effect of K promoter on the reducibility of Cu–Mg–Fe–Zn catalyst. Reproduced with permission from Ref. [19]. **b** H_2-TPD profiles for clarifying the effect of K promoter on the H_2 activation on modified Cu–Mg–Fe–Zn catalyst. Reproduced with permission from Ref. [19]. **c** Schematic of the tailoring of product distribution by varying K contents on modified Cu–Mg–Fe–Zn catalyst. Reproduced with permission from Ref. [19]. **d** Electronic interaction at K–ZrO_2 interface for tuning the CO^* dissociative and non-dissociative activation. Reproduced with permission from Ref. [20]. **e** K–CuMgZnFe and Cu–Zn–Al composite catalyst for enhancing the formation and migration of CO^* intermediate. Reproduced with permission from Ref. [21]

other metallic and non-metallic promoters, including Na [23], S [23, 24], Cr [25], and In [26], also exhibit good promotion via the electronic or geometric effect on FeC_x-based catalysts for the synthesis of ethanol.

Co-based catalysts for the synthesis of ethanol can be divided into two classes: Co_2C catalysts (for fixed-bed reactors) and partially reduced CoO_x catalysts (for fixed-bed and tank reactors). Co_2C nanoprism has been reported for the synthesis of light olefins and higher alcohols in CO-Fischer Tropsch synthesis (FTS) [27]. However, its utilization in CO_2 hydrogenation is more difficult considering the high CO_2/H_2 concentration would induce its decomposition [28]. The metal-support interaction could be introduced to improve the catalyst structure stability. Zhang et al. [29] prepared several supported Co_2C catalysts and found that only the Si-based supports (SiO_2/Si_3N_4) strongly interacted with Co_2C via the Co–O–Si bonds could protect the Co_2C phase from the decomposition to metallic Co while the generated CO in RWGS reaction would induce a surface reconstruction of Co toward Co_2C (Fig. 3.4a). Ethanol is generated following the CO^* insertion mechanism and it occupies 62.8% in total alcohols while the total selectivity of alcohols is 9.2% with a CO_2 conversion of 18.8% on Na–Co_2C/SiO_2 catalyst at 250 °C and 5 MPa. They further conducted extended X-ray absorption fine structure (EXAFS) spectroscopy analysis and DFT calculations to elucidate the effect of Na promoter [30]. It evidenced the Na–Co_2C interaction and according to DFT calculations on Na–Co_2C (111) surface, whether the direct dissociation or the H-assisted dissociation of CO^* are inhibited (Fig. 3.4b), thereby resulting in a high-surface CO^*/CH_x^* ratio. Similar with the FeC_x-based catalysts, Co_2C could be also coupled with the CZA component, while the results of in situ chemical transient kinetics (CTK) experiments demonstrate that the generated olefins diffuse from upstream Co_2C to downstream CZA and then be hydrogenated to form the R-CH_x^* species [31]. This species is further coupled with CH_xO^* to form the precursor of ethanol and other C_{2+} alcohols. Notably, the predicted energy barrier of the key coupling between CH_2^* and CHO^* is just 0.35 eV, preferring the formation of ethanol.

For partially reduced CoO_x catalysts, the existence of Co^0–$Co^{\delta+}$ pair generally plays a pivotal role. In fixed-bed reactors, Witoon et al. [32] evaluated several K-modified CoO_x–In_2O_3 catalysts and a $KCoO_2$ species has been determined according to the results of Co–K edge EXAFS spectroscopy (Fig. 3.4c), while the strongly adsorbed H_2 on its surface would react with CO^* to form the CH_x^*. The CO^* insertion reaction would occur at the interface of $KCoO_2$ and CoO (the CO^* is adsorbed on CoO), resulting in enhanced ethanol selectivity of ~ 40% (in total alcohols) and yield of ~ 75 mg g_{cat}^{-1} h^{-1}, respectively, while the total yield of C_{2+} alcohols achieves ~ 170 g kg_{cat}^{-1} h^{-1} at 380 ºC, 4 MPa. The Co^0–$Co^{\delta+}$ active pairs can be also obtained from the reduction of Co–Ga–Al–O spinel in H_2 at 600 ºC [33]. As displayed in Fig. 3.4d, the CO^* species is generated on $Co(GaAl)_2O_4$ spinel while the CH_x^* species accumulates on metallic Co sites. The strong interaction between GaO_x and Co confirms the configuration of Co^0–$Co^{\delta+}$ pairs and increases the ethanol selectivity

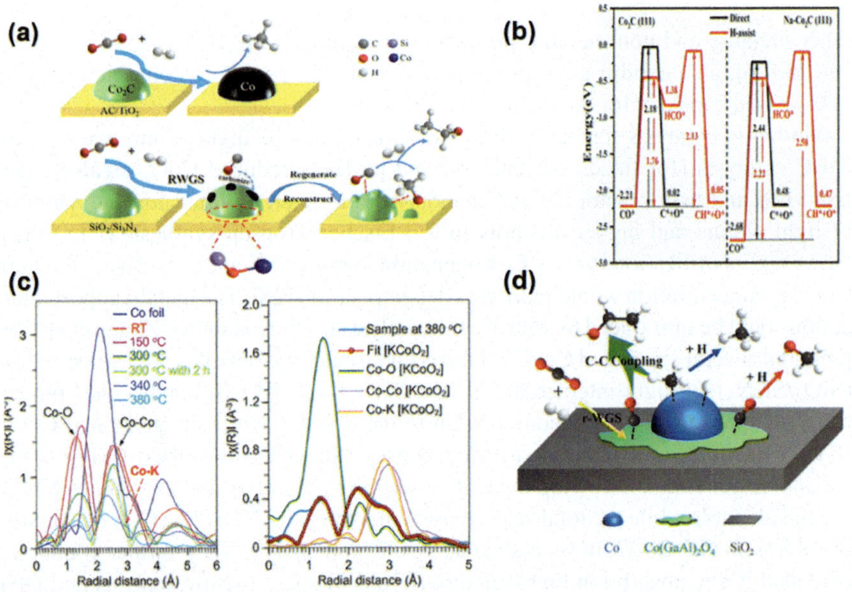

Fig. 3.4 CO_2 hydrogenation to ethanol on Co-based catalysts. **a** Reconstruction of supported Co_2C catalysts in the reaction. Reproduced with permission from Ref. [29]. **b** Energetic information for direct or H-assisted dissociation of CO^* on Co_2C (111) and Na–Co_2C (111) surface. Reproduced with permission from Ref. [30]. **c** EXAFS spectra for evidencing the formation of $KCoO_2$ species. Reproduced with permission from Ref. [32]. **d** Reaction mechanism for CO_2 hydrogenation to ethanol on Ga-modified Co–Al–O spinel catalyst. Reproduced with permission from Ref. [33]

to 20.2% at 270 °C and 3 MPa. For tank reactors, the modified $CoAl_2O_4$ spinel exhibits an excellent ethanol selectivity (~90%) at 140 ~ 200 °C. The spinel would transform to a partially reduced $CoAlO_x$ while the reducibility is quite important. The low reducibility would lead to the formation of methanol while the addition of Ni could enhance the catalyst reduction and increase the surface concentration of $^*CH_x$ species [34, 35]. The ethanol yields achieve 0.444 mmol g_{cat}^{-1} h^{-1} and 15.8 mmol g_{cat}^{-1} h^{-1} on $CoAlO_x$ and Ni-doped $CoAlO_x$ catalysts, respectively.

Moreover, besides Fe–/Co-carbide catalysts, a recent report by Wang et al. [36] shows that an active K-modified Ni–Zn carbide (K–$Ni_3Zn_1C_{0.7}$) is in situ generated and stabilized in CO_2 hydrogenation. Compared with metallic Ni or Ni–Zn alloy, the composite carbide boosts the C–C coupling with the selectivity shifting from CH_4 to C_{2+} alcohols. The selectivity and yield of ethanol reach ~ 55% (in alcoholic products) and ~ 22 mg g_{cat}^{-1} h^{-1} with a catalyst lifetime of ~ 220 h at 350 °C and 3 MPa. Notably, the mechanism investigations show that CH_3^* is directly coupled with CO_2^* instead of CO^* to form the CH_3COO^* intermediate. It inspires us to explore more transition metal and carbide catalysts for CO_2 hydrogenation to ethanol while the control of surface species is essential. However, we also note that the selectivity of ethanol is relatively low on metal carbide catalysts compared with that on noble metal catalysts, and thus how to break the selectivity-conversion trade-off and inhibit the side-reactions toward C_1 products (CO and CH_4) are important research fields in the future.

3.1.3 Modified Cu-Based Catalysts

Cu-based catalysts are one of the most important methanol synthesis catalysts in both CO and CO_2 hydrogenation owing to its intrinsic ability on the activation of C=O bond and the formation of CH_3O^* species. However, the production of ethanol on Cu-based catalysts is limited owing to the weak C–C coupling. Recent reports show that the rational design of modified Cu-based catalysts could enhance the coupling reaction at unique interfaces and therefore increases the ethanol selectivity. Among them, Co–Cu-based catalysts are promising for the synthesis of ethanol because the incorporation of Co could promote the C–C coupling for Cu-based catalysts. Yang et al. [37] developed a supported Co–Cu catalyst, and Co_3O_4 with different morphologies is applied as the support. Compared with rhombic dodecahedra, the Co_3O_4 nanorod prefers to be partially reduced to CoO_x instead of complete reduction to Co^0, while it exhibits a circa 95% selectivity of CO via the decomposition of CH_3O^* intermediate. The Cu/Co_3O_4 nanorod catalyst (the mass fraction of Cu is 10 wt.%) was further prepared using the method of rotary evaporation. In situ DRIFTS results show that in comparison with Cu/Co_3O_4 rhombic dodecahedra, more $^*CH_3O$ species is formed on Cu/Co_3O_4 nanorod, resulting in nearly nine-fold higher ethanol yield of 1.87 mmol g_{cat}^{-1} h^{-1} at 250 °C and 3 MPa.

For Co–Cu bimetallic alloy catalyst, Co could also act as a geometric and electronic promoter. Besides, considering the interaction between CO^* intermediate with

Co is stronger than that with Cu, the Co atoms could diffuse to the catalyst surface at high CO pressures. As illustrated in Fig. 3.5a, Liu et al. adjusted the CO partial pressure during the catalyst preparation and a series of Co–Cu alloys with various surface compositions were obtained [38]. XPS results demonstrate that the high CO concentration leads to the increased surface Co/Cu ratio from trace amount (0.1% CO) to 2.23 (100% CO). For the catalytic tests at 200 °C and 4 MPa (Fig. 3.5b), it is clear to see that the selectivity of ethanol exhibits a volcano-like trend with respect to the CO partial pressure during the preparation, while the optimum ethanol selectivity achieves 61% at treated conditions of 1% CO and the surface Co/Cu ratio of 1. It highlights the feasibility of tuning surface Co/Cu ratio to optimize the performance of CO_2 hydrogenation to ethanol. Theoretical calculations further provide more mechanistic understanding. Figure 3.5c shows a crossroad in the conversion of key CH_2O^* intermediate to CH_2^*, CH_3O^*, or CH_2OH^* species. The surface composition and CO concentration strongly influence the reaction energy barrier (Fig. 3.5d). Figure 3.5e shows the transition states of C–O bond scission of CH_2O^* species on three surfaces with different Co/Cu ratios while Figs. 3.5f and g show the hydrogenation to CH_3O^* and CH_2OH^* species, respectively. In brief, the adsorption of CH_2O^* is weaker than that of CO^* and it is mainly adsorbed on Cu surface, which is disadvantageous to the cleavage of C–O bond. While on moderate Co-segregated surfaces, the CH_2O^* would co-adsorb on the Co–Cu boundary, thus accelerating its dissociation. For following hydrogenation, the Co perhaps acts as a geometric promoter while the high surface Co/Cu ratio and the CO concentration significantly increase the energy barrier of CH_2O^* to CH_3O^* which is the key intermediate for methanol formation. The Ga-modified Co–Cu-based catalyst also shows the potential for the synthesis of ethanol, while the interfacial Cu^0/Cu^+–$CoGaO_x$ is evidenced to be active for the CH_x–CO^* coupling and ethanol generation [39]. The control of surface Co/Cu composition is the most important concept and the formation and stabilization of Co–Cu interface would lead to the high performance for CO_2 hydrogenation to ethanol.

3.2 CO_2 Hydrogenation to Liquid Hydrocarbons

Despite the rapid development of new energy technology nowadays (hydrogen energy, solar energy, etc.), liquid hydrocarbons (C_{5+} HCs) from fossil energy still show an irreplaceable role as transportation fuels and chemical feedstocks. Researchers urgently need to find a way to produce liquid hydrocarbons from non-fossil energy. CO_2 hydrogenation to liquid hydrocarbons can improve our energy structure, reduce fossil energy consumption, and achieve effective carbon resource cycling with reduced CO_2 emissions.

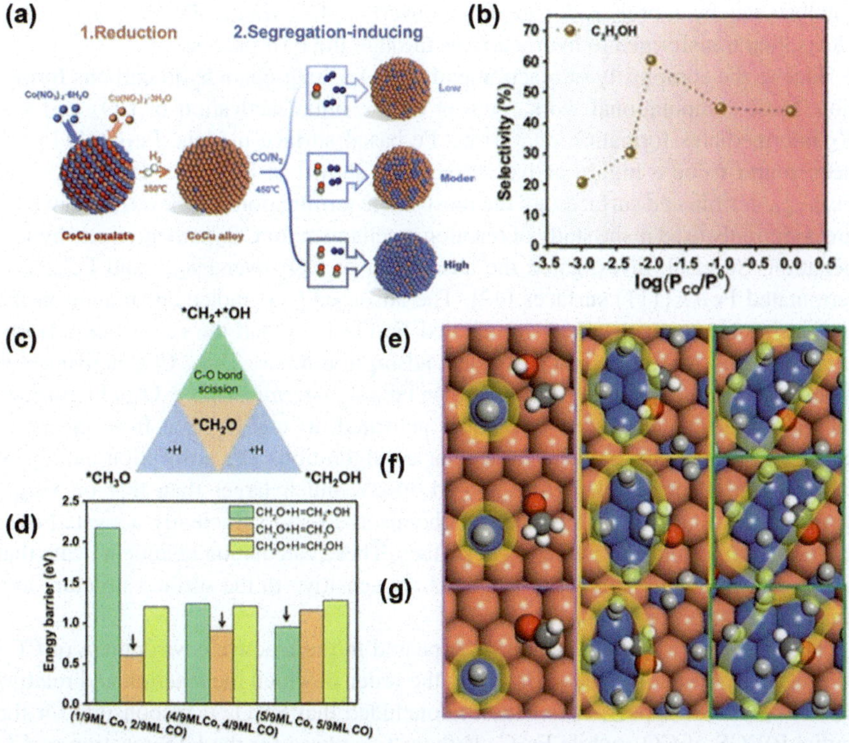

Fig. 3.5 CO_2 hydrogenation on Co–Cu alloy catalysts. [38] Reproduced with permission from ref. **a** Synthesis of Co–Cu alloy with various surface compositions. **b** Relationship of ethanol selectivity in CO_2 hydrogenation and CO partial pressure during the catalysts preparation. **c** Potential conversion route of CH_2O^* intermediate. **d** Energy profile for C–O bond scission and hydrogenation of CH_2O^* on three surfaces with different Co proportion and CO concentration. **e** Transition states of C–O bond scission of CH_2O^* on three surfaces. **f** and **g** Transition states of the hydrogenation of CH_2O^* to CH_3O* and CH_2OH^*. The Co atom is pink while the Cu atom is blue. The C, O, and H atoms are gray, red, and white, respectively

3.2.1 Fe-Based Catalysts

Fe-based catalysts have attracted increasing interests in the field of catalytic conversion of CO_2 to liquid fuels due to their excellent performance in the RWGS reaction and Fischer–Tropsch synthesis (FTS). The precursors of Fe-based catalysts for CO_2 hydrogenation usually include oxides such as Fe_2O_3, Fe_3O_4, $Fe_2C_2O_4$, etc. After pre-treatment (e.g., reduction by H_2 or carbonization by CO) and CO_2 hydrogenation reaction, the catalyst may contain multiple phases such as metallic iron, iron oxide, and iron carbide. Zhu et al. evidenced that χ-Fe_5C_2 is the main active phase for carbon chain growth compared with Fe_3O_4 and Fe_3C by combining (quasi) in situ structural characterizations and DFT calculations [40, 41]. In terms of the mechanism of CO_2 hydrogenation to long-chain hydrocarbons, a two-step process over Fe-based catalysts has been proposed: CO_2 first converts to CO via the RWGS reaction and CO is then transformed to hydrocarbons through the FTS process.

Due to the complexity of reaction pathways for long-chain hydrocarbons formation, most computational work focused on the initial activation of CO_2 and key C_1 intermediates formation on various Fe-based surface models. $Fe_3O_4(111)$ has been found to be a highly active surface. As shown in Fig. 3.6a, the Fe_{tet1}- and $Fe_{oct2\text{-}tet1}$-terminated surfaces are the most stable terminations for $Fe_3O_4(111)$ which are commonly used in the study of reaction mechanism for CO_2 activation and hydrogenation. Su et al. investigated the adsorption of CO_2 over Fe_{tet1}- and $Fe_{oct2\text{-}tet1}$-terminated $Fe_3O_4(111)$ surfaces [42]. The adsorbed CO_2 molecules remain in the linear structure over the Fe_{tet1}-terminated $Fe_3O_4(111)$, and the calculated adsorption energy is – 0.43 eV via a physical adsorption mode. Thus, CO_2 activation is insufficient on this surface. By contrast, on $Fe_{oct2\text{–}tet1}$-terminated $Fe_3O_4(111)$ surface the CO_2 molecule has been effectively activated, as can be seen from the elongated C=O bond and the bent molecular configuration. The adsorption energy of most stable adsorption structure is – 1.36 eV, much larger than that on Fe_{tet1}-terminated $Fe_3O_4(111)$. Thus, CO_2 is chemisorbed and sufficiently activated over the $Fe_{oct2\text{-}tet1}$-terminated $Fe_3O_4(111)$ surface. These calculation results indicate that CO_2 adsorption and activation behaviors are sensitive to the surface structure and facet of Fe-based catalysts.

Fe_5C_2 is also a main active phase appeared in the reactive environment of CO_2 hydrogenation to liquid fuels. Based on the order in which the reaction and relative Fe active phases appear, some studies concluded that Fe_3O_4 is responsible for the initial RWGS of CO_2, while Fe_5C_2 is the active phase for the FTS reaction of CO generated from CO_2. Computational studies could give a clear explanation of the active sites for initial CO_2 activation and reaction at the molecular level. As shown in Fig. 3.6b, Nie et al. investigated the adsorption of CO_2 over six crystal planes of χ-Fe_5C_2 including (510), (111), $(11\bar{1})$, (100), (110), and $(10\bar{1})$ [43]. The adsorption strength of CO_2 on these facets ranks as $(11\bar{1})$ > (510) > $(10\bar{1})$ > (110) > (111) > (100), and the most stable CO_2 adsorption occurs over the $(11\bar{1})$ plane with an adsorption energy of – 0.95 eV. Different from Fe_3O_4, Fe atoms instead of C atoms interact with the CO_2 molecule and activate it on Fe_5C_2. As shown in Fig. 3.6c, Bai et al. calculated CO_2 adsorption on seven crystal surfaces of χ-Fe_5C_2 [44]. The (010) surface has the strongest adsorption of CO_2, with an adsorption energy of – 1.28 eV. They concluded that the surface structure of Fe_5C_2 has a great influence on the adsorption of CO_2, and Fe-rich surfaces of Fe_5C_2 show better adsorption for CO_2 molecules. Compared with the adsorption energies of CO_2 on Fe_3O_4 and χ-Fe_5C_2, it is found that CO_2 is more strongly adsorbed on Fe_3O_4, indicating that CO_2 is preferentially adsorbed on the Fe_3O_4 phase for activation and further reaction. Furthermore, Liu et al. believed that the adsorbed CO_2 over the surface of Fe_5C_2 is facile to spill to Fe_3O_4, and the reaction heat is – 2.14 eV [44].

The experimental studies proposed that the Fe_3O_4 phase promotes the reduction of CO_2 to CO, which subsequently converts to the targeted liquid fuels over the χ-Fe_5C_2 phase. Specifically, after the adsorption and activation of CO_2, three kinds of reactions can occur over the Fe_3O_4 surface, including the direct dissociation of CO_2 ($CO_2^* \rightarrow$

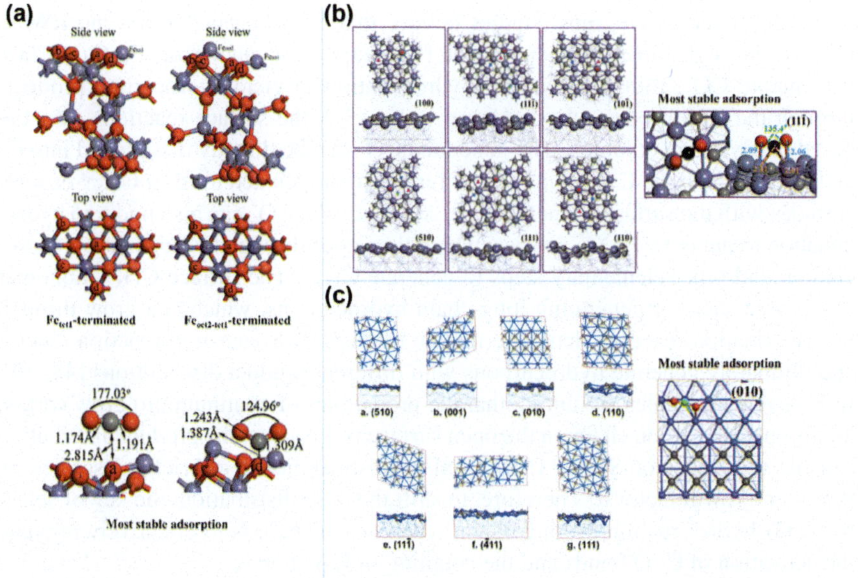

Fig. 3.6 Top and side views of optimized structures of **a** Fe_{tet1}- and $Fe_{oct2\text{-}tet1}$-terminated $Fe_3O_4(111)$ surfaces [42]. **b** Optimized structures of χ-$Fe_5C_2(100)$, $(11\bar{1})$, $(10\bar{1})$, (510), (111), and (110) facets. Reproduced with permission from Ref. [43]. **c** Optimized structures of χ-$Fe_5C_2(510)$, (001), (010), (110), $(11\bar{1})$, $(\bar{4}11)$, and (111) facets. Reproduced with permission from Ref. [44]. The most stable adsorption structures of CO_2 are provided for each work

$CO^* + O^*$), and the hydrogen(H)-assisted pathway of CO_2 conversion along with the generation of HCOO* or COOH* intermediate ($CO_2^* + H^* \rightarrow COOH^*$ or $CO_2^* + H^* \rightarrow HCOO^*$). Relevant computational studies clarified the favorable pathways for CO_2 initial transformation on Fe-based catalysts. Nie et al. revealed that the surface of $Fe_3O_4(111)$ hindered the direct dissociation of adsorbed CO_2 molecules [45]. They believed that CO_2 is adsorbed and bonded to surface O to form CO_3^- species, which then reacts with surface H^* to form O_2COH (i.e., (O)–$COOH^*$) or O_3CH (i.e., $HCOO^*$–(O)) intermediate for subsequent conversion. Liu et al. also suggested that the surface of $Fe_3O_4(111)$ catalyzes the formation of $COOH^*$ intermediate from CO_2, followed by its further dissociation to generate CO^* for subsequent reactions [46]. However, Han et al. reported an opposite opinion on the basis of a comprehensive study of the reaction network [47]. They calculated the free energy diagrams of CO_2-RWGS through the above three pathways over $Fe_3O_4(111)$ at 300 ºC and 2.5 MPa. Although it is easier for the hydrogenation of CO_2 to generate $HCOO^*$ or $COOH^*$ intermediate in the early stage of the reaction compared with CO_2 direct dissociation, it is more difficult to generate HCO^* or CO^* intermediate through further dissociation of the C–O bond in the H-assisted pathway. The overall comparison showed that the direct dissociation of CO_2 to generate CO and H_2O is energetically more favorable, indicating that the direct dissociation of CO_2 dominated the RWGS reaction over $Fe_3O_4(111)$ among the three reaction routes.

As discussed above, most studies believe that CO_2 first undergoes the RWGS reaction through direct dissociation or H-assisted dissociation on Fe_3O_4 surface to generate CO^*, followed by CO^* hydrogenation to yield hydrocarbon products through the FTS process. As displayed in Fig. 3.7, the surface reactions over Fe-based catalysts are complex. The generated CO^* can be directly dissociated into C^* and O^*, in which the C^* species can be further hydrogenated to form $CH_x{}^*$ or form carbides with the surface of metallic Fe. Alternatively, CO^* can also undergo hydrogenation to generate HCO^* species, which then goes through hydrogenation, dissociation, and other elementary steps to generate $CH_x{}^*$. The surface $CH_x{}^*$ monomer is a key species for generating long-chain hydrocarbons, which can grow through oligomerization reaction chains, terminate through hydrogenation reaction chains, and ultimately generate hydrocarbons with different product distributions [48, 49]. In FTS-related studies, assuming that the probability of continuing to grow carbon atoms on each carbon chain on the open surface is the same, the product distribution satisfies the Anderson Schulz Flory (ASF) statistical model, as briefly described by the chain growth factor α. Therefore, to shift the ASF distribution, the key of selective CO_2 hydrogenation to long-chain hydrocarbons here is to effectively regulate the activation of C=O bonds and the coupling of C–C bonds.

Monometallic iron does not have sufficient ability of C=O activation and C–C coupling, resulting in the majority of C_1 products (i.e., CO and CH_4) in CO_2 hydrogenation. Alkali metals (Na, K, etc.) are generally applied as electronic promoters to improve the catalytic performance. For instance, Wang et al. have reported that K-modified Fe–Cu can significantly increase the selectivity to C_{5+} HCs to over 50%, which can be used as ultra-clean liquid fuels [50, 51]. K can effectively promote the chemisorption and dissociation of CO_2 and weaken the adsorption of H_2 to create a C-rich and H-lean surface environment [52]. It enhances the carburization of iron species and suppresses excessive hydrogenation, facilitating the chain growth. As a result, compared with Fe–Cu catalyst, the C_{5+} HCs yield on Fe–Cu–K catalyst increases 7.8-fold from 2.3 to 18.1% [53]. Besides, the type of alkali promoter also has an important impact. The kinetic experiments performed by Yang et al. demonstrated that compared with Li and Na, the K, Rb, and Cs with larger ionic radius

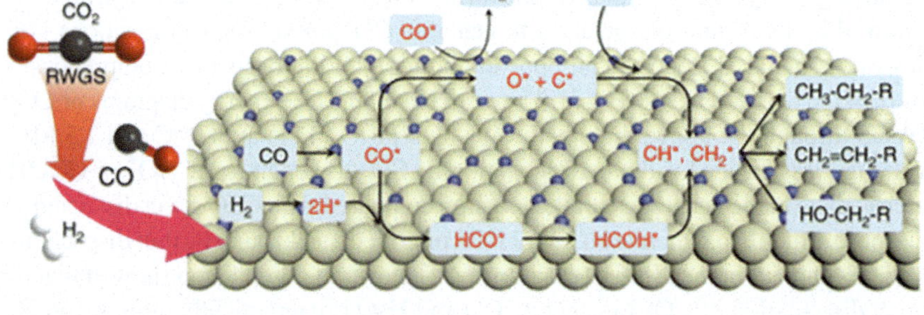

Fig. 3.7 CO_2-FTS reaction to C_{2+} products. Reproduced with permission from Ref. [49]

could better inhibit the generation of side product CH_4 and increase the proportion of long-chain hydrocarbons [54].

Transition and main group metals (Cu, Zn, Mn, etc.) are also commonly used as promoters for Fe-based catalysts. Satthawong et al. have demonstrated that adding a small amount of transition metals (M=Cu, Co, Ni, Pd) can significantly promote the activity of Fe catalyst and the selectivity to C_{2+} HCs thus making the Fe–M catalysts potential candidates for further development [55]. On Al_2O_3-supported Fe-based catalyst, the incorporation of Cu increases the CO_2 conversion and C_{5+} selectivity because it promotes the adsorption of CO_2 and intermediates of light olefins thus facilitating further carbon chain growth to C_{5+} HCs [56]. Choi et al. reported that the delafossite-$CuFeO_2$ after reduction and carburization exhibits excellent activity with ~65% liquid hydrocarbons selectivity and only 2 ~ 3% CH_4 selectivity [57]. Cu doping and K promoter have a synergistic promoting effect on the C–C bond coupling. For example, the Fe_2O_3 sheet modified by Cu and K prepared using the facet control method exhibits high CO_2 conversion (46.5%) and C_{5+} hydrocarbons selectivity (53.6%), accompanied with low CO (5.3%) and CH_4 selectivity (12.2%) at 300°C [58].

Zn is a good structural promoter which can improve the dispersion of iron species. In situ structural investigations revealed that the migration of Zn to the Fe surface and following formation of ZnO effectively inhibit the oxidation of FeC_x [59]. Chen et al. plumbed the effect of different metal promoters on activated carbon (AC) supported Fe-based catalysts and found that the Zn–KFe/AC catalyst possesses excellent performances and a good stability for the generation of C_{5+} hydrocarbons [60]. On the Na-promoted $ZnFe_2O_4$ catalyst, Fe species is uniformly dispersed. The combination of Zn and Na with Fe increases the CO_2 adsorption and promotes the formation of Fe_5C_2 for boosting the selectivity to C_{5+} HCs of 51.6% with a CO_2 conversion of 34.0% [61].

Mn, as both structural and electronic promoters, can improve the dispersion of Fe, promote the reduction of iron oxide, and facilitate the formation of iron carbide [62]. Zhao et al. reported a series of modified Fe-based catalysts, while the Mn–Fe_5C_2 achieves a C_{5+} yield as high as 1282.7 $g_{fuel}kg_{cat}^{-1}\,h^{-1}$ at CO_2 conversion of 44.9% and maintains low selectivity of CO (9.6%) and CH_4 (12.8%), respectively [63]. Multistructural investigation demonstrated that the introduction of Mn promotes the formation of iron carbide, and also inhibits its agglomeration via the geometric effect.

Some noble metal promoters also have positive effects on CO_2 hydrogenation to C_{5+} HCs. The introduction of Pd promotes the reduction of Fe species and the ability of activating H_2 and CO_2. In addition, the formation of Pd–Fe alloy accelerates dissociated H from active H pool spillover to Fe active site to form CH_x and promote the carbon chain growth [64]. The K and Pd co-promoted $K_{0.016}Pd_{0.12}Fe$ catalyst possesses high C_{5+} HCs selectivity of 70.6% (CO-free) and CO_2 conversion of 41.8%. The effect of rare Earth metals (La, Ce, Y) on the CO_2 hydrogenation to C_{5+} HCs over Al_2O_3-supported Fe catalysts has also been explored [65]. With the addition of rare Earth metal, the selectivity of CH_4 and C_{2-4} HCs in hydrocarbons decreases, while the selectivity of CO and C_{5+} HCs increases. The addition of rare Earth metals significantly regulates the adsorption of CO_2 and H_2 on the catalyst surface, thereby decreases the surface H/C ratio to improve the chain growth ability.

The supports also play a crucial role in promoting CO_2 hydrogenation. The interaction between metals and supports has an impact on the catalyst structure and electronic properties, and some supports can improve the adsorption of reactant molecules. Liu et al. physically mixed Fe_5C_2 nanoparticles with K-modified supports such as SBA-15, B-ZSM-5, TiO_2, activated carbon, and Al_2O_3 for CO_2 hydrogenation [66]. The Al_2O_3 carrier significantly reduces the generation of CO and CH_4, promoting the generation of C_{5+} long-chain hydrocarbons. The acidic sites of Al_2O_3 can also participate in CO_2 hydrogenation. As illustrated in Fig. 3.8a, Khan et al. prepared a $FeAlO_x$ catalyst containing Fe_3O_4 and amorphous AlO_x using the co-precipitation method [67]. AlO_x can enhance CO_2 and H_2 adsorption, and promote RWGS, while its acidic site can facilitate CH_x^* coupling. The $FeAlO_x$ catalyst with moderate acidity possesses 20% CO_2 conversion with high C_{5+} HCs selectivity of 62.4%. Kim et al. reported a K-promoted $Fe/CuAl_2O_4$ catalyst with high C_{5+} HCs productivity of 429.4 mL g_{cat}^{-1} h^{-1}, while during the reduction process, $CuAl_2O_4$ reconstructs to form Cu^0 and AlO_x support [68]. The interface with Fe_3O_4, Fe_xC_y, and AlO_x ensures its high catalytic performance.

In addition, carbon materials have the advantages of high specific surface area and promoting the adsorption of CO_2. The weak interactions with Fe facilitate the carbonization of Fe species during the reaction. Compared with multi-walled carbon nanotubes (MWNTs), single-walled carbon nanotubes (SWNTs) can better facilitate the generation of long-chain hydrocarbons [69]. SWNTs with larger curvature are conducive to electron transfer, promoting C=O bond breakage and CH_x^* monomer formation. Additionally, the residence time of short-chain hydrocarbons on FeK/SWNTs surface is longer to further chain growth, and thereby the FeK/SWNTs exhibits a higher C_{5+} HCs selectivity of 50% with high CO_2 conversion of 52.7%. As shown in Fig. 3.8b, Hwang et al. reported a Fe–Co alloy catalyst derived from N-coordinated Co single atom mesoporous carbon for CO_2 hydrogenation [70]. It exhibits excellent liquid fuels selectivity of 42.4% and CO_2 conversion of 51.7%. Mesoporous carbon materials ensure the stable formation of Co–Fe alloy and carbide during the reaction.

Fe-based catalysts can be combined in series with zeolites to further undergo oligomerization, isomerization, aromatization, etc., on the acidic sites to generating liquid fuel components (Fig. 3.9). The 10MR zeolites favored the formation of $C_{5–11}$ HCs, while both 8MR and 12MR zeolites could not promote the C–C chain growth [71]. The series combination method also has a significant impact on the catalyst. The intimate contact between Fe-based catalysts and zeolites can lead to the migration of alkali metal promoters to acidic sites, resulting in the deactivation of zeolite catalysts. The FeMnK + H-ZSM-5 catalysts offer a 70% selectivity of $C_{5–11}$ hydrocarbons together with a 17% selectivity of $C_{2–4}$ olefins. Wei et al. reported the Na–Fe_3O_4/H-ZSM-5 catalysts for directly converting CO_2 into a gasoline fuel [72]. It exhibits 78% selectivity to $C_{5–11}$ as well as low CH_4 and CO selectivity, and gasoline fractions are mainly iso-paraffins and aromatics thus favoring the octane number. The iso-paraffin/aromatic ratio in gasoline-range hydrocarbons can be tuned by altering

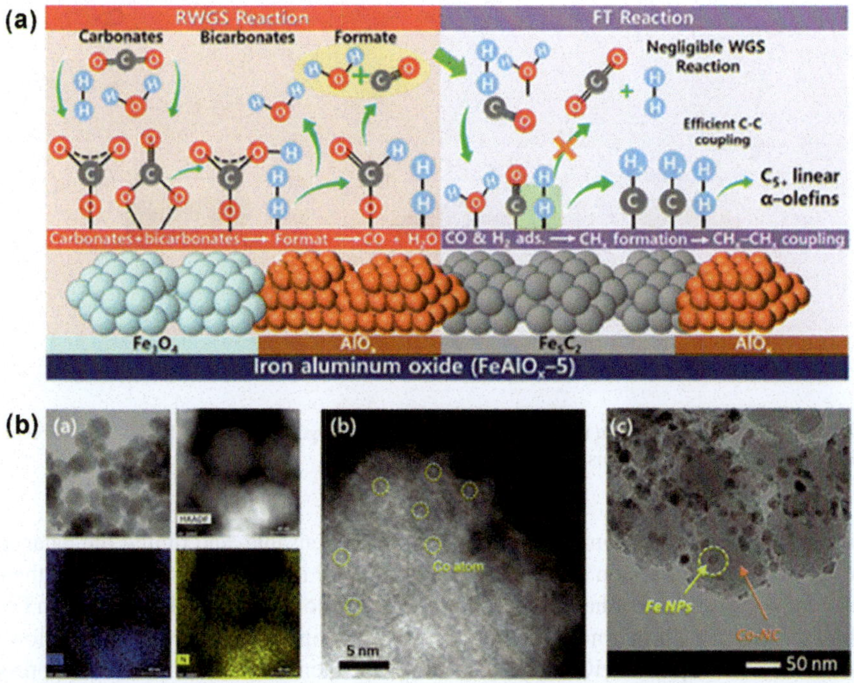

Fig. 3.8 Supported Fe-based catalysts for CO_2 hydrogenation to liquid C_{5+} hydrocarbons. **a** Proposed reaction mechanism over the $FeAlO_x$ composite catalyst. Reproduced with permission from Ref. [67]. **b** Morphological information of K-modified Fe–Co alloy supported on N-coordinated carbon material. Reproduced with permission from Ref. [70]

zeolite type and the integration manner of multifunctional catalyst. Additionally, this catalyst has been applied in a 1000 ton/year unit for CO_2 hydrogenation to gasoline. The recycle CO_2 conversion has reached 85.1% with gasoline selectivity to 76.1%, and the octane number, iso-paraffins, and aromatics content of gasoline products have reached the standard. Connecting iron-based catalysts in series with various zeolites can improve the composition of iso-paraffins in CO_2 hydrogenation products [73]. NaFe, SAPO-11, and H-ZSM-5 form the triple-bed system, which shows the highest selectivity of gasoline to 71.7% in hydrocarbons with a maximum of 38.2% iso-paraffins possessing a research octane number value of 91.7 at CO_2 conversion of 31.2%. SAPO-11 located on the second bed promotes the isomerization of the short-chain hydrocarbons generating from NaFe, resulting in improved isomerization activity with reduced aromatization.

The main catalytic conditions and data of representative Fe-based and corresponding tandem catalysts are listed in Table 3.1. To sum up, we can see that the alkali metal promoters, especially as Na and K, play important roles in increasing the CO_2 conversion and C_{5+} selectivity. However, most of Fe-based catalysts are limited by the high CO selectivity for their practical applications. It indicates that

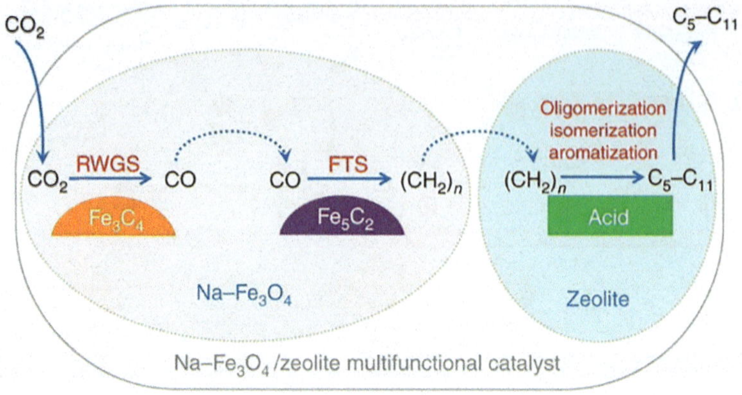

Fig. 3.9 Reaction scheme on tandem catalyst for CO_2 hydrogenation to gasoline-range hydrocarbons. Reproduced with permission from Ref. [72]

more active sites for promoting CO-FTS are indispensable, and tuning the balance of various active sites on catalyst surfaces are quite important for optimizing their synergistic effect and enhancing CO_2 conversion to long-chain hydrocarbons on Fe-based catalysts. Furthermore, optimizing the reaction route which directly follows the intermediate path of $HCOO^*$ to form CH_x^* and then to hydrocarbons can suppress the side-reaction toward CO, which is also beneficial for designing better Fe-based catalysts.

3.2.2 *Cobalt-Based Catalysts*

Cobalt-based catalyst is well-responsible for the conversion of syngas to liquid fuels at relatively low temperatures owing to its promoting effect on C–C coupling [74, 75]. Metallic Co is considered as the active site while its phase transformation to CoO_x or CoC_x undergoing the oxidation by H_2O or carbonization would result in the irreversible catalyst deactivation [76–78]. However, for CO_2 hydrogenation, the selectivity shifts to CH_4 and other short-chain $C_2 \sim C_4$ alkanes because the hydrogenation reactions dominant on the surface with high H_2/CO_2 coverage and low CO coverage [79–81]. Considering exorbitant CO concentration limits the CO_2 conversion, the surface reaction environment with appropriate $CO_2/H_2/CO$ partial pressures is indispensable for the synthesis of liquid fuels [82]. Hence, how to tune the adsorption/activation of reactants and the concentrations of some key intermediates are of great importance for enhancing CO_2 hydrogenation to liquid fuels on Co-based catalysts.

Computational studies have shed light on the adsorption and activation of CO_2 over Co-based catalysts. O'Shea et al. investigated the adsorption and activation of CO_2 over Co(111), Co(100), and Co(110) surfaces, and concluded that Co(110) had

Table 3.1 Catalytic performance of Fe-based catalysts for the synthesis of liquid C_{5+} hydrocarbons in CO_2 hydrogenation

Catalyst	Reaction conditions					C_{5+} HCs			References
	T (°C)	p (MPa)	H_2/CO_2 ratio	CO_2 Conv. (%)	CO Sel. (%)	Sel. (%)[a]	Yield (%)[b]	Stability (h)	
Fe–Cu–K	300	2.5	3	35.7	–	50.7[c]	18.1	~ 70	[53]
10Fe3Cu1K/Al_2O_3	400	3	3	41.7	26.5	18.7	5.8	–	[56]
$CuFeO_2$-24	300	1	3	16.7	31.4	~ 65	7.4	–	[57]
3 K/6Cu–Fe–S	300	2	3	46.5	5.3	53.6	23.6	205	[58]
$ZnFe_2O_4$	340	1	3	34.0	11.7	58.5	17.5	–	[61]
$K_3/FeMn_{10}Ti_{20}$	320	5	2.8	34.9	9.7	51.0	16.0	–	[63]
$K_{0.016}Pd_{0.12}Fe$	340	3	3	41.8	33.8	70.6	20.0	100	[64]
$Pd_{0.12}/K_{0.01}Fe$	340	3	3	46.6	13.7	48.2	18.6	–	[64]
Fe–K/La–Al_2O_3	300	1	3	24.5	37.5	57.9	8.8	–	[65]
Fe–K/Ce–Al_2O_3	300	1	3	21.2	34.2	60.6	8.4	–	[65]
Fe–K/Y–Al_2O_3	300	1	3	22.7	40.6	59.8	8.0	–	[65]
Fe_5C_2-10 K/a-Al_2O_3	400	3	3	44.8	23.2	33.9	11.6	~ 35	[66]
$FeAlO_x$-5	330	3.5	3	36.8	30.5	77.0	19.7	450	[67]
22Fe3K/$CuAl_2O_4$	320	3	3	41.9	10.2	54.5	20.5	~ 240	[68
FeK/SWNTs	340	2	3	52.7	9.6	55.4	26.4	120	[69]
FeK/Co-NC	300	2.5	3	51.7	–	–	21.9	100	[70]
FeMnK/H-ZSM-5	320	3	3	32.0	25.0	61.5	14.7	~ 25	[71]
Na–Fe_3O_4/H-ZSM-5	320	3	3	22.0	20.1	79.4	13.9	1000	[72]
NaFe/SAPO-11/ZSM-5	320	3	3	31.2	13.2	72.5	19.6	50	[74]

[a] C_{5+} HCs selectivity excluding CO
[b] Yield = CO_2 Conv. × (1-CO Sel.) × C_{5+} HCs Sel
[c] C_{5+} HCs selectivity including CO

the strongest adsorption ability of CO_2 with an adsorption energy of – 0.61 eV [83]. Have et al. concluded that the degree of adsorption activation of CO_2 over Co(110) is more significant than that over CoO(100), with stronger adsorption energy (– 0.66 vs. – 0.35 eV), bigger bent angle of O=C=O bond and longer elongated C=O bond of CO_2 molecule [84]. The adsorption properties of CO_2 over the mainly exposed (101) and (020) surfaces of Co_2C nanoprisms and (111) surface of Co_2C nanospheres were investigated by Zhang et al. [85, 86] The calculation results showed that the adsorption energy of CO_2 over $Co_2C(111)$ is – 0.23 eV, while that obtained over the (101) and (020) facet of Co_2C nanoprisms are – 0.52 and – 0.76 eV, respectively, indicating that the catalyst morphology can impact the dominantly exposed crystal planes of Co_2C catalyst, then influencing CO_2 adsorption and activation.

Recent studies highlight that the well-controlled Co-based bimetallic catalysts with good geometric or electronic interactions are quite promising in CO_2 hydrogenation. He et al. discovered a Co_6/MnO_x catalyst with a high selectivity to liquid hydrocarbons (C_{5+} HCs) in a batch reactor [87]. The obtained C_5~C_{26} hydrocarbons (mostly *n*-paraffin) achieve 53.2% at 200 °C, 8 MPa. CO_2 and H_2-TPD results reveal that the MnO_x incorporation boosts the adsorption of CO_2 but inhibits that of H_2. The C-rich and H-lean surface environment suppresses the rapid hydrogenation and enables the carbon chain growth to C_{5+} products. Mechanistic investigations including the in situ IR and isotopic tracing experiments using ^{13}CO suggest that initially activated CO_2, which exists as the forms of carboxylate ($CO_2^{\delta-}$), formate ($HCOO^*$), and methoxy (CH_3O^*), converts to CH_x^* ($x = 2/3$) species without the formation of CO^* intermediate. This inferred reaction pathway is different from the present mainstream CO_2-FTS catalyst, while it provides an opportunity for the synthesis of valuable fuels directly from CO_2 without undergoing the CO-mediated tandem route considering the Co-based catalyst does not well response to the RWGS reaction.

For fixed-bed reactor, which is more common in the practical applications, the direct CO_2 hydrogenation to C_{5+} hydrocarbons on Co-based catalyst is extremely challenging. Compared with batch reactors, the C–C coupling process is limited here due to the short residence time. It demands for a second component or promoter to increase the concentration of CO intermediate [88]. Among them, Fe-based components are well-responsible for the RWGS reaction and therefore its introduction into Co-based material to form the Co–Fe bimetallic catalyst is considered as a potential strategy to increase the CO partial pressure and create a C-rich surface environment. For instance, Satthawong et al. reported a strong synergy of Co-Fe bimetallic catalysts for CO_2 hydrogenation to higher hydrocarbons [89]. As shown in Fig. 3.10a, we can see that the CO_2 conversion increases with the Co content, while the optimum STY of $C_2 \sim C_7$ products is obtained at the Co/(Co + Fe) atomic ratio of 0.5, indicating that the excessive Co increases the selectivity of C_1 products. The addition of K promoter further improves the CO_2 conversion and the selectivity of multi-carbon products. Noritatsu Tsubaki and co-workers compared the catalytic properties of Na or K-modified $ZnFe_2O_4$ and K–$Zn(FeCo)_2O_4$ catalysts, and interestingly, they found that the introduction of Co inhibits the formation of long-chain hydrocarbons (Fig. 3.10b) [90]. As displayed in Fig. 3.10c and d, the mechanistic investigations show that the initial CO_2 conversion to CO^* intermediate occurs on Fe_3O_4 sites, and it would further transfer to neighboring Co sites for the cleavage of C=O bond, providing more C_1 building blocks for chain growth. However, excessive Co sites also accelerate the formation of CH_4 and suppress that of long-chain products. It gives a sign that the proportion of Co in Co–Fe catalysts plays an important role for the synthesis of liquid hydrocarbons. However, clearly clarifying the active Co–Fe phases is still challenging because they are highly sensitive to the air, and the continuous development of in situ or operando characterization technologies would provide more opportunities.

Considering Cu and Mo components are also active for promoting the RWGS reaction, several Co–Cu or Co–Mo bimetallic catalysts have been applied to facilitate the generation of liquid hydrocarbons [91–93]. For example, Shi et al. studied a series of alkali metal (Li, Na, K, Rb, and Cs) modified Co–Cu/TiO_2 catalysts. XRD results and corresponding catalytic performance indicate that the metallic Co possibly cooperates with metallic Cu to promote the highly effective conversion to C_{5+} HCs [92]. The properties and contents of the promoters have important impacts on the surface CO_2/H_2 adsorption and activation. The Na-modified Co–Cu/TiO_2 catalyst exhibits an optimum selectivity of C_{5+} HCs up to 42.1% with a CO_2 conversion of 18.4% at 250 °C and 5 MPa. The catalytic performance remains basically stable within ~ 200 h. Owen et al. evaluated the supported Na–Co–Mo catalyst in CO_2 hydrogenation and

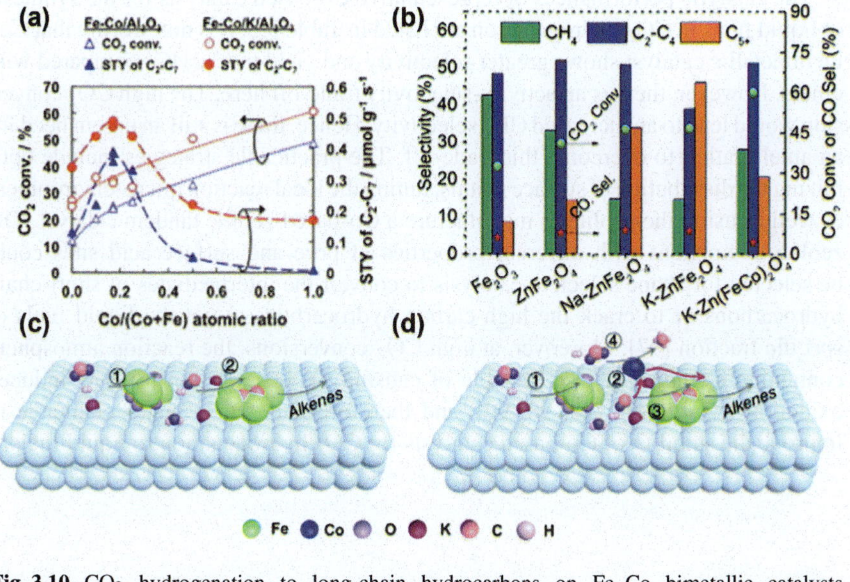

Fig. 3.10 CO_2 hydrogenation to long-chain hydrocarbons on Fe–Co bimetallic catalysts. **a** Catalytic performance with respect to Co/(Co + Fe) atomic ratio on Fe–Co/Al_2O_3 and K-modified Fe–Co/Al_2O_3 catalysts. Reproduced with permission from Ref [89]. **b** Catalytic performance on a series of $ZnFe_2O_4$, modified $ZnFe_2O_4$, and Co-doped $ZnFe_2O_4$ catalysts. Reproduced with permission from Ref. [90] [. **c** and **d** Proposed reaction scheme for the formation of multi-carbon products on K–$ZnFe_2O_4$ and K–$Zn(FeCo)_2O_4$ catalysts. Reproduced with permission from Ref. [90]

clarified the effect of metal-support interaction [93]. The interaction influences the average size of Co_3O_4 nanoparticles (NPs) and the catalytic results reveal that long-chain products prefer to be generated on TiO_2 and CeO_2-supported catalysts (C_{5+} HCs selectivity as 24.2% and 31.2%) while the size of Co_3O_4 NPs mainly distributes in the range of 20–30 nm. As staged above, it is noted that C_{5+} HCs is generated undergoing the CO-mediated route on most reported Co-based bimetallic catalysts. The interaction and synergistic effect between various components are essential for the formation and following conversion of CO intermediate, and it requires more fundamental investigations and scientific understanding.

Apart from metallic Co, cobalt carbide (Co_2C) is also one of the most important catalysts for CO_x hydrogenation [94, 95]. Compared with metallic Co, the chain growth is relatively weakened on Co_2C, resulting in a more intensive product distribution. Wang et al. designed an in situ synthesized K-modified Co_2C catalyst mainly exposed with (101) and (020) facets, and the optimum selectivity of C_{5+} HCs reaches 28.7% with the co-production of gaseous $C_2 \sim C_4$ olefins of 31.6% [96]. Interestingly, the in situ XRD patterns and quasi in situ XPS spectra evidence that adding moderate amount of H_2O is instrumental in the formation and stabilization of Co_2C phase. As a result, co-feeding 0.5% H_2O into the feed gas inhibits the decomposition of Co_2C to Co and prolongs the lifetime of this K–Co_2C catalyst from ~ 35 h to over 200 h. These results offer important insights into tuning the reaction atmosphere to improve the production of target multi-carbon products on metallic Co– or Co_2C-based catalysts.

The catalytic performances of representative Co-based catalysts for the synthesis of liquid fuels in CO_2 hydrogenation are listed in Table 3.2. It is noteworthy that Co–Fe bimetallic catalyst shows greater selectivity and yield to C_{5+} HCs compared with others. However, there is an activity-selectivity trade-off here. The high CO_2 conversion would lead to an increased CH_4 selectivity. Hence, there is still an urgent need for an amelioration to overcome this trade-off. The practicable strategies include optimizing the distribution of surface metals, tuning the local reaction micro-environment as well as using the zeolite to manufacture a Co-based/zeolite tandem catalyst. The zeolite component with different properties of pore and surface acid sites could be selected for shape selective catalysis to convert the intermediates of short-chain hydrocarbons or to crack the high-carbon hydrocarbons to get the liquid fuels of specific fraction [97]. Moreover, at high CO_2 conversions, the reaction atmosphere containing abundant H_2O is capable of causing the deactivation by H_2O-induced oxidation or sintering [78, 98, 99], and therefore it is still a major challenge to improve the stability of Co-based catalysts for CO_2 hydrogenation to liquid fuels.

Table 3.2 Catalytic performance of representative Co-based catalysts for the synthesis of liquid C_{5+} HCs in CO_2 hydrogenation

Catalyst	Reaction conditions					C_{5+} HCs			References
	T (°C)	p (MPa)	H_2/CO_2 ratio	CO_2 Conv. (%)	CO Sel. (%)	Sel. (%) [a]	Yield (%) [b]	Stability (h)	
Co_6/MnO_x [c]	200	8	1	–	0.4	53.2	15.1 [d]	18	[87]
Na–Co–(Fe_3O_4 + Fe_5C_2)	320	2	3	46.7	9.2	34.0	14.4	36	[90]
Na–Co–Cu/TiO_2	250	5	3	18.4	30.2	42.1	5.4	200	[92]
K–Co–Cu/TiO_2	250	5	3	13.0	35.1	35.1	3.0	200	[91]
Na–Co–Mo/CeO_2	200	0.1	3	15.1	70.2	31.2	1.4	–	[93]
K–Co_2C	300	3	3	21.6	23.0	28.7	4.8	–	[96]

[a] C_{5+} HCs selectivity refers to its proportion in total hydrocarbons based on C-mol.%
[b] C_{5+} HCs yield is calculated according to the equation: Yield = CO_2 Conv. × (1-CO Sel.) × C_{5+} HCs Sel
[c] Co_6/MnO_x catalyst was evaluated in the batch reactor
[d] C_{5+} HCs yield is 15.1 mmol CO_2 g_{cat}^{-1} h^{-1}

3.3 Challenges and Opportunities for CO_2 Hydrogenation to Liquid Fuels

There are several challenges and opportunities for realizing CO_2 hydrogenation to liquid fuels (ethanol and liquid hydrocarbons). First, the complex reaction network, including various competitive reactions (C–C coupling, insertion, hydrogenation, and others), limits the highly selective conversion of CO_2 and H_2 to target liquid fuels. Moreover, the catalyst structure is dependent on the reaction atmosphere and its transformation influences the catalytic performance. How to clarify the structure-performance relationship and control the structural evolution during the reaction is another key challenge. However, considering the separation cost of liquid is considerable, how to get a concentrated product distribution is also of great importance for reducing the overall cost of production process.

Designing effective catalysts for tuning the surface reaction is the most hopeful strategy. In recent years, several advanced technologies such as high-throughput catalyst synthesis and evaluation, 3D printing and "machine learning" are used to screen, discover, and design the novel catalysts [100]. Besides, the reaction can be simulated in the in situ/operando chamber and it significantly improves the understanding of real catalyst structure and catalytic mechanisms. Based on this understanding, a series of modified strategies, including adding the promoters and supports, tuning the surface hydrophilicity/hydrophobicity properties, and introducing the second components are applied to further concentrate the product distribution. These strategies can also be employed for the similar catalyst design for CO_2 hydrogenation to other highly valued chemicals.

References

1. Xu D, Wang Y, Ding M, Hong X, Liu G, Tsang SCE (2021) Advances in higher alcohol synthesis from CO_2 hydrogenation. Chem 7:849–881
2. Zeng F et al (2021) Catalysts design for higher alcohols synthesis by CO_2 hydrogenation: trends and future perspectives. Appl Catal B 291:120073
3. Chandel AK et al (2014) Techno-economic analysis of secondgeneration ethanol in Brazil: competitive, complementary aspects with firstgeneration ethanol, biofuels in Brazil edn. Sprigner
4. Ali SS, Ali SS, Tabassum N (2022) A review on CO_2 hydrogenation to ethanol: reaction mechanism and experimental studies. J Environ Chem Eng 10:106962
5. Zhang F et al (2021) Selective hydrogenation of CO_2 to ethanol over sodium-modified rhodium nanoparticles embedded in zeolite silicalite-1. J Phys Chem C 125:24429–24439
6. Caparrós FJ et al (2018) Remarkable carbon dioxide hydrogenation to ethanol on a palladium/iron oxide single-atom catalyst. ChemCatChem 10:2365–2369
7. Lou Y et al (2021) CeO_2 supported Pd dimers boosting CO_2 hydrogenation to ethanol. Appl Catal B 291:120122
8. Chen J, Zha Y, Liu B, Li Y, Xu Y, Liu X (2023) Rationally designed water enriched nano reactor for stable CO_2 hydrogenation with near 100% ethanol selectivity over diatomic palladium active sites. ACS Catal 13:7110–7121

9. Liu B et al (2018) Effects of mesoporous structure and Pt promoter on the activity of Co-based catalysts in low-temperature CO_2 hydrogenation for higher alcohol synthesis. J Catal 366:91–97
10. Ouyang B, Xiong S, Zhang Y, Liu B, Li J (2017) The study of morphology effect of Pt/Co_3O_4 catalysts for higher alcohol synthesis from CO_2 hydrogenation. Appl Catal A 543:189–195
11. Ye X et al (2020) Highly selective hydrogenation of CO_2 to ethanol via designed bifunctional Ir_1–In_2O_3 single-atom catalyst. J Am Chem Soc 142:19001–19005
12. Zheng K, Li Y, Liu B, Jiang F, Xu Y, Liu X (2022) Ti-doped CeO_2 stabilized single-atom rhodium catalyst for selective and stable CO_2 hydrogenation to ethanol. Angew Chem Int Ed 61:e202210991
13. Bai S, Shao Q, Wang P, Dai Q, Wang X, Huang X (2017) Highly active and selective hydrogenation of CO_2 to ethanol by ordered Pd–Cu nanoparticles. J Am Chem Soc 139:6827–6830
14. Guo H, Shi T-T, Chen J-T, Chen Z-X (2023) Theoretical investigation of the properties of Pd_2Cu intermetallics and CO_2 hydrogenation to ethanol catalyzed on Pd_2Cu (110). J Phys Chem C 127:10097–10106
15. Zhai P et al (2021) Development of direct conversion of syngas to unsaturated hydrocarbons based on Fischer-Tropsch route. Chem 7:3027–3051
16. Huang J et al (2022) Boosting the production of higher alcohols from CO_2 and H_2 over Mn- and K-modified iron carbide. Ind Eng Chem Res 61:7266–7274
17. Huang J et al (2023) The synthesis of higher alcohols from CO_2 hydrogenation over Mn-Cu-K modified Fe_5C_2 and CuZnAlZr tandem catalysts. Front Energy Res 10:1–6
18. Xu D, Ding M, Hong X, Liu G, Tsang SCE (2020) Selective C_{2+} alcohol synthesis from direct CO_2 hydrogenation over a Cs-promoted Cu–Fe–Zn catalyst. ACS Catal 10:5250–5260
19. Xu D, Ding M, Hong X, Liu G (2020) Mechanistic aspects of the role of K promotion on Cu–Fe-based catalysts for higher alcohol synthesis from CO_2 hydrogenation. ACS Catal 10:14516–14526
20. Liu T, Xu D, Song M, Hong X, Liu G (2023) K–ZrO_2 interfaces boost CO_2 hydrogenation to higher alcohols. ACS Catal 13:4667–4674
21. Xu D, Yang H, Hong X, Liu G, Tsang SCE (2021) Tandem catalysis of direct CO_2 hydrogenation to higher alcohols. ACS Catal 11:8978–8984
22. Wang Y et al (2021) Direct conversion of CO_2 to ethanol boosted by intimacy-sensitive multifunctional catalysts. ACS Catal 11:11742–11753
23. Si Z et al (2022) Synthesis of alkene and ethanol in CO_2 hydrogenation on a highly active sputtering CuNaFe catalyst. ACS Sustainable Chem. Eng. 10:14972–14979
24. Yao R et al (2021) Monometallic iron catalysts with synergistic Na and S for higher alcohols synthesis via CO_2 hydrogenation. Appl Catal B 298:120556
25. Zhang Q et al (2023) Hydrogenation of CO_2 to higher alcohols on an efficient Cr-modified CuFe catalyst. Appl Catal B 337:123013
26. Xi X et al (2021) CO_2 Hydrogenation to higher alcohols over K-promoted bimetallic Fe–In catalysts on a Ce–ZrO_2 support. ACS Sustain Chem Eng 5:6235–6249
27. Lin T et al (2021) Cobalt carbide nanocatalysts for efficient syngas conversion to value-added chemicals with high selectivity. Acc Chem Res 54:1961–1971
28. Wang M et al (2023) Stabilizing Co_2C with H_2O and K promoter for CO_2 hydrogenation to C2+ hydrocarbons. Sci Adv 9:eadg0167
29. Zhang S, Liu X, Shao Z, Wang H, Sun Y (2020) Direct CO_2 hydrogenation to ethanol over supported Co_2C catalysts: studies on support effects and mechanism. J Catal 38:86–96
30. Zhang S et al (2021) Tuning the interaction between Na and Co_2C to promote selective CO_2 hydrogenation to ethanol. Appl Catal B 293:120207
31. Zhang S et al (2023) Revealing and regulating the complex reaction mechanism of CO_2 hydrogenation to higher alcohols on multifunctional tandem catalysts. ACS Catal 13:3055–3065
32. Witoon T et al (2022) Enhanced CO_2 hydrogenation to higher alcohols over K-Co promoted In_2O_3 catalysts. Chem Eng J 431:133211

33. An K, Zhang S, Wang H, Li N, Zhang Z, Liu Y (2022) $Co^0 - Co^{\delta+}$ active pairs tailored by Ga-Al-O spinel for CO_2-to-ethanol synthesis. Chem Eng J 433:134606
34. Wang L et al (2018) Selective hydrogenation of CO_2 to ethanol over cobalt catalysts. Angew Chem Int Ed 57:6104–6108
35. Wang L, He S, Wang L, Lei Y, Meng X, Xiao F-S (2019) Cobalt-nickel catalysts for selective hydrogenation of carbon dioxide into ethanol. ACS Catal 9:11335–11340
36. Wang J, Wang T, Xi Y, Gao G, Sun P, Li F (2023) In-situ-formed potassium-modified nickel-zinc carbide boosts production of higher alcohols beyond CH_4 in CO_2 hydrogenation. Angew Chem Int Ed e202311335
37. Yang C et al (2019) The interplay between structure and product selectivity of CO_2 hydrogenation. Angew Chem Int Ed 58:11242–11247
38. Liu S et al (2022) Moderate surface segregation promotes selective ethanol production in CO_2 hydrogenation reaction over CoCu catalysts. Angew Chem Int Ed 61:e202109027
39. Zhang G, Fan G, Zheng L, Li F (2022) Ga-promoted CuCo-based catalysts for efficient CO_2 hydrogenation to ethanol: the key synergistic role of Cu-$CoGaO_x$ interfacial sites. ACS Appl Mater Interfaces 14:35569–35580
40. Zhu J et al (2023) Unveiling the promoting effect of potassium on the structural evolution of iron catalysts during CO_2 hydrogenation. Chem Eng Sci 282:119228
41. Zhu J et al (2022) Dynamic structural evolution of iron catalysts involving competitive oxidation and carburization during CO_2 hydrogenation. Sci Adv 8:eabm3629
42. Su T et al (2016) Density functional theory study on the interaction of CO_2 with $Fe_3O_4(111)$ surface. Appl Surf Sci 378:270–276
43. Nie X, Han G, Song C, Guo X (2022) Computational identification of facet-dependent CO_2 initial activation and hydrogenation over iron carbide catalyst. J CO_2 Util 59:101967
44. Bai Y et al (2022) Theoretical study about adsorbed oxygen reduction over χ-Fe_5C_2: formation of H_2O and CO_2. Mol Catal 524:112236
45. Nie X et al (2018) DFT insight into the effect of potassium on the adsorption, activation and dissociation of CO_2 over Fe-based catalysts. Phys Chem Chem Phys 20:14694–14707
46. Liu B et al (2018) CO_2 formation mechanism in Fischer-Tropsch synthesis over iron-based catalysts: a combined experimental and theoretical study. Catal Sci Technol 8:5288–5301
47. Han SJ et al (2020) Identification of active sites for CO_2 hydrogenation in Fe catalysts by first-principles microkinetic modelling. J. Mater. Chem. A 8:13014–13023
48. Wei J, Yao R, Han Y, Ge Q, Sun J (2021) Towards the development of the emerging process of CO_2 heterogenous hydrogenation into high-value unsaturated heavy hydrocarbons. Chem Soc Rev 50:10764–10805
49. Ye RP et al (2019) CO_2 hydrogenation to high-value products via heterogeneous catalysis. Nat Commun 10:5698
50. Wang W, Jiang X, Wang X, Song C (2018) Fe–Cu bimetallic catalysts for selective CO_2 hydrogenation to Olefin-Rich C_2^+ hydrocarbons. Ind Eng Chem Res 57:4535–4542
51. Wang W et al (2021) CO_2 Hydrogenation to Olefin-Rich hydrocarbons over Fe-Cu bimetallic catalysts: an investigation of Fe-Cu interaction and surface species. Front Chem Eng 3:708014
52. Boreriboon N, Jiang X, Song C, Prasassarakich P (2018) Higher hydrocarbons synthesis from CO_2 hydrogenation over K- and La-promoted Fe–Cu/TiO_2 catalysts. Top Catal 61:1551–1562
53. Hwang S-M et al (2019) Mechanistic insights into Cu and K promoted Fe-catalyzed production of liquid hydrocarbons via CO_2 hydrogenation. J CO_2 Util 34:522–532
54. Yang Q et al (2022) Identifying performance descriptors in CO_2 hydrogenation over iron-based catalysts promoted with alkali metals. Angew Chem-Int Edit 61:e202116517
55. Satthawong R, Koizumi N, Song C, Prasassarakich P (2013) Comparative study on CO_2 hydrogenation to higher hydrocarbons over Fe-based bimetallic catalysts. Top Catal 57:588–594
56. Liu J et al (2018) Selective CO_2 hydrogenation to hydrocarbons on Cu-promoted Fe-based catalysts: dependence on Cu–Fe interaction. ACS Sustain Chem Eng 6:10182–10190
57. Choi YH et al (2017) Carbon dioxide Fischer-Tropsch synthesis: a new path to carbon-neutral fuels. Appl Catal B 202:605–610

58. Qin C et al (2023) Facet-controlled Cu-doped and K-promoted Fe_2O_3 nanosheets for efficient CO_2 hydrogenation to liquid hydrocarbons. Chem Eng J 467:143403
59. Zhang C et al (2021) Unraveling the role of zinc on bimetallic Fe_5C_2–ZnO catalysts for highly selective carbon dioxide hydrogenation to high carbon α-olefins. ACS Catal 11:2121–2133
60. Chen J et al (2023) Benchmarking promoters of Fe/activated carbon catalyst for stable hydrogenation of CO_2 to liquid hydrocarbons. Appl Catal B 325:122370
61. Choi YH et al (2017) Sodium-containing spinel zinc ferrite as a catalyst precursor for the selective synthesis of liquid hydrocarbon fuels. Chemsuschem 10:4764–4770
62. Zhang Z et al (2020) Insights into the regulation of FeNa catalysts modified by Mn promoter and their tuning effect on the hydrogenation of CO_2 to light olefins. J Catal 390:12–22
63. Zhao H et al (2021) Multi-promoters regulated iron catalyst with well-matching reverse water-gas shift and chain propagation for boosting CO_2 hydrogenation. J CO_2 Util 52:101700
64. Xiong L et al (2022) Highly selective hydrogenation of CO_2 to C_5+ hydrocarbons over Fe catalysts copromoted by K with Pd. J Environ Chem Eng 10:108407
65. Zhang L et al (2020) The rare earth elements modified FeK/Al_2O_3 catalysts for direct CO_2 hydrogenation to liquid hydrocarbons. Catal Today 356:613–621
66. Liu J et al (2018) Direct transformation of carbon dioxide to value-added hydrocarbons by physical mixtures of Fe_5C_2 and K-modified Al_2O_3. Ind Eng Chem Res 57:9120–9126
67. Khan MK et al (2020) Selective conversion of carbon dioxide into liquid hydrocarbons and long-chain α-olefins over Fe-amorphous AlO_x bifunctional catalysts. ACS Catal 10:10325–10338
68. Kim Y, Song Y, Kim Y, Na K (2022) Multifunctional long-lived catalysts for direct hydrogenative conversion of CO_2 to liquid hydrocarbons with upscaling C_{5+} productivity. J Mater Chem A 10:21862–21873
69. Wang S et al (2020) Iron-potassium on single-walled carbon nanotubes as efficient catalyst for CO_2 hydrogenation to heavy olefins. ACS Catal 10:6389–6401
70. Hwang S-M et al (2021) Atomically alloyed Fe–Co catalyst derived from a N-coordinated Co single-atom structure for CO_2 hydrogenation. ACS Catal 11:2267–2278
71. Li Y et al (2023) Direct conversion of carbon dioxide into liquid fuels and chemicals by coupling green hydrogen at high temperature. Appl Catal B 324:122299
72. Wei J et al (2017) Directly converting CO_2 into a gasoline fuel. Nat Commun 8:15174
73. Noreen A et al (2020) One-pass hydrogenation of CO_2 to multibranched isoparaffins over bifunctional zeolite-based catalysts. ACS Catal 10:14186–14194
74. Lin T et al (2022) Advances in selectivity control for Fischer-Tropsch synthesis to fuels and chemicals with high carbon efficiency. ACS Catal 12:12092–12112
75. Rommens KT, Saeys M (2023) Molecular views on Fischer-Tropsch synthesis. Chem Rev 123:5798–5858
76. Wolf M, Fischer N, Claeys M (2020) Water-induced deactivation of cobalt-based Fischer-Tropsch catalysts. Nat Catal 3:962–965
77. ten Have IC, Weckhuysen BM (2021) The active phase in cobalt-based Fischer-Tropsch synthesis. Chem Catal 1:339–363
78. Rahmati M, Safdari MS, Fletcher TH, Argyle MD, Bartholomew CH (2020) Chemical and thermal sintering of supported metals with emphasis on cobalt catalysts during Fischer-Tropsch synthesis. Chem Rev 120:4455–4533
79. Li W et al (2019) CO_2 hydrogenation on unpromoted and M-promoted Co/TiO_2 catalysts (M = Zr, K, Cs): effects of crystal phase of supports and metal-support interaction on tuning product distribution. ACS Catal 9:2739–2751
80. Fan T et al (2021) Cobalt catalysts enable selective hydrogenation of CO_2 toward diverse products: recent progress and perspective. J Phys Chem Lett 12:10486–10496
81. Zhou Z, Gao P (2022) Direct carbon dioxide hydrogenation to produce bulk chemicals and liquid fuels via heterogeneous catalysis. Chin J Catal 43:2045–2056
82. Liu R et al (2023) Reactive separations of CO/CO_2 mixtures over Ru–Co single atom alloys. ACS Catal 13:2449–2461

83. de la Peña O'Shea VA, González S, Illas F, Fierro JLG (2008) Evidence for spontaneous CO_2 activation on cobalt surfaces. Chem Phys Lett 454:262–268
84. Have ICT et al (2022) Uncovering the reaction mechanism behind CoO as active phase for CO_2 hydrogenation. Nat Commun 13:324
85. Zhang M et al (2022) Theoretical and experimental insights into CO_2 formation on Co_2C catalysts in syngas conversion to value-added chemicals. Appl Surf Sci 602:154379
86. Zhang S et al (2021) Tuning the interaction between Na and Co_2C to promote selective CO_2 hydrogenation to ethanol. Appl Catal B 293:120207
87. He Z et al (2019) Synthesis of liquid fuel via direct hydrogenation of CO_2. Proc Natl Acad Sci 116:12654–12659
88. Khangale PR, Meijboom R, Jalama K (2020) CO_2 hydrogenation to liquid hydrocarbons via modified Fischer–Tropsch over alumina-supported cobalt catalysts: Effect of operating temperature, pressure and potassium loading. J CO_2 Util 41:101268
89. Satthawong R, Koizumi N, Song C, Prasassarakich P (2013) Bimetallic Fe–Co catalysts for CO_2 hydrogenation to higher hydrocarbons. J CO_2 Util 3–4:102–106
90. Guo L et al (2020) Spinel-structure catalyst catalyzing CO_2 hydrogenation to full spectrum alkenes with an ultra-high yield. Chem Commun 56:9372–9375
91. Shi Z et al (2018) Direct conversion of CO_2 to long-chain hydrocarbon fuels over K–promoted $CoCu/TiO_2$ catalysts. Catal Today 311:65–73
92. Shi Z et al (2018) Effect of alkali metals on the performance of $CoCu/TiO_2$ catalysts for CO_2 hydrogenation to long-chain hydrocarbons. Chinese J. Catal. 39:1294–1302
93. Owen RE et al (2016) Effect of support of Co-Na-Mo catalysts on the direct conversion of CO_2 to hydrocarbons. J CO_2 Util 16:97–103
94. Zhong L et al (2016) Cobalt carbide nanoprisms for direct production of lower olefins from syngas. Nature 538:84–87
95. Lin T et al (2021) Cobalt carbide nanocatalysts for efficient syngas conversion to value-added chemicals with high selectivity. Acc Chem Res 54:1961–1971
96. Wang M et al (2023) Stabilizing Co_2C with H_2O and K promoter for CO_2 hydrogenation to C_{2+} hydrocarbons. Sci Adv 9:eadg0167
97. Li J et al (2018) Integrated tuneable synthesis of liquid fuels via Fischer-Tropsch technology. Nat Catal 1:787–793
98. Claeys M et al (2015) Impact of process conditions on the sintering behavior of an alumina-supported cobalt Fischer-Tropsch catalyst studied with an in situ magnetometer. ACS Catal 5:841–852
99. Bezemer GL, Remans TJ, Bavel APV, Dugulan AI (2010) Direct evidence of water-assisted sintering of cobalt on carbon nanofiber catalysts during simulated Fischer-Tropsch conditions revealed with in Situ Mössbauer spectroscopy. J Am Chem Soc 132:8540–8541
100. Gao P, Zhang L, Li S, Zhou Z, Sun Y (2020) Novel heterogeneous catalysts for CO_2 hydrogenation to liquid fuels. ACS Cent Sci 6:1657–1670

Chapter 4
Status of Catalyst Development for CO_2 Hydrogenation to Platform Chemicals CH_3OH and CO

Qingxin Yang and Evgenii V. Kondratenko

Abstract To ensure sustainable development in the chemical industry, new catalytic materials and large-scale processes are required for minimizing the carbon footprint. Many efforts have been made in the valorization of carbon dioxide (CO_2), the primary greenhouse gas, through thermo-, electro-, and photocatalysis over the past decades. Within this concept, the direct hydrogenation of CO_2 with hydrogen (H_2) produced on a CO_2-free basis has been attracting a special attention due to its high efficiency to close carbon cycles. Among various products formed through CO_2 hydrogenation, methanol (CH_3OH) and carbon monoxide (CO) stand out due to their multiple applications as platform chemicals. The former is also used as a fuel with high energy density. This contribution critically analyzes the up-to now developed catalysts for CO_2 hydrogenation to methanol or CO via thermo catalytic processes, with a special focus being put on the discussion of active phases/sites and structure–activity relationships reported in previous studies. The suggested reaction mechanisms considering both selective and side reactions over different catalysts have also been briefly reviewed.

4.1 Introduction

The use of CO_2 as a feedstock to produce value-added chemicals offers an opportunity for reducing the environmental impact of this greenhouse gas, realizing the goal of carbon neutrality in our society, and lowering our dependence on fossil-based products. As CO_2 is chemically inert, a reducing agent or/and input of energy are required. Thanks to the advances in the production of green H_2 generated by renewable energies, various thermocatalytic CO_2 hydrogenation reactions can become promising for large-scale industrially relevant applications [1, 2]. Methanol (Eq. 4.1) and carbon monoxide (Eq. 4.2) are the most prominent products formed from CO_2 considering

Q. Yang · E. V. Kondratenko (✉)
Leibniz-Institut für Katalyse e. V., Albert-Einstein-Straße 29a, 18059 Rostock, Germany
e-mail: evgenii.kondratenko@catalysis.de

G. Zhang et al. (eds.), *Advances in CO_2 Utilization*, Green Chemistry and Sustainable Technology, https://doi.org/10.1007/978-981-99-8822-8_4

their versatile applications in the chemical industry. For example, the methanol-to-olefin reaction is the basis of various large-scale technologies to produce lower olefins [3]. The most profound examples of CO utilization include Fischer–Tropsch synthesis [4] and various hydroformylation reactions [5].

$$CO_2 + 3H_2 \leftrightarrows CH_3OH + H_2O \quad \Delta H_{298\,K} = -49.3\ kJ\ mol^{-1} \tag{4.1}$$

$$CO_2 + H_2 \leftrightarrows CO + H_2O \quad \Delta H_{298\,K} = +41.2\ kJ\ mol^{-1} \tag{4.2}$$

One of the most challenging issues in the synthesis of methanol or carbon monoxide from CO_2 is the control of the selectivity due to the presence of side competing reactions [6, 7]. In addition, the non-carbon-containing product, H_2O, of these reactions can aggravate catalyst deactivation due to restructuring/sintering of active species [8, 9]. The latter structural changes also are caused by high reaction temperatures required for achieving high CO_2 conversion in the hydrogenation of CO_2 to CO called as the reverse water gas shift (RWGS) reaction due to its endothermicity. Thus, the ongoing research is focused on (i) identifying physicochemical catalyst properties including catalytically active site/species that are relevant to controlling product selectivity and (ii) elucidating pathways leading to the desired and side products. Such fundamental knowledge is the basis for purposeful catalyst design/preparation. Against this background, we review the progress in the field of heterogeneously catalyzed CO_2 conversion into methanol or carbon monoxide. We also provide our personal view on possible future developments.

4.2 CO_2 Hydrogenation to Methanol

Although CO and CO_2 coexist in industrially used reaction feeds to produce methanol over Cu-based catalysts, the latter is hydrogenated to this alcohol as proven by mechanistic studies with isotopic tracers in 1987 [10]. CO contributes to the production of CO_2 through the water gas shift (WGS) reaction. Since that time various catalysts based on Cu and other metals have been developed and tested for the direct CO_2 hydrogenation to methanol [6–9, 11]. The scope of this section is to describe and to analyze the progress achieved with extensively studied Cu-based catalysts as well as attractive alternatives based on In_2O_3, $ZnZrO_x$ solid solutions, noble metals, or Ni_xGa_y intermetallic materials. Especially we focus on mechanistic aspects relevant for catalyst development.

4.2.1 Cu-Based Catalysts

Cu-based catalysts are the most widely studied materials for CO_2 hydrogenation to methanol due to their high reactivity and low price. To improve their onstream stability and selectivity in methanol synthesis from CO_2 toward a level relevant for industrial applications, various oxidic promoters, e.g., ZnO, Al_2O_3, SiO_2, TiO_2, CeO_2, or their combinations, have been used [6, 9]. A typical commercial methanol synthesis Cu/ZnO/Al_2O_3 catalyst contains the metals in the molar ratio of 6:3:1, respectively, where Al_2O_3 is generally regarded as a structural promoter [12]. Although such catalysts are applied on a large scale, the nature of active sites and the structure-performance relationships are still matters of debates.

4.2.1.1 Active Sites

The debate on the active sites of industrial Cu-based catalysts has been in the literature for a long history, partially due to the complexity of catalyst composition and its dynamic nature in the reaction atmosphere [13, 14]. Although metallic Cu alone converts CO_2 into methanol, the addition of ZnO significantly increases the activity both of Cu single crystals and powder catalysts [15–17]. In their pioneering work in 1991, Clausen et al. [18] concluded on the basis of in situ XRD analysis that metallic Cu is the only crystalline copper phase in an industrial Cu/ZnO/Al_2O_3 catalyst under methanol synthesis conditions. Later, many researchers focused on elucidating the role of ZnO. Several models were proposed. For example, Fujitani et al. [17] reported that the promotional effect of Zn depends on the coverage of this metal. The Cu-related turnover frequency (TOF) of methanol formation over Zn/Cu(111) increased with Zn loading and reached its highest value at the Zn coverage of 0.2. A further increase in the coverage resulted in a decrease in the activity. Thus, a Cu–Zn alloy was proposed to be active for methanol synthesis (Fig. 4.1a). This mechanistic concept was later confirmed for a powder Cu/ZnO catalyst [13]. Formate species, an important intermediate, were concluded to be directly bound to Zn in the alloy and initially hydrogenated to methoxy species followed by the conversion of the latter into methanol. In addition to being integrated into the active sites, Zn was also reported to improve dispersion of metallic Cu [13].

Another important work published in 2000 by Grunwaldt et al. [19] emphasized the importance of dynamic changes in the morphology of Cu particles over a Cu/ZnO catalyst. Applying *operando* extended X-ray absorption fine structure (EXAFS) analysis, those authors determined relatively low Cu–Cu coordination numbers under reducing conditions. On this basis, they suggested the presence of a disk-like Cu particles on the surface of ZnO. The Cu–Cu coordination numbers increased under more oxidizing conditions due to the formation of spherically shaped Cu particles. Such morphology changes were assumed to affect methanol production. Due to a larger abundance of the active Cu(110) plane exposed on flat particles than on spherical particles, the catalyst displayed higher activity under reducing conditions.

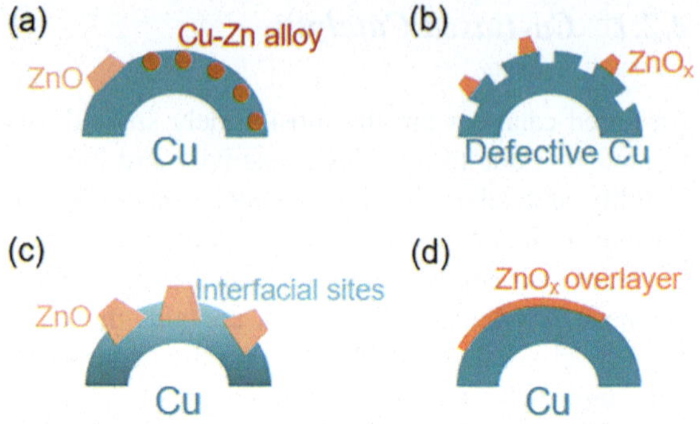

Fig. 4.1 Models reflecting the nature of active sites and the chemical state of Zn in Cu/ZnO-based catalysts. **a** Cu–Zn surface alloy [13, 15]; **b** metallic Cu species with defects (such as step sites) close to partially oxidized $Zn^{\delta+}$ species [14]; **c** Cu–ZnO interfacial sites [21]; **d** ZnO_x overlayer on metallic Cu particles [20]

The changes were proved to be reversible when periodically switching between more reducing and more oxidizing feeds. However, the Cu content in this catalyst is only 4.5 wt%, that is much lower than 40–60 wt% in commercial catalysts.

As ZnO can be reduced under reducing conditions, strong metal-support interactions (SMSI) between partially reduced ZnO and metallic Cu were suggested to play an important role. In this regard, Behrens et al. [14] determined the highest methanol formation activity over the catalysts consisting of Cu step sites close to $Zn^{\delta+}$ species (Fig. 4.1b). Lunkenbein et al. [20] observed a "graphitic-like" ZnO_x overlayer on metallic Cu particles (Fig. 4.1d). Both works emphasized the importance of interfacial sites between oxidized Zn species and metallic Cu for ensuring of high activity of Cu/ZnO-based catalysts.

To contribute to elucidating the role and the state of Zn species, model ZnCu(111) and ZnO/Cu(111) catalysts representing Cu–Zn alloy sites (Fig. 4.1a) and ZnO–Cu interfacial sites (Fig. 4.1c), respectively, were applied for CO_2 hydrogenation to methanol [21]. While no ZnO reduction was observed in the ZnO/Cu(111) catalyst at 550 K, 4.5 bar of H_2, and 0.5 bar of CO_2 as concluded from XPS results, metallic Zn in the ZnCu(111) catalyst underwent oxidation and eventually formed ZnO under the same conditions. DFT calculations suggested that Cu–ZnO interfacial sites contribute to lowering the energy barriers of key elementary reaction steps in CO_2 hydrogenation to methanol.

4.2.1.2 Structure–Activity Relationships

In addition to the nature of active sites/species, another frequently discussed topic is structural sensitivity, i.e., the activity of methanol synthesis may vary significantly depending on the shape and the size of particles as well as on the exposed crystal facets of Cu. For the latter case, the activity was reported to follow the order of Cu(110) > Cu(100) > Cu(111) [22, 23]. However, there is no significant difference in the activation energy of methanol synthesis from CO_2 on single crystalline Cu with different surface structures as reported by Yoshihara et al. [22] and Nakamura et al. [13, 23].

Chinchen et al. [24] established, in 1986, a linear relationship between metallic Cu surface area and methanol formation rate per Cu for various Cu-based catalysts regardless of the kind of support material. Note that this correlation was made based on the catalysts possessing Cu particles larger than 10 nm. In 2016, the group of de Jong [16] reported that decreasing Cu particle size has a negative effect on the Cu-related TOF of methanol formation. This conclusion is valid for catalysts with Cu particle size ranging from 2 to 10 nm on SiO_2 and C supports. The particle-size-effect did not depend on the presence or the absence of Zn species. Those authors further claimed that small Cu particles could not accommodate active sites required for methanol synthesis, e.g., step-edge Cu sites, which are usually stabilized on particles above 8 nm. However, a recent study led by de Jongh [25] suggested that the Cu-related TOF of methanol formation is insensitive to Cu particle size in the range of 5–20 nm, while methanol selectivity increased with a decrease in the size in the expense of CO. To explain these results, a cuboctahedral model of Cu particles was constructed and the fraction of corner, edge, and terrace sites on differently sized Cu particles was determined by DFT calculations. The portion of low-coordinated edge sites was found to increase from 5 to 20% with decreasing Cu species size from 30 to 3 nm. Computational results showed that the adsorption free energy of formate, key species for CH_3OH formation, on the edge Cu sites is higher than that on the terraces sites. Therefore, the abundance of active sites favoring CH_3OH formation increases with decreasing Cu particle size. On the other hand, the nature of active sites for CH_3OH and CO formation did not change on Cu particles with different sizes, as indicated by the size-independent activation energy of products formation. Therefore, a higher ratio of CH_3OH formation rate to that of CO was achieved on smaller particles and consequently the selectivity to methanol increased.

4.2.1.3 The Nature/Role of Oxide Component in Cu-Based Catalysts

The investigation of the role of metal oxide promoters, the metal oxidation state, and the real active sites of Cu-based catalysts has never stopped and is recently promoted by the availability of advanced in situ/*operando* characterization techniques. In 2021, Beck et al. [26] discussed the pressure-gap-effects existing in various previous works dealing with methanol synthesis over Cu/ZnO-based catalysts and systematically studied structural changes of Zn species under catalyst reduction and CO_2 hydrogenation conditions using *operando* XAS technique at H_2 partial pressure ranging from 1 mbar to 10 bar without changing the H_2/CO_2 feed ratio that was 3. This study ruled out the presence of any Cu–Zn alloy under reaction conditions independent on the pressure. Those authors also demonstrated that metallic Zn formed during catalyst reduction at high H_2 pressures underwent oxidation to ZnO under methanol synthesis conditions. More recently, Nilsson' group employed an ambient pressure XPS technique (up to 500 mbar) with tunable surface sensitivity to monitor the oxidation state of zinc under different conditions [27]. It was concluded that co-feeding of CO to a CO_2/H_2 feed enhanced the stability of surface Cu–Zn alloy sites efficiently hydrogenating CO_2 to methanol.

In addition to ZnO and Al_2O_3, other oxide components, e.g., MgO, ZrO_2, CeO_2, and Ga_2O_3 are used for the preparation of Cu-based catalysts. Zander et al. [28] compared Cu/ZnO and Cu/MgO catalysts (Cu:M molar ratio of 80:20) in terms of their structural properties and performance in CO_2 hydrogenation to methanol. Although the use of MgO resulted in a higher Cu dispersion, Cu/MgO showed lower activity for methanol formation than Cu/ZnO. This is due to the unreducible nature of MgO, which results in the lack of strong interactions between Cu and MgO, highlighting the essential role of ZnO in Cu/ZnO-based catalysts. Graciani et al. [29] constructed model catalysts CeO_x/Cu(111) and Cu/CeO_x/TiO_2(110) which showed much higher rates of methanol formation than Cu/ZnO(0001). The reason was attributed to the presence of interfacial Cu and CeO_2 sites.

A lot of efforts were devoted to understanding of the synergy induced by ZrO_2 in Cu-based catalysts with or without ZnO. In 1997, Fisher et al. [30] prepared a ZrO_2 (30.5 wt%)-promoted Cu/SiO_2 catalyst that showed a 25-fold increase in the rate of methanol formation from CO_2 in comparison with Cu/SiO_2. It was suggested that ZrO_2 could be involved in CO_2 adsorption and intermediate formation. Later, the enhancing effect of ZrO_2 was associated with the interface between ZrO_2 and Cu, as originally suggested for ZnO and Cu, which contributes to the stabilization of partially oxidized Cu species that is relevant for CO_2 adsorption [31]. Combining DFT calculations and experimental studies, Kattel et al. [32] concluded that metal/oxide interfaces consisting of Cu and Zr^{3+} are suitable to bind reaction intermediates (*CO, *HCO, and *H_2CO) and thus the reaction proceeds via the RWGS reaction followed by the hydrogenation of CO into methanol, rather than the formate pathway proposed over Cu/ZrO_2 catalyst in another study [33]. Molecular-level details about the reaction mechanism are separately discussed in Sect. 4.2.5.

4.2.2 *In_2O_3-Based Catalysts*

Due to the ability of Cu-based materials to catalyze the competing RWGS reaction, the selectivity to methanol over Cu/ZnO/Al_2O_3 catalysts is typically below 100% even at low CO_2 conversion degrees [6]. Inspired by the very low RWGS activity of indium oxide, Ye et al. [34] introduced this metal oxide for methanol synthesis in their computational study in 2013. The calculations revealed that In_2O_3(110) with defective oxygen sites should be highly selective for methanol synthesis because the formation of formate (*HCOO) species from adsorbed CO_2 is kinetically and thermodynamically favored. This finding has stimulated both theoretical and experimental studies with In_2O_3-based catalysts.

4.2.2.1 Relevance of In_2O_3 for Methanol Synthesis

In 2015, Liu et al. [35] verified experimentally the ability of bulk In_2O_3 to produce methanol. The selectivity of 55% was achieved at 1.1% CO_2 conversion at 270 °C and

40 bar. By engineering the surface structure of In_2O_3 and optimizing reaction conditions, Pérez-Ramírez and co-workers [36] achieved a near 100% methanol selectivity at 5.2% CO_2 conversion over In_2O_3 (9 wt%)/ZrO_2 at 300 °C and 50 bar. The catalyst showed high stability during 1000 h on stream. Note that the Cu/ZnO/Al_2O_3 catalyst exhibited a methanol selectivity of only 10% under the same conditions [36]. Two types of active oxygen vacancies, i.e., thermally and hydrogen-induced vacancies, were identified, with the former being more active for CO_2 activation. Methanol space time yield (STY(CH_3OH)) over bulk In_2O_3 was enhanced when CO was co-fed due to the beneficial role of this reactant to generate oxygen vacancies on the surface of In_2O_3.

An *operando* study combined XAS with XRD revealed structural changes of In_2O_3 [37]. This oxide was gradually reduced to metallic In^0 with rising time on reaction feed. When relating these results to the catalytic data, partially reduced In_2O_{3-x} should be the active phase in the desired reaction, while In^0 causes catalyst deactivation. Dang et al. [38] calculated energy barriers of the rate-determining step in CO_2 hydrogenation to methanol on different In_2O_3 surfaces and predicted on this basis the most active In_2O_3 phase and surface. In line with the computational results, the hexagonal In_2O_3 phase with the predominantly exposed (104) surface displayed the highest methanol selectivity of 99.5% at 6.7% CO_2 conversion at 300 °C and 50 bar.

Considering the importance of ZrO_2 support for In_2O_3, some research studies have been devoted to understanding of the synergistic effect between these two components in catalyst [39, 40]. ZrO_2 was initially considered to interact with In_2O_3 electronically, providing a new kind of oxygen vacancies [36]. A later study dealing with differently loaded In_2O_3/ZrO_2 catalysts suggested that the methanol selectivity strongly depends on the proximity between In species and ZrO_2 [41]. H_2 dissociation was enhanced with increasing In content, resulting in the facile hydrogenation of *HCOO and *H_3CO intermediates. When the loading was below 1 wt%, highly dispersed InO_x species were stabilized forming substantial interface sites (InO_x/ZrO_2) on which the decomposition of *HCOO into CO was favored due to the lack of active hydrogen species.

However, the interfacial sites on the surface of In_2O_3/m-ZrO_2 were used in other studies to explain higher methanol productivity in comparison with those supported on tetragonal ZrO_2 (t-ZrO_2), Al_2O_3, and CeO_2, as well as bulk In_2O_3 (Fig. 4.2a) [39]. The unique role of m-ZrO_2 as support was attributed to the more obvious lattice mismatch between In_2O_3 and m-ZrO_2, in comparison with t-ZrO_2, which brings geometry distortion of In_2O_3 supported on m-ZrO_2 and results in high concentration oxygen vacancies on In_2O_3 (Fig. 4.2b). Additionally, a DFT study suggested the importance of electronic interactions between In_2O_3 and m-ZrO_2, i.e., electron donation from m-ZrO_2 to In_2O_3, for methanol synthesis [42]. This conclusion was further supported by *quasi* in situ XPS results. Thus, the improved catalyst activity and methanol selectivity were attributed to the enhanced H_2 activation on electron-enriched In-O-In sites.

On the basis of the above discussions, there are still contradictive conclusions about the indispensable effect of m-ZrO_2 on In_2O_3 on methanol synthesis from

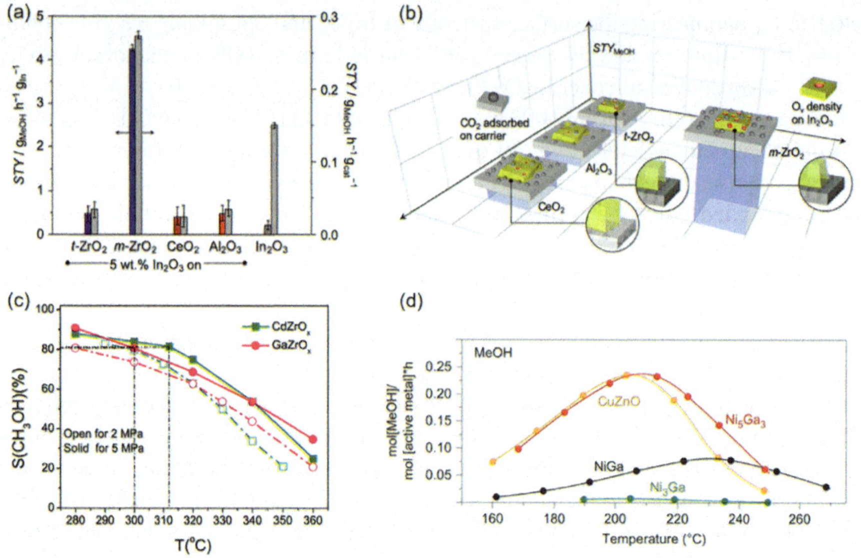

Fig. 4.2 **a** Space time yield of methanol formation (STY) with respect to the indium content (left *Y* axis) or the total catalyst mass (right *Y* axis) as determined over differently supported In_2O_3 catalysts and bulk In_2O_3. **b** Schematic representation of the interplay between oxygen vacancy density on In_2O_3 and the CO_2 adsorption capacity of the support for achieving high STY of supported In_2O_3 catalysts. **c** Methanol selectivity (S(CH_3OH)) over $CdZrO_x$ and $GaZrO_x$ catalysts as a function of reaction temperature. **d** Metal-related STY of Ni_xGa_y catalysts and the reference catalyst Cu/ZnO/Al_2O_3 as a function of reaction temperature at atmospheric pressure. **a–b** Reprinted with permission from Ref. [39]. Copyright 2020, American Chemical Society; **c** reprinted from Ref. [56]. Copyright 2019, American Chemical Society; **d** reprinted with permission from Ref. [64]. Copyright 2014, Springer Nature

CO_2. In summary, the former material functioning as support, promoter, and/or, even catalytically active component has been claimed to contribute to (i) maintaining the high dispersion of In_2O_3 [36], (ii) changing the electronic properties of In_2O_3 entities [42], (iii) increasing the density of oxygen vacancies via electronic interactions with In_2O_3 or geometric effect [36, 39, 43], (iv) affecting the direct CO_2 adsorption/activation on itself [39], and (v) influencing the adsorption/desorption configurations of reaction intermediates and thereby altering reaction pathways [40, 41]. On this issue, the wide applications of ZrO_2 as support and/or promoter for catalysts for methanol synthesis from CO_2 and the induced synergistic effects have been summarized by Li and Chen [44] in 2019, but not limited within In_2O_3-based catalysts. Further studies are still required to provide deeper fundamental insights (especially at an atomic level) needed to optimize the current In_2O_3-ZrO_2-based systems with the purpose to reach industrial relevance.

4.2.2.2 Effect of Promoters

Although In_2O_3-based catalysts generally deliver high methanol selectivity, their activity is not high enough due to the insufficient ability to activate H_2. To improve this property, such catalysts are typically modified by different promoters, e.g., Ni, Ru, Rh, Pd, Re, Ir, Pt, or Au [40, 45–52]. The Pt-group metals may also enhance the activity to the RWGS reaction and correspondingly lower the selectivity to methanol. Thus, the precise control of geometric and electronic catalyst properties caused by the promoter(s) is necessary to minimize their detrimental effects.

A theoretical and microkinetic study with a model catalyst $Pd_4/In_2O_3(110)$ suggested that methanol formation follows through a reaction route involving surface *HCOO species [53]. To check experimentally the potential of promotion with Pd, Rui et al. [54] prepared In_2O_3-supported Pd nanoparticles with an average size of 3.6 nm through mixing In_2O_3 and a Pd/peptide composite. Both fresh and spent samples contained monometallic Pd species on the surface of In_2O_3. No bimetallic PdIn particles could be detected. The developed catalyst displayed a methanol selectivity of 72% at a CO_2 conversion of around 21%, at 300 °C and 50 bar. It also showed STY(CH_3OH) of 0.89 g_{MeOH} h^{-1} $g_{cat}{}^{-1}$. The excellent performance was attributed to (i) that improved H_2 dissociation on Pd provides sufficient concentration of H species for hydrogenation steps and (ii) that the spillover effect of H species from Pd nanoparticles to In_2O_3 facilitates the generation of oxygen vacancies on the surface of the latter which are involved in CO_2 adsorption/activation. Frei et al. [48] prepared Pd/In_2O_3 through co-precipitation (Pd/In_2O_3-CP) or impregnation (Pd/In_2O_3-DI) methods. Pd atoms were embedded into the In_2O_3 matrix in the fresh Pd/In_2O_3-CP material. Upon exposure to the reaction feed, $Pd_xIn_{4\text{-}x}$ and/or Pd_2In_{5-x} clusters with 1-2 Pd atoms were formed on the surface of In_2O_3 and act as active centers for methanol synthesis. To improve the utilization efficiency of palladium and indium, the same group later prepared ternary Pd-In_2O_3-ZrO_2 catalysts by a flame spray pyrolysis (FSP) method resulting in low-nuclearity Pd species and the high abundance of oxygen vacancies [47]. Benefiting from these structural characteristics, the best-performing catalyst showed an exceptional methanol productivity of 1.3 g_{CH_3OH} h^{-1} g_{cat}^{-1}. A positive correlation between the STY(CH_3OH) and the concentration of oxygen vacancies was established.

In_2O_3-based catalysts with incorporated positively charged platinum or rhenium atoms were prepared through a co-precipitation method [49, 51]. They enhanced the heterolytic dissociation of H_2 and thus improved both CO_2 conversion and methanol selectivity, while Pt nanoparticles deposited on In_2O_3 by an impregnation method increased the activity of the RWGS reaction. However, Sun et al. [50] prepared In_2O_3 catalyst with Pt nanoparticles over which CO_2 conversion was enhanced by two times, in comparison with pure In_2O_3, without sacrificing methanol selectivity. Rui et al. [52] proposed that the $Au^{\delta+}$-$In_2O_{3\text{-}x}$ interfacial sites enabled efficient CO_2 and H_2 activation, and therefore, boosted STY(CH_3OH) by 1.5 times in comparison with pure In_2O_3.

Besides the above-discussed noble metals, nickel was also employed as a promotor for In_2O_3-based catalysts [45, 46]. The accordingly prepared catalysts showed high

CO_2 conversion but exhibited distinct product distribution that was strongly dependent on Ni loading. This is not surprising as metallic Ni is very active for CO_2 methanation [7]. It is possible to simultaneously improve overall catalyst activity and methanol productivity through the control of catalyst nanostructure and electronic properties, e.g., the metal-support interaction and the nuclearity of supported Ni species. Frei et al. [46] reported that the In_2O_3 catalyst containing 1 wt% of Ni showed more than two times higher STY(CH_3OH) compared with pure In_2O_3, due to the presence of a thin $InNi_3$ layer on which H_2 dissociates homolytically and the formed H species spilled over to In_2O_3. The selectivity to methanol over the promoted catalyst was slightly lower (70 vs. 80%). Increasing Ni content to 20 wt% shifted the product spectrum to CO (57% selectivity) and minor products CH_4 (13%) and methanol (30%) because $InNi_3$ clusters grew into nanoparticles and Ni metal was formed when the Ni loading was increased. CO_2 could be activated and hydrogenated to CO over the former structure and to CH_4 over the latter as suggested by DFT calculations.

4.2.3 *Solid Solution Catalysts*

Although a significant progress has been made in pursuing efficient methanol synthesis catalysts, methanol selectivity above 90% at CO_2 conversion higher than 10% has been barely reported over Cu- or In-based catalysts and remains a grand challenge. In 2017, Wang et al. [55] reported a new kind of catalysts based on a binary ZnO-ZrO_2 solid solution. The catalyst with 13 wt% Zn resulted in the methanol selectivity of 91% at CO_2 conversion of about 10% at 315 °C and 50 bar. Such outstanding performance was attributed to a synergistic effect between zinc and zirconium sites which could enhance both CO_2 and H_2 activation due to higher concentration of oxygen vacancies as probed by titration and spectroscopic experiments. The formation of solid solution from single phases of ZnO and ZrO_2 seems to be indispensable for methanol synthesis, as the counterpart prepared through physically mixing is not active for this reaction [55]. The concept of binary solid solution catalysts was extended to other materials, i.e., $CdZrO_x$ and $GaZrO_x$ with molar ratio of Cd/(Cd + Zr) and Ga/(Ga + Zr) being 11.5% [56]. CH_3OH selectivity over these two selectivity was found to be higher than 80% at reaction temperatures below 300 °C (Fig. 4.2c).

Several strategies have been developed to further improve the performance of ZnO-ZrO_2 solid solution catalysts, for example by using a metal promoter [57–59]. The promotional effect is generally associated with an increase in the catalyst ability to generate surface hydrogen species from gas-phase H_2 that is relevant for the creation of anion vacancies. Sha et al. [57] promoted $ZnZrO_x$ catalysts with different Ga amounts. A volcano-type curve in terms of CO_2 conversion or STY(CH_3OH) was obtained at 320 °C and 50 bar when varying the Ga molar fraction from 0 to 10%. The authors did not, however, explain this dependence in detail. The selectivity to methanol was not affected and was about 85%. The optimal catalyst was determined over 5%$GaZnZrO_x$ which delivered STY(CH_3OH) of 0.63 $g_{CH_3OH}\,h^{-1}\,g_{cat}^{-1}$.

Promoting of $ZnZrO_x$ with only 0.5 mol% Cu resulted in an increase in CO_2 conversion from 2 to 9% without any changes in the selectivity to methanol (ca. 80%) at 290 °C and 45 bar [60]. It was proposed that the hydrogen spillover effect induced by Cu contributes to the fast hydrogenation of surface intermediates. When the Cu content was increased to 2 mol% the conversion increased further to 12% but the selectivity decreased to only 55%. A similar improvement in STY(CH_3OH) was also observed when $ZnZrO_x$ was promoted with 0.1 wt% Pd [61]. Those authors claimed that a PdZn alloy formed after reductive catalyst treatment boosts the concentration of oxygen vacancies in ZrO_2. However, Lee et al. [58] observed atomically dispersed Pd species on the surface of $ZnZrO_x$ up to Pd molar fraction of 0.8%. DFT calculations explained the promotional effect of Pd by (i) facilitating the formation of anion vacancies and (ii) accelerating H_2 activation/dissociation. Considering the importance of Pd for H_2 activation, the same authors introduced, very recently, a highly active catalyst system (Pd/CNT + $ZnZrO_x$) by integrating $ZnZrO_x$ with Pd/CNT (carbon nanotube) where the spillover effect of H species activated on Pd could be intensified due to the presence of CNT [59]. Thus, a very high STY(CH_3OH) of 0.90 $g_{CH_3OH}\,h^{-1}\,g_{cat}^{-1}$ was achieved.

4.2.4 Other Kinds of Catalysts

In addition to the above-discussed methanol synthesis catalysts, some other catalysts have also been reported to be active in the desired reaction. They include noble metals [62, 63] and intermetallic catalysts [64, 65]. In this section, we focus on the well-studied Pd-based monometallic/bimetallic catalysts and Ni_xGa_y intermetallic catalysts. In addition, it is worth mentioning that very recently a MoS_2 material was reported as an efficient low-temperature catalyst for CO_2 hydrogenation to methanol [66, 67]. In the work of Hu et al. [66], MoS_2 nanosheet displayed methanol selectivity of 94.3% with a CO_2 conversion of 12.5% at 50 bar and only 180 °C. The active sites were suggested to be in-plane S vacancies where CO_2 dissociation and hydrogenation to CH_3OH were favored instead of CH_4 formation.

4.2.4.1 Pd-Based Catalysts

The use of supported palladium-containing catalysts for methanol synthesis from CO and H_2 began in the late 1970s [68]. In 1995, such catalysts have been, for the first time, introduced by Fujitani et al. [62] for CO_2 hydrogenation to methanol. The kind of support is a key parameter for CO_2 conversion at 250 °C and 50 bar. The following activity order was reported Pd/Ga_2O_3 > Pd/ZnO > Pd/ZrO_2 > Pd/La_2O_3 > Pd/CeO_2 > Pd/Al_2O_3. The Pd (10 wt%)/Ga_2O_3 catalyst showed methanol selectivity of 51.5% at 19.6% CO_2 conversion. Under the same conditions, the selectivity of 36.1% at the conversion of 11.7% was achieved over the classical Cu/ZnO catalyst. The same authors later claimed that the high methanol formation activity of Pd/Ga_2O_3 and Pd/

ZnO should be attributed to PdGa or PdZn alloys [69]. A controversial result was reported by Kim et al. [70] who concluded that the alloy causes catalyst deactivation, while Pd nanoparticles modified by ZnO, i.e., Pd-ZnO interfaces provide active sites for the desired reaction. Xu et al. [71] found that the ratio of PdZn to metallic Pd increased with the temperature of the reduction of Pd/ZnO/Al_2O_3 catalyst prior to CO_2 hydrogenation reaction. The selectivity to methanol augmented, too. Additionally, ZnO_x-modified Pd species were identified by EXAFS analysis in the catalysts possessing less than 2 wt% Pd. However, the selectivity did not change in comparison with the counterparts with PdZn species. Therefore, those authors stated that both the alloy and the ZnO_x-modified Pd species are active for methanol synthesis.

A mechanistic study using FTIR spectroscopy revealed that Ga_2O_3 chemically adsorbed CO_2 and the formed polydentate carbonates were hydrogenated to formate species and finally to methanol [72]. These hydrogenation reactions could be accelerated by the spillover effect of H species formed from gas-phase H_2 on Pd. Thus, the function of oxide supports/promoters, i.e., ZnO and Ga_2O_3 is still a matter of debates. Over the past decade, several research groups have proposed that Pd may present as Pd_xZn_y and Pd_xGa_y intermetallic phases in Pd/ZnO- and Pd/Ga_2O_3-based catalysts [73]. In 2012, Collins et al. [73] identified bimetallic Pd_xGa_y on the surface of the Pd/Ga_2O_3 catalyst by means of HRTEM and FTIR spectra of adsorbed CO. However, those authors claimed that the exact composition of Pd_xGa_y could not be unambiguously determined due to the structural similarity among Pd_5Ga_2, Pd_5Ga_3, and Pd_2Ga. Similarly, the catalysts containing Pd_2Ga or PdZn intermetallic compound showed higher activity toward methanol synthesis than the catalyst with monometallic Pd [74]. The Pd_2Ga intermetallic phase could also be generated through a simple deposition of nitrates of Pd and Ga on SiO_2 followed by reduction at 550 °C as proven by in situ XRD and environmental TEM studies [63]. Importantly, Pd_2Ga did not experience any changes in the phase composition and in the dispersion/particle size during the reaction.

4.2.4.2 Ni-Ga Intermetallic Catalyst

In 2014, Studt et al. [64] predicted on the basis of DFT calculations that Ni_xGa_y intermetallic catalysts can be active for methanol synthesis from CO_2. The prepared Ni_5Ga_3 intermetallic catalyst displayed a comparable methanol formation rate with an industrially relevant Cu/ZnO/Al_2O_3 catalyst in CO_2 hydrogenation at 1 bar and showed a lower CO/methanol molar ratio (Fig. 4.2d). The latter property of the Ni_5Ga_3 catalyst was attributed to Ga-rich surface sites. This work provided a basis for designing alternative catalysts for low-pressure methanol synthesis and stimulated several subsequent studies aiming to the optimization of the Ni-Ga catalyst systems [65, 75–77].

The formation of the Ni_5Ga_3 phase requires a high-temperature reduction of Ni and Ga oxides or hydrotalcite structures [65, 76]. The reduction temperature is an important parameter as it influences both the catalyst composition and the structure [75]. Experimental and theoretical results revealed a synergistic effect between

Ni_5Ga_3 and Ga_2O_3, with the latter contributing to CO_2 adsorption and activation. Hejral et al. [77] combined *operando* XAS, XRD, and near ambient pressure XPS to monitor the phase evolution of Ni-Ga nanocatalysts under reduction and reaction conditions. They found that the Ni_5Ga_3 phase provides active sites for methanol formation, while metallic Ni hydrogenates CO_2 to CH_4.

4.2.5 Reaction Mechanisms

Mechanistic studies of CO_2 hydrogenation into methanol are mainly devoted to the identification of surface intermediates. To this end, FTIR spectroscopy and DFT calculations are the two primary approaches [8, 9, 78]. However, too many controversial results have been reported in the literature, pointing out that there are two possible reaction pathways for methanol synthesis from CO_2: (i) the RWGS reaction followed by CO hydrogenation to methanol and (ii) the direct hydrogenation of CO_2 to methanol via formate (*HCOO) species, as summarized in Fig. 4.3. In the first pathway, CO could be formed through the direct CO_2 dissociation or through the decomposition of carboxyl (*COOH) [33, 79]. Dating back to the 1980s, isotopic labeling experiments confirmed that CO_2 is the primary carbon source for methanol synthesis over a commercial $Cu/ZnO/Al_2O_3$ catalyst using $CO_2/CO/H_2$ as a reaction feed [10]. Later studies showed different activation energies of CO formation (100–130 kJ mol^{-1}) and methanol formation (70–100 kJ mol^{-1}) in CO_2 hydrogenation over Cu-based catalysts [13, 17, 25].

Formate species were experimentally observed by XPS and FTIR [8, 9, 23, 27, 29]. However, a question arises if they are real intermediates or spectators. The formate and carboxyl routes of methanol formation over Cu(111) were compared with DFT calculations [79]. Although the formate formation is thermodynamically favored than the formation of carboxyl, it has been suggested that the hydrogenation of the former into $*H_2COO$ is not energetically feasible [80]. A subsequent study experimentally proved that formate species pre-adsorbed over Cu/SiO_2 catalyst could not be hydrogenated to methanol [80]. Kattel et al. [21] also performed DFT calculations and reported the potential energy profiles of CO_2 hydrogenation into methanol over ZnCu(211) and ZnO/Cu(111). In these models, the formate pathway was identified to dominate methanol formation. However, different rate-determining steps were predicted for these systems. The hydrogenation of $*H_3CO$ to $*CH_3OH$ limits the formation of methanol over ZnCu(211) with an energy barrier of 1.49 eV. The hydrogenation of *HCOO to *HCOOH with a lower energy barrier (0.85 eV) is the RDS over ZnO/Cu(111).

Graciani et al. [29], however, considered formate species as spectators in CO_2 hydrogenation to methanol over CeO_x/Cu(111). Those authors also performed DFT calculations and claimed that methanol synthesis follows a RWGS + CO hydrogenation pathway in which carboxyl is a key intermediate. Larmier and co-workers [33] studied the methanol synthesis from CO_2 over a Cu/ZrO_2 catalyst by means of in situ spectroscopic tools and DFT calculations and concluded that formate species

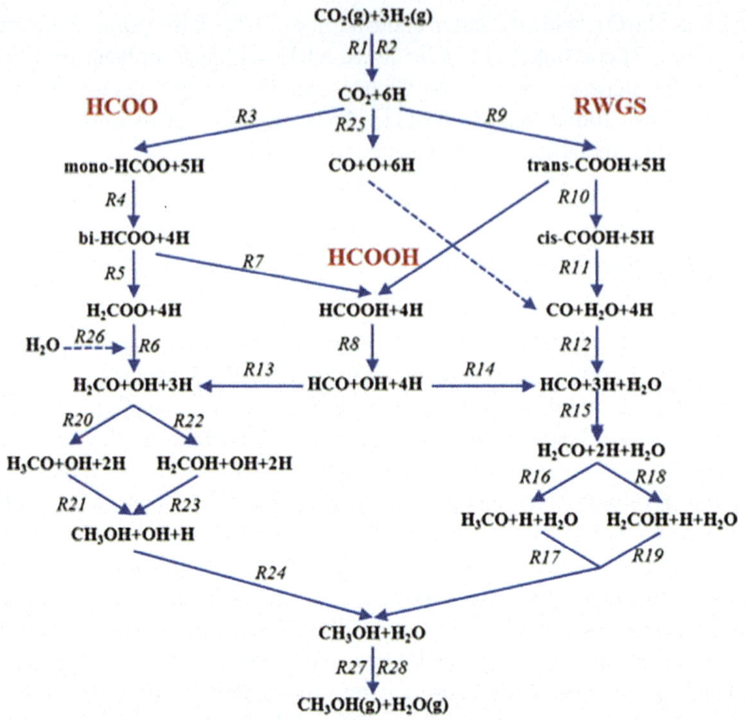

Fig. 4.3 Reaction network of methanol synthesis from CO_2 and H_2 on Pd_4/In_2O_3. Reprinted with permission from Ref. [53]. Copyright 2014, Elsevier

can be hydrogenated to methoxy species which are methanol precursors. They also analyzed different ways of CO_2 dissociation to *CO and *O on Cu-ZrO_2 interfacial sites as well as hydrogenation to carboxyl intermediates, and hydrogenation into formate species. The latter option was found to be the most thermodynamically and kinetically favorable route. Other DFT calculations also suggested that adsorbed CO_2 prefers hydrogenation to *HCOO on the surface of In_2O_3 and $ZnZrO_x$ [34, 55]. Nevertheless, the consensus on the major active intermediates involved in methanol formation has not been reached. A unified approach to compare and evaluate the reaction pathway of CO_2 hydrogenation into methanol over different catalysts is still needed.

It is necessary to mention that a major part of experimental and theoretical studies concluded that CO formation through the RWGS reaction is the only origin of not 100% methanol selectivity. This reaction cannot, however, explain why the selectivity to methanol decreases with rising CO_2 conversion under isothermal conditions, while the selectivity to carbon monoxide increases over Cu-containing catalysts [33, 81, 82]. Such dependence can be understood when considering consecutive transformations of primarily formed methanol to CO. It is well-known that this alcohol can be directly decomposed to CO and H_2 or be involved in steam reforming reactions

[83]. On the basis of sophisticated kinetic studies of CO_2 hydrogenation to methanol and methanol decomposition in the presence of H_2 or H_2O over a $CuZnAlO_x$ catalyst, Ortner et al. [81] concluded that methanol synthesis and decomposition occur on different sites. Those involved in the undesired reaction(s) are saturated above certain partial pressures of water and methanol, while the sites responsible for the preferred reaction are not saturated and can produce methanol in downstream located catalyst bed segments. Further experimental and theoretical studies would significantly advance our knowledge in this regard and provide fundamentals needed for improving catalyst efficiency for CO_2 conversion into methanol.

4.3 CO_2 Hydrogenation to CO

In the context of CO_2 valorization, another attractive topic is the selective formation of CO through the RGWS reaction (Eq. 4.2). The challenge is associated with the presence of competitive pathways of CO_2/CO conversion into CH_4 and CH_3OH. In the past decades, numerous studies have been dedicated to supported catalysts with noble metals, e.g., Pt, Ru, Ir, and Rh as well as understanding of principles for selectivity control [6, 7, 78]. The most developed strategy for this purpose includes the modification of metal-support interactions typically through the size of supported metal, the kind of support material, and the procedure of catalyst treatment. Therefore, we herein do not extend this part to noble metal-based catalysts but focus on the progress made on the cost-effective catalysts consisting of non-noble metals or metal oxides.

4.3.1 Co-Based Catalysts

An earlier study on CO_2 hydrogenation over Co-based catalysts reported that methane is the main product [4]. However, when using a mixture of CO and H_2, liquid/wax products are formed. These distinctive results were explained by the different strengths of adsorption of CO_2 and CO on reduced Co species, with the former being adsorbed weaker. As a result, the local C/H atomic ratio is low in the case of CO_2 hydrogenation that is favorable for complete hydrogenation to methane. However, later works revealed that the product spectrum in CO_2 hydrogenation over Co-based catalysts could be shifted from CH_4 to CO through improving the dispersion of Co species and tuning the properties of metal-support interfacial sites toward adsorption of reactants/intermediates [84–86]. CO selectivity of 100% and CO_2 conversion of about 24% could be achieved at 500 °C over Co/CeO_2 catalysts with Co loading below 2 wt% [84]. Increasing the loading resulted in larger Co_3O_4 particles and the formation of CH_4 instead of CO. Although these catalytic results were interesting and provided inspiration for the design of selective RWGS catalysts, those authors did not report any in-depth characterization results.

Co/ZrO_2 catalysts with atomically dispersed Co^{2+} in the lattice of ZrO_2 were prepared in Ref. [86]. The speciation did not change in the range of Co loading between 1 and 15 atom%. Due to the charge imbalance between Co^{2+} and Zr^{4+}, oxygen vacancies were formed in ZrO_2. They play a crucial role in CO_2 adsorption but not in CO adsorption as evidenced by DFT calculations and FTIR spectroscopic measurements. Consequently, CO easily desorbs thus avoiding its further hydrogenation to CH_4. The importance of the strength of CO adsorption for product selectivity was supported in another study dealing with Co-containing catalysts based on different TiO_2 support (anatase (a-TiO_2) and rutile (r-TiO_2)) [85]. CO was established to adsorb weaker on Co/a-TiO_2 than on Co/r-TiO_2. As a result, CO is the main product over the former catalyst without any obvious subsequent hydrogenation to CH_4.

4.3.2 Ni-Based Catalysts

Supported catalysts with high Ni loading are well-known for their excellent performance in the methanation of CO and CO_2 [6]. Similar to Co-based catalysts, lowering Ni loading results in the selectivity change from CH_4 to CO [87]. This phenomenon is typically related to the size of Ni particles which is an important parameter affecting H_2, CO_2, and CO adsorption/activation. Ni/SBA-15 catalysts with Ni loading of 5.9, 14.2, and 21.7 wt% possess Ni particles of 2.6, 3.7, and 5.1 nm, respectively [87]. The catalyst with the smallest Ni particles displayed the highest CO selectivity of about 92% at CO_2 conversion of 20% at 500 °C.

As mentioned above, low Ni loadings are required for selective CO production, but are detrimental for CO_2 conversion. To realize both high selectivity and conversion, Fe species were added to 1.5 wt% Ni/CeO_2 with a Ni/Fe molar ratio of 3 [88]. A CO/CH_4 ratio of 14 was achieved over this catalyst that is higher than 7.4 obtained over its Fe-free counterpart. Detailed experimental and theoretical studies revealed that the high CO selectivity is attributed to the fast CO desorption rate from Ni-FeO_x interfacial sites [89]. Rodrigues et al. [90] supported Ni on an Mg(Al)O solid solution. Both the concentration of oxygen vacancies in the support and the content of metallic Ni were concluded to be crucial for the selectivity to CO. The resulting catalyst displayed CO selectivity of 94% with 60% CO_2 conversion at 700 °C. A non-oxidic support, nitrogen-doped carbon, was used to prepare a catalyst with $Ni_3ZnC_{0.7}$ species through calcination of Ni-containing zeolitic imidazolate framework material (ZIF-8) [91]. A metal carbide ($Ni_3ZnC_{0.7}$) was formed as evidenced by XRD and TEM analyses. The support is responsible for CO_2 adsorption, while the DFT study revealed that the metal carbide species provide active sites for CO_2 hydrogenation reaction.

4.3.3 Fe- or Cu-Based Catalysts

Fe- or Cu-based catalysts are also widely applied for the RWGS reaction. Earlier studies dealing with this reaction suggest that Fe_3O_4 is involved in RWGS reaction through a redox mechanism [92]. In other words, CO_2 is reduced by Fe_3O_4 to CO followed by the reduction of oxidized iron oxide with H_2. To improve the reactivity of Fe-based catalysts, alkali metals (K or Cs) or Cu are typically used as promotors [93]. However, the promotional effects of K and Cu differ strongly. The presence of K increases the basicity of the catalyst surface, resulting in an enhanced CO_2 adsorption and the formation of carbon-containing surface intermediates, such as formate and carboxyl [92, 94]. However, a direct observation of these intermediates is very difficult, therefore only limited spectroscopic evidence was reported in the literature [94]. On the contrary, metallic Cu increases the reaction rate of CO formation mainly due to its ability to activate H_2 [93]. Dissociated H species could be transferred through the spillover effect to iron oxide and facilitate the removal of lattice oxygen. The major challenge for Fe-based catalysts lies in the uncertainty of structure under reaction conditions [95]. It is well-known that iron oxides can be reduced and carburized into iron carbides. For example, Chen et al. [95] prepared a Pt-promoted Fe_3O_4 catalyst (0.2 wt% Pt/Fe_3O_4) and tested its performance in the RWGS reaction. Compared with pure Fe_3O_4, Pt/Fe_3O_4 showed higher CO_2 conversion without a decrease in CO selectivity. The increase in the activity was attributed to a larger fraction of iron carbides in catalyst composition. Very recently, Yang et al. [96] prepared a series of Fe_2O_3 materials differing in their reducibility and planar defects in the sublattice of iron. Such defects are preserved in the structure of Fe_5C_2 formed in situ from Fe_2O_3 and determine product selectivity. The selectivity to CO of above 98% was achieved at about 86% equilibrium CO_2 conversion over Fe_5C_2 with the low defectiveness.

Over the past two decades, numerous studies dealt with Cu-based catalysts [6, 7]. Analogous to the requirements of methanol synthesis catalysts, Zn species seems to be indispensable for Cu-based catalysts used for the RWGS reaction [97, 98]. In their pioneering work, Stone and Waller [97] varied the atomic ratio of Cu/Zn and determined the maximum of CO_2 conversion obtained over the catalyst with the ratio of 2. The presence of Al_2O_3 in the catalyst changed, however, this activity trend. CO_2 conversion over Cu-ZnO/Al_2O_3 catalysts increased monotonically as the Cu/Zn ratio augmented from 1 to 4. Regardless of the presence or the absence of Al_2O_3, the catalyst activity is positively affected by the Cu metal surface area. However, some researchers stated the importance of the Cu-ZnO interface but not metallic Cu [98].

A drawback of Cu-based catalysts is their deactivation above 400 °C because of sintering of Cu nanoparticles [99]. Thus, recent studies are devoted to the design of stable catalysts over which highly dispersed Cu species could be maintained during RWGS reaction [100]. By utilizing the classical SMSI effect, Cu nanoparticles confined onto the overlayer of $LaTiO_2$ support did not show any decrease in CO formation rate over 500 h on reaction stream even at 800 °C [100].

4.3.4 Reaction Mechanisms

From a mechanistic viewpoint, there are two different concepts describing CO formation from CO_2, which are called as redox and associated mechanisms [7, 78]. According to the former, adsorbed CO_2 directly dissociates on reduced sites to yield adsorbed CO and oxygen species ($*CO_2 \rightarrow *CO + *O$). *CO gives gas-phase CO, while *O reacts with H_2 to regenerate the active sites. On the contrary, the associated mechanism suggests the formation of various intermediates (e.g., carbonate, carboxyl, and formate) from CO_2 without direct dissociation.

Gas switch experiments, spectroscopic techniques, temperature-programmed desorption/surface reaction, and computational studies have been used to elucidate the reaction mechanism. For example, earlier works with Cu-based catalysts claimed different pathways leading to CO [101, 102]. When pre-reduced Cu/ZnO catalyst was subjected to CO_2 at 150 °C, CO was detected and its formation rate decreased with raising time on CO_2 stream due to the oxidation of metallic Cu species to Cu_2O as evidenced by XPS results [101]. Subsequent H_2 treatment of the oxidized catalyst produced H_2O. Its amount corresponded to the amount of formed CO from CO_2. Importantly, the initial formation rate of CO was independent of the presence of H_2 in the reaction feed. These results provided strong evidence for the redox mechanism. By performing CO_2 pulse experiments with or without H_2 at low coverages of oxygen species, the authors from Ref. [102] found, however, that the CO formation rate at 500 °C after CO_2/H_2 pulsing was higher than that after CO_2 pulsing. On this basis, it was suggested that formate intermediates contributed to CO formation.

Although formate species were observed in various FTIR studies, there is still uncertainty about their real function [92, 103]. In this regard, combining steady-state isotopic transient kinetic analysis (SSITKA) with FTIR measurements, Meunier and co-worker [103] investigated the reactivity of different intermediates presented on the surface of Pt/CeO_2 in the course of the RWGS reaction. When switching from $^{12}CO_2/H_2$ (1:4) to $^{13}CO_2/H_2$ (1:4), the intensity of bands related to ^{12}C-labeled surface intermediates decreased with time on $^{13}CO_2/H_2$ stream. The time required to reach a steady-state concentration of carbonyl, carbonate, and formate species was determined to be 54, 48, and 660 s, respectively. Thus, carbonate species were proposed as the main active surface intermediates which were mainly formed on the interfacial sites between Pt nanoparticles and CeO_2. Later, Bobadilla et al. [104] used *operando* DRIFTS and UV–Vis spectroscopy to study the reaction mechanism of the RWGS reaction over Au/TiO_2 catalyst and identified hydroxycarbonyl species (*HOCO) as the main intermediates (Fig. 4.4). They are formed through a reaction of CO_2 and H activated on Au nanoparticle. However, the use of Al_2O_3 as support for Au changed the reaction pathway to the formate-dominant mechanism. Formate species are generated from bicarbonates (*OOCOH) formed from CO_2 reacting with hydroxyl groups. Therefore, the dominant reaction mechanism could be strong influenced by the type of support and the kind of active sites.

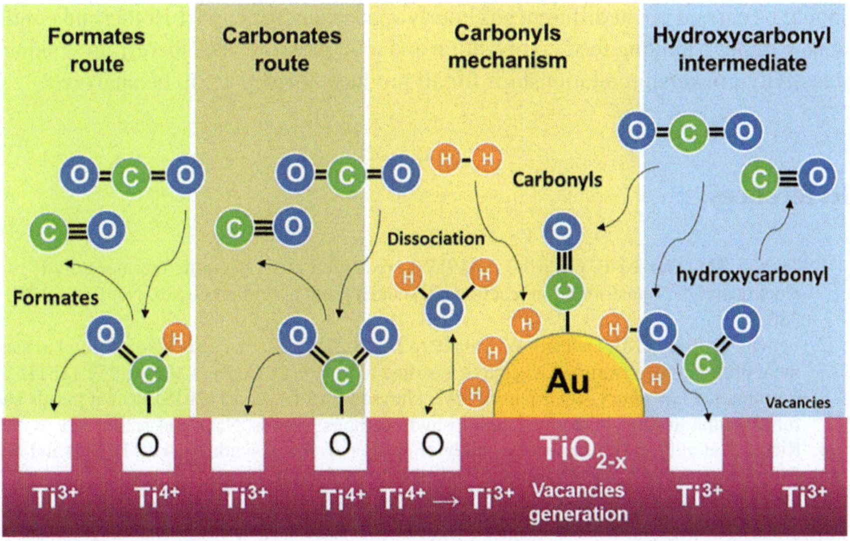

Fig. 4.4 Proposed mechanisms of the RWGS reaction over Au/TiO_2. Reprinted with permission from Ref. [104]. Copyright 2018, American Chemical Society

4.4 Conclusions and Outlook

Direct production of methanol or CO from CO_2 with green H_2 is an attractive option for the development of commercial processes with a focus on their sustainability and environmental friendliness. Significant progresses in the development of catalysts for these reactions and disentangling their mechanisms have been made over the past decades. However, several major challenges are still hindering their implementation on industrially relevant scales. Although Cu-based catalysts display high activity in methanol synthesis, the competing RWGS and CH_3OH decomposition reactions and reaction-induced changes in the structure of Cu nanoparticles are the major reasons for the relatively low methanol selectivity and severe catalyst deactivation, respectively. Alternative methanol synthesis catalysts afford high methanol selectivity but with unsatisfied methanol productivity. Similar shortcomings are also valid for the RWGS reaction to generate CO. Therefore, breaking the trade-off between catalyst activity and product selectivity over affordable catalysts should be a direction in ongoing studies. To establish descriptors required for tailored catalyst design, *operando* catalyst characterization techniques should be combined with kinetic studies. As CO_2 conversion tests are typically carried out in fixed-bed reactors, spatially resolved characterization studies are needed to check if the active species/sites/phases change along the catalyst bed. For CO_2 hydrogenation to CH_3OH, it is highly desirable to check if CH_3OH is involved in side reactions that result in loss of selectivity at high degrees of CO_2 conversion. To this end, catalytic tests should be carried out at different gas hourly space velocities under isothermal conditions without changing feed composition and total pressure. The therefrom obtained selectivity-conversion relationships for all products formed are to be analyzed.

References

1. Tackett BM, Gomez E, Chen JG (2019) Net reduction of CO_2 *via* its thermocatalytic and electrocatalytic transformation reactions in standard and hybrid processes. Nat Catal 2:381–386
2. Fernández-Torres MJ, Dednam W, Caballero JA (2022) Economic and environmental assessment of directly converting CO_2 into a gasoline fuel. Energy Convers Manag 252:115115
3. Yarulina I, Chowdhury AD, Meirer F, Weckhuysen BM, Gascon J (2018) Recent trends and fundamental insights in the methanol-to-hydrocarbons process. Nat Catal 1:398–411
4. Riedel T et al (1999) Comparative study of Fischer–Tropsch synthesis with H_2/CO and H_2/CO_2 syngas using Fe- and Co-based catalysts. Appl Catal A 186:201–213
5. Franke R, Selent D, Börner A (2012) Applied hydroformylation. Chem Rev 112:5675–5732
6. Wang W, Wang S, Ma X, Gong J (2011) Recent advances in catalytic hydrogenation of carbon dioxide. Chem Soc Rev 40:3703–3727
7. De S, Dokania A, Ramirez A, Gascon J (2020) Advances in the design of heterogeneous catalysts and thermocatalytic processes for CO_2 utilization. ACS Catal 10:14147–14185
8. Jiang X, Nie X, Guo X, Song C, Chen JG (2020) Recent advances in carbon dioxide hydrogenation to methanol *via* heterogeneous catalysis. Chem Rev 120:7984–8034
9. Zhong J et al (2020) State of the art and perspectives in heterogeneous catalysis of CO_2 hydrogenation to methanol. Chem Soc Rev 49:1385–1413
10. Chinchen GC, Denny PJ, Parker DG, Spencer MS, Whan DA (1987) Mechanism of methanol synthesis from CO_2/CO/H_2 mixtures over copper/zinc oxide/alumina catalysts: use of ^{14}C-labelled reactants. Appl Catal 30:333–338
11. Kondratenko EV, Mul G, Baltrusaitis J, Larrazábal GO, Pérez-Ramírez J (2013) Status and perspectives of CO_2 conversion into fuels and chemicals by catalytic, photocatalytic and electrocatalytic processes. Energy Environ Sci 6:3112–3135
12. Hansen JB, Højlund Nielsen PE (2008) Methanol synthesis. In: Ertl G, Knözinger H, Schüth F, Weitkamp J (eds) Handbook of heterogeneous catalysis. Wiley-VCH, pp 2920–2949
13. Nakamura J, Choi Y, Fujitani T (2003) On the issue of the active site and the role of ZnO in Cu/ZnO methanol synthesis catalysts. Top Catal 22:277–285
14. Behrens M et al (2012) The active site of methanol synthesis over Cu/ZnO/Al_2O_3 industrial catalysts. Science 336:893–897
15. Kuld S, Conradsen C, Moses PG, Chorkendorff I, Sehested J (2014) Quantification of zinc atoms in a surface alloy on copper in an industrial-type methanol synthesis catalyst. Angew Chem Int Ed 53:5941–5945
16. van den Berg R et al (2016) Structure sensitivity of Cu and CuZn catalysts relevant to industrial methanol synthesis. Nat Commun 7:13057
17. Fujitani T, Nakamura I, Uchijima T, Nakamura J (1997) The kinetics and mechanism of methanol synthesis by hydrogenation of CO_2 over a Zn-deposited Cu(111) surface. Surf Sci 383:285–298
18. Clausen BS et al (1991) *In situ* cell for combined XRD and on-line catalysis tests: studies of Cu-based water gas shift and methanol catalysts. J Catal 132:524–535
19. Grunwaldt JD, Molenbroek AM, Topsøe NY, Topsøe H, Clausen BS (2000) *In Situ* investigations of structural changes in Cu/ZnO catalysts. J Catal 194:452–460
20. Lunkenbein T, Schumann J, Behrens M, Schlögl R, Willinger MG (2015) Formation of a ZnO overlayer in Industrial Cu/ZnO/Al_2O_3 catalysts induced by strong metal-support interactions. Angew Chem Int Ed 54:4544–4548

21. Kattel S, Ramírez PJ, Chen JG, Rodriguez JA, Liu P (2017) Active sites for CO_2 hydrogenation to methanol on Cu/ZnO catalysts. Science 355:1296–1299
22. Yoshihara J, Campbell CT (1996) Methanol synthesis and reverse water-gas shift kinetics over Cu(110) model catalysts: structural sensitivity. J Catal 161:776–782
23. Nakamura I, Fujitani T, Uchijima T, Nakamura J (1996) A model catalyst for methanol synthesis: Zn-deposited and Zn-free Cu surfaces. J Vac Sci Technol A 14:1464–1468
24. Chinchen GC, Waugh KC, Whan DA (1986) The activity and state of the copper surface in methanol synthesis catalysts. Appl Catal 25:101–107
25. Barberis L et al (2022) Competition between reverse water gas shift reaction and methanol synthesis from CO_2: influence of copper particle size. Nanoscale 14:13551–13560
26. Beck A et al (2021) Following the structure of copper-zinc-alumina across the pressure gap in carbon dioxide hydrogenation. Nat Catal 4:488–497
27. Amann P et al (2022) The state of zinc in methanol synthesis over a Zn/ZnO/Cu(211) model catalyst. Science 376:603–608
28. Zander S et al (2013) The role of the oxide component in the development of copper composite catalysts for methanol synthesis. Angew Chem Int Ed 52:6536–6540
29. Graciani J et al (2014) Highly active copper-ceria and copper-ceria-titania catalysts for methanol synthesis from CO_2. Science 345:546–550
30. Fisher IA, Woo HC, Bell AT (1997) Effects of zirconia promotion on the activity of Cu/SiO_2 for methanol synthesis from CO/H_2 and CO_2/H_2. Catal Lett 44:11–17
31. Arena F et al (2008) Solid-state interactions, adsorption sites and functionality of Cu-ZnO/ZrO_2 catalysts in the CO_2 hydrogenation to CH_3OH. Appl Catal A 350:16–23
32. Kattel S, Yan B, Yang Y, Chen JG, Liu P (2016) Optimizing binding energies of key intermediates for CO_2 hydrogenation to methanol over oxide-supported copper. J Am Chem Soc 138:12440–12450
33. Larmier K et al (2017) CO_2-to-methanol hydrogenation on zirconia-supported copper nanoparticles: reaction intermediates and the role of the metal-support interface. Angew Chem Int Ed 56:2318–2323
34. Ye J, Liu C, Mei D, Ge Q (2013) Active oxygen vacancy site for methanol synthesis from CO_2 hydrogenation on In_2O_3(110): a DFT study. ACS Catal 3:1296–1306
35. Sun K et al (2015) Hydrogenation of CO_2 to methanol over In_2O_3 catalyst. J CO_2 Util 12:1–6
36. Martin O et al (2016) Indium oxide as a superior catalyst for methanol synthesis by CO_2 hydrogenation. Angew Chem Int Ed 55:6261–6265
37. Tsoukalou A et al (2019) Structural evolution and dynamics of an In_2O_3 catalyst for CO_2 hydrogenation to methanol: an *operando* XAS-XRD and *in situ* TEM study. J Am Chem Soc 141:13497–13505
38. Dang S et al (2020) Rationally designed indium oxide catalysts for CO_2 hydrogenation to methanol with high activity and selectivity. Sci Adv 6:eaaz2060
39. Frei MS et al (2020) Role of zirconia in indium oxide-catalyzed CO_2 hydrogenation to methanol. ACS Catal 10:1133–1145
40. Sun K, Shen C, Zou R, Liu C (2023) Highly active Pt/In_2O_3-ZrO_2 catalyst for CO_2 hydrogenation to methanol with enhanced CO tolerance: the effects of ZrO_2. Appl Catal B 320:122018
41. Chen T-Y et al (2019) Unraveling highly tunable selectivity in CO_2 hydrogenation over bimetallic In-Zr oxide catalysts. ACS Catal 9:8785–8797
42. Yang C et al (2020) Strong electronic oxide-support interaction over In_2O_3/ZrO_2 for highly selective CO_2 hydrogenation to methanol. J Am Chem Soc 142:19523–19531
43. Tsoukalou A et al (2020) Operando X-ray absorption spectroscopy identifies a monoclinic ZrO_2:In solid solution as the active phase for the hydrogenation of CO_2 to methanol. ACS Catal 10:10060–10067
44. Li K, Chen JG (2019) CO_2 hydrogenation to methanol over ZrO_2-containing catalysts: insights into ZrO_2 induced synergy. ACS Catal 9:7840–7861
45. Zhu J et al (2021) Ni–In synergy in CO_2 hydrogenation to methanol. ACS Catal 11:11371–11384

46. Frei MS et al (2021) Nanostructure of nickel-promoted indium oxide catalysts drives selectivity in CO_2 hydrogenation. Nat Commun 12:1960
47. Pinheiro Araújo T et al (2022) Flame-made ternary Pd-In_2O_3-ZrO_2 catalyst with enhanced oxygen vacancy generation for CO_2 hydrogenation to methanol. Nat Commun 13:5610
48. Frei MS et al (2019) Atomic-scale engineering of indium oxide promotion by palladium for methanol production *via* CO_2 hydrogenation. Nat Commun 10:3377
49. Tang S et al (2023) Mononuclear Re sites on In_2O_3 catalyst for highly efficient CO_2 hydrogenation to methanol. J Catal 417:462–472
50. Sun K et al (2020) A highly active Pt/In_2O_3 catalyst for CO_2 hydrogenation to methanol with enhanced stability. Green Chem 22:5059–5066
51. Han Z, Tang C, Wang J, Li L, Li C (2021) Atomically dispersed Pt^{n+} species as highly active sites in Pt/In_2O_3 catalysts for methanol synthesis from CO_2 hydrogenation. J Catal 394:236–244
52. Rui N et al (2020) Hydrogenation of CO_2 to methanol on a $Au^{\delta+}$–In_2O_{3-x} catalyst. ACS Catal 10:11307–11317
53. Ye J, Liu C-J, Mei D, Ge Q (2014) Methanol synthesis from CO_2 hydrogenation over a Pd_4/In_2O_3 model catalyst: a combined DFT and kinetic study. J Catal 317:44–53
54. Rui N et al (2017) CO_2 hydrogenation to methanol over Pd/In_2O_3: effects of Pd and oxygen vacancy. Appl Catal B 218:488–497
55. Wang J et al (2017) A highly selective and stable ZnO-ZrO_2 solid solution catalyst for CO_2 hydrogenation to methanol. Sci Adv 3:e1701290
56. Wang J et al (2019) High-performance $MaZrO_x$ (Ma = Cd, Ga) solid-solution catalysts for CO_2 hydrogenation to methanol. ACS Catal 9:10253–10259
57. Sha F et al (2021) The promoting role of Ga in $ZnZrO_x$ solid solution catalyst for CO_2 hydrogenation to methanol. J Catal 404:383–392
58. Lee K et al (2022) Atomic Pd-promoted $ZnZrO_x$ solid solution catalyst for CO_2 hydrogenation to methanol. Appl Catal B 304:120994
59. Lee K et al (2023) Engineering nanoscale H supply chain to accelerate methanol synthesis on $ZnZrO_x$. Nat Commun 14:819
60. Xu D, Hong X, Liu G (2021) Highly dispersed metal doping to ZnZr oxide catalyst for CO_2 hydrogenation to methanol: insight into hydrogen spillover. J Catal 393:207–214
61. Huang C et al (2021) CO_2 hydrogenation to methanol over PdZnZr solid solution: effects of the PdZn alloy and oxygen vacancy. ACS Appl Energy Mater 4:9258–9266
62. Fujitani T et al (1995) Development of an active Ga_2O_3 supported palladium catalyst for the synthesis of methanol from carbon dioxide and hydrogen. Appl Catal A 125:L199–L202
63. Fiordaliso EM et al (2015) Intermetallic $GaPd_2$ nanoparticles on SiO_2 for low-pressure CO_2 hydrogenation to methanol: catalytic performance and *in situ* characterization. ACS Catal 5:5827–5836
64. Studt F et al (2014) Discovery of a Ni-Ga catalyst for carbon dioxide reduction to methanol. Nat Chem 6:320–324
65. Sharafutdinov I et al (2014) Intermetallic compounds of Ni and Ga as catalysts for the synthesis of methanol. J Catal 320:77–88
66. Hu J et al (2021) Sulfur vacancy-rich MoS_2 as a catalyst for the hydrogenation of CO_2 to methanol. Nat Catal 4:242–250
67. Zhou S, Zeng HC (2022) Boxlike assemblages of few-layer MoS_2 nanosheets with edge blockage for high-efficiency hydrogenation of CO_2 to methanol. ACS Catal 12:9872–9886
68. Poutsma ML, Elek LF, Ibarbia PA, Risch AP, Rabo JA (1978) Selective formation of methanol from synthesis gas over palladium catalysts. J Catal 52:157–168
69. Fujitani T, Nakamura I (2002) Methanol synthesis from CO and CO_2 hydrogenations over supported palladium catalysts. Bull Chem Soc Jpn 75:1393–1398
70. Kim C-H, Lee JS, Trimm DL (2003) The preparation and characterisation of Pd–ZnO catalysts for methanol synthesis. Top Catal 22:319–324
71. Xu J et al (2016) Methanol synthesis from CO_2 and H_2 over Pd/ZnO/Al_2O_3: catalyst structure dependence of methanol selectivity. Appl Catal A 514:51–59

72. Collins SE, Baltanás MA, Bonivardi AL (2004) An infrared study of the intermediates of methanol synthesis from carbon dioxide over Pd/β-Ga_2O_3. J Catal 226:410–421
73. Collins SE et al (2012) The role of Pd–Ga bimetallic particles in the bifunctional mechanism of selective methanol synthesis *via* CO_2 hydrogenation on a Pd/Ga_2O_3 catalyst. J Catal 292:90–98
74. Ota A et al (2012) Comparative study of hydrotalcite-derived supported Pd_2Ga and PdZn intermetallic nanoparticles as methanol synthesis and methanol steam reforming catalysts. J Catal 293:27–38
75. Gallo A et al (2020) Ni_5Ga_3 catalysts for CO_2 reduction to methanol: exploring the role of Ga surface oxidation/reduction on catalytic activity. Appl Catal B 267:118369
76. Men Y et al (2020) Synthesis of Ni_5Ga_3 catalyst by hydrotalcite-like compound (HTlc) precursors for CO_2 hydrogenation to methanol. Appl Catal B 275:119067
77. Hejral U et al (2022) Tracking the phase changes in micelle-based NiGa nanocatalysts for methanol synthesis under activation and working conditions. J Catal 405:183–198
78. Alam MI, Cheula R, Moroni G, Nardi L, Maestri M (2021) Mechanistic and multiscale aspects of thermo-catalytic CO_2 conversion to C_1 products. Catal Sci Technol 11:6601–6629
79. Zhao Y-F et al (2011) Insight into methanol synthesis from CO_2 hydrogenation on Cu(111): complex reaction network and the effects of H_2O. J Catal 281:199–211
80. Yang Y, Mei D, Peden CHF, Campbell CT, Mims CA (2015) Surface-bound intermediates in low-temperature methanol synthesis on copper: participants and spectators. ACS Catal 5:7328–7337
81. Ortner N et al (2023) Revealing origins of methanol selectivity loss in CO_2 hydrogenation over CuZn-containing catalysts. ACS Catal 13:60–71
82. Ortner N, Lund H, Armbruster U, Wohlrab S, Kondratenko EV (2022) Factors affecting primary and secondary pathways in CO_2 hydrogenation to methanol over CuZnIn/$MZrO_x$ (La, Ti or Y). Catal Today 387:47–53
83. Iulianelli A, Ribeirinha P, Mendes A, Basile A (2014) Methanol steam reforming for hydrogen generation *via* conventional and membrane reactors: a review. Renew Sustain Energy Rev 29:355–368
84. Wang L, Liu H, Chen Y, Yang S (2017) Reverse water–gas shift reaction over co-precipitated Co–CeO_2 catalysts: effect of Co content on selectivity and carbon formation. Int J Hydrogen Energy 42:3682–3689
85. Li W et al (2019) CO_2 hydrogenation on unpromoted and M-promoted Co/TiO_2 catalysts (M = Zr, K, Cs): effects of crystal phase of supports and metal-support interaction on tuning product distribution. ACS Catal 9:2739–2751
86. Dostagir NHMD et al (2021) Co single atoms in ZrO_2 with inherent oxygen vacancies for selective hydrogenation of CO_2 to CO. ACS Catal 11:9450–9461
87. Chen C-S, Budi CS, Wu H-C, Saikia D, Kao H-M (2017) Size-tunable Ni nanoparticles supported on surface-modified, cage-type mesoporous silica as highly active catalysts for CO_2 hydrogenation. ACS Catal 7:8367–8381
88. Winter LR, Gomez E, Yan B, Yao S, Chen JG (2018) Tuning Ni-catalyzed CO_2 hydrogenation selectivity *via* Ni-ceria support interactions and Ni-Fe bimetallic formation. Appl Catal B 224:442–450
89. Yan B et al (2019) Tuning CO_2 hydrogenation selectivity *via* metal-oxide interfacial sites. J Catal 374:60–71
90. Rodrigues MT et al (2017) RWGS reaction employing Ni/Mg(Al,Ni)O—the role of the O vacancies. Appl Catal A 543:98–103
91. Li Y et al (2018) Highly dispersed metal carbide on ZIF-derived pyridinic-N-doped carbon for CO_2 enrichment and selective hydrogenation. ChemSusChem 11:1040–1047
92. Loiland JA, Wulfers MJ, Marinkovic NS, Lobo RF (2016) Fe/γ-Al_2O_3 and Fe–K/γ-Al_2O_3 as reverse water-gas shift catalysts. Catal Sci Technol 6:5267–5279
93. Yang L et al (2020) CO_2 valorisation *via* reverse water-gas shift reaction using promoted Fe/CeO_2-Al_2O_3 catalysts: showcasing the potential of advanced catalysts to explore new processes design. Appl Catal A 593:117442

94. Gu M et al (2021) Structure-activity relationships of copper- and potassium-modified iron oxide catalysts during reverse water-gas shift reaction. ACS Catal 11:12609–12619
95. Chen H et al (2021) Dynamic phase transition of iron oxycarbide facilitated by Pt nanoparticles for promoting the reverse water gas shift reaction. ACS Catal 11:14586–14595
96. Yang Q et al (2023) Activity and selectivity descriptors for iron carbides in CO_2 hydrogenation. Appl Catal B 327:122450
97. Stone FS, Waller D (2003) Cu–ZnO and Cu–ZnO/Al_2O_3 catalysts for the reverse water-gas shift reaction. The effect of the Cu/Zn ratio on precursor characteristics and on the activity of the derived catalysts. Top Catal 22:305–318
98. Álvarez Galván C et al (2016) Reverse water-gas shift reaction at the Cu/ZnO interface: influence of the Cu/Zn ratio on structure-activity correlations. Appl Catal B 195:104–111
99. Yang S-C et al (2018) Synergy between ceria oxygen vacancies and Cu nanoparticles facilitates the catalytic conversion of CO_2 to CO under mild conditions. ACS Catal 8:12056–12066
100. Yu J et al (2021) Ultra-high thermal stability of sputtering reconstructed Cu-based catalysts. Nat Commun 12:7209
101. Fujita S-I, Usui M, Takezawa N (1992) Mechanism of the reverse water gas shift reaction over Cu/ZnO catalyst. J Catal 134:220–225
102. Chen C-S, Cheng W-H, Lin S-S (2000) Mechanism of CO formation in reverse water–gas shift reaction over Cu/Al_2O_3 catalyst. Catal Lett 68:45–48
103. Goguet A, Meunier FC, Tibiletti D, Breen JP, Burch R (2004) Spectrokinetic investigation of reverse water-gas-shift reaction intermediates over a Pt/CeO_2 catalyst. J Phys Chem B 108:20240–20246
104. Bobadilla LF, Santos JL, Ivanova S, Odriozola JA, Urakawa A (2018) Unravelling the role of oxygen vacancies in the mechanism of the reverse water-gas shift reaction by *Operando* DRIFTS and ultraviolet-visible spectroscopy. ACS Catal 8:7455–7467

Chapter 5
CO_2 Methanation over the Supported Ni Catalysts: The Structural Effect

Chenyang Shen and Chang-jun Liu

Abstract Increasing research interests can be found in supported Ni catalyst for CO_2 methanation because of the low price of the metal with comparable activity to the noble metal catalysts. CO_2 methanation on Ni catalyst shows high CO_2 conversion with high selectivity of methane at temperatures around 300 °C at atmospheric pressure. The established pipelines for transportation of natural gas can be directly used for the methane product. CO_2 methanation on Ni catalyst has a big potential in the large-scale utilization of carbon dioxide. CO_2 methanation over the Ni catalyst is a structure-sensitive reaction, with which the mechanism is still in debate. It remains a big challenge in the structural controllable preparation of the supported Ni catalyst. In this chapter, recent advances in the studies of structural effect of the supported Ni catalyst on CO_2 methanation were summarized. Future development is also discussed.

5.1 Introduction

Hydrogenation of CO_2 into methane (i.e., CO_2 methanation), known as Sabatier reaction, attracts increasing attention worldwide because the methane product has broad applications as clean fuel and feedstock for syntheses of various chemicals. It is promising for utilization of carbon dioxide in a large scale. It has been also considered as a potential way for energy storage. The established pipelines for natural gas transportation can be directly used for the methane product from utilization of CO_2 at the remote site.

Table 5.1 shows the principal reactions in CO_2 methanation. CO_2 methanation is an exothermal reaction. It is thermodynamically favorable ($\Delta G^{\circ}_{298\,K} = -114$ kJ/mol). The low temperature operation is theoretically favored for this reaction [1]. However, the CO_2 methanation reaction is an eight-electron one with kinetic limitation. An active catalyst is needed to achieve high reaction rate at relatively low temperatures. Moreover, there exists a competitive side reaction to carbon monoxide,

C. Shen · C. Liu (✉)
School of Chemical Engineering and Technology, Tianjin University, Tianjin 300072, China
e-mail: cjL@tju.edu.cn

G. Zhang et al. (eds.), *Advances in CO_2 Utilization*, Green Chemistry and Sustainable Technology, https://doi.org/10.1007/978-981-99-8822-8_5

which is called as the reverse water gas shift (RWGS) reaction. The RWGS reaction is an endothermic one, leading to formation of carbon monoxide at temperatures higher than 450 °C. It remains as a significant challenge in the development of catalysts with high activity at temperatures at and below 200 °C for CO_2 methanation.

From Table 5.1, one can find that several reactions may cause carbon formation at suitable reaction condition like elevated reaction temperatures. These may cause the deactivation of the catalyst in some cases. Special care is required for the design and preparation of the catalyst and for the design of the reactor [1].

Various catalysts have been exploited for CO_2 methanation, including noble metal catalysts and non-noble metal-based ones. The supported noble metal catalysts, mostly Rh, Rh and Pd [2], normally show good activity with high methane selectivity for CO_2 methanation even at elevated temperatures [3, 4]. The investigation of the noble metal catalysts is important for the understanding of the structure-performance relationship, which further helps the design and preparation of non-noble metal catalysts for CO_2 methanation. Rh, Ru and other noble metals have also been applied as the promoter to improve the performance of the noble metal catalysts. Regarding the non-noble metal catalysts, the nickel-based catalysts have been extensively investigated for CO_2 methanation. Few papers can be found in the literature with the single Fe catalysts, supported on the oxides, for CO_2 methanation. Fe has been more used as a promoter of Ni catalysts [5] and noble metal catalysts [6] for CO_2 methanation. The supported cobalt catalysts can be active ones for CO_2 methanation if appropriate metal-support interaction is achieved [7, 8]. Specific catalyst design is required.

There are rapidly increasing publications on Ni-based catalysts for CO_2 methanation in the literature. However, the reaction mechanism is still unclear. The support, the nickel-support interaction and the catalyst preparation method have significant effects on the activity of the catalyst. A different support or different catalyst preparation method may cause a totally different activity for CO_2 methanation. The low temperature activity of the supported nickel catalysts for CO_2 methanation still needs to be improved. To prepare the catalyst in the structure or performance controllable way remains a big challenge. In this chapter, we attempt to review the recent

Table 5.1 Major reactions involved in CO_2 methanation

	Reaction	$\Delta H°_{298\ K}$ (kJ/mol)	$\Delta G°_{298\ K}$ (kJ/mol)
1	$CO_2 + 4H_2 \leftrightharpoons CH_4 + 2H_2O$ (g)	– 165.0	– 114.0
2	$CO_2 + H_2 \leftrightharpoons CO + H_2O$ (g)	41.2	28.6
3	$CO + 3H_2 \leftrightharpoons CH_4 + H_2O$	– 206.1	– 141.8
4	$2CO + 2H_2 \leftrightharpoons CH_4 + CO_2$	– 247.3	– 170.4
5	$CH_4 \leftrightharpoons C + 2H_2$	74.8	50.7
6	$2CO \leftrightharpoons C + CO_2$	– 172.4	– 119.7
7	$CO + H_2 \leftrightharpoons C + H_2O$ (g)	– 131.3	– 91.1
8	$CO_2 + 2H_2 \leftrightharpoons C + 2H_2O$ (g)	– 90.1	– 62.5

progresses in the structural effect of the supported Ni catalyst for CO_2 methanation. The future developments will be discussed as well.

5.2 The Structure Effect

The structure of the nickel catalyst has significant effect on the activity and selectivity of CO_2 methanation [9, 10]. However, a structural controllable preparation method of the supported nickel catalyst is still not available. The present structural effect study of the supported Ni catalyst has been normally carried out via the size effect or via the strong Ni-support interaction in empirical ways. A small size Ni or a highly dispersed Ni catalyst possesses high proportion of surface species, surface defects and heterogeneity of the atomic structure, improving the hydrogen activation, enhancing the Ni-support interaction and promoting the quantum size effects with changes in the electronic nature.

The structural effect via the catalyst size is still a very complex issue. From the literature, the Ni catalyst of 1–10 nm shows higher activity than the one with catalyst size higher than 10 nm. However, in some cases, the increasing nickel catalyst size may increase the turnover frequency (TOF) of methane [11, 12]. This strongly suggests the structural sensitivity of the supported nickel catalysts for CO_2 methanation, with which the support plays extremely important roles. The support can provide sites for CO_2 adsorption. The unique support enhances the Ni-support interaction and forms the interfacial active sites. These also lead to high-active nickel catalyst for CO_2 methanation.

Han et al. [13] prepared a supported Ni catalyst on sepiolite, a kind of clay mineral materials, via wet impregnation with 10 wt.% Ni loading. The size of the obtained Ni catalyst is 6.8 nm with a specific surface area of 73 m^2/g. The small catalyst size is caused by intense interaction between Ni and the support, which is resulted from the electron transfer from Mg and Si in the magnesium-containing clay mineral support to Ni. The catalyst shows high methane selectivity (close to 100%) and enhanced coke resistance for CO_2 methanation at reaction temperatures higher than 350 °C. Ni(111) is clearly identified in the high-resolution transmission electron microscopical image. The presence of Ni(111) would cause the enhanced coke resistance at elevated reaction temperatures [9]. The sepiolite-supported Ni catalyst exhibits excellent stability at 400 °C for 70 h. The strong interaction between Ni and the support was considered to be the reason for the enhanced stability [13].

Prof. Kawi and his co-workers prepared a catalyst with Ni nanoparticles encapsulated in mesoporous ceria for CO_2 methanation [14]. The average Ni catalyst size is 3.0 nm with 10 wt.% of the Ni loading. The catalyst shows a high low-temperature activity. At a high gas hourly space velocity (GHSV) of 60,000 mL g^{-1} h^{-1}, CO_2 conversion increases from 26% at 225 °C to 70% at 250 °C, and to the highest of 81% at 350 °C.

Xie et al. [15] compared the activity of CeO_2-nanorod-supported Ni (Ni/CeO_2-R), CeO_2-nanocube-supported Ni (Ni/CeO_2-C) and CeO_2-nanooctahedra-supported Ni

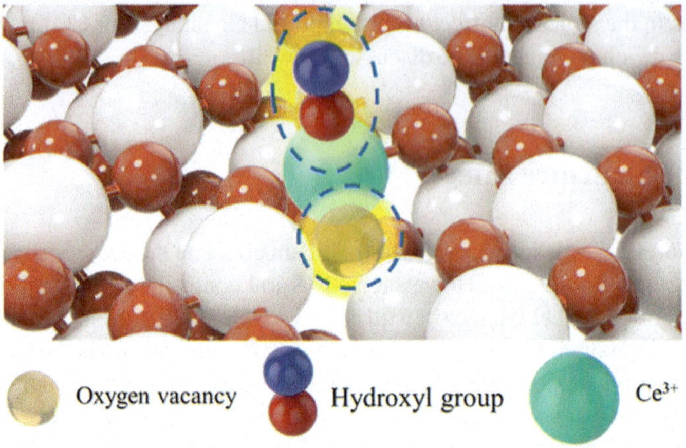

Fig. 5.1 FLP structure (with yellow light) on the Ni/CeO_2 (110) facet (red–white ball). Adapted with permission from Ref. [15], ACS

(Ni/CeO_2-O) catalysts for CO_2 methanation. The average size of Ni catalyst is 7.1 nm, 8.2 nm and 22.8 nm, respectively, for Ni/CeO_2-R, Ni/CeO_2-C and Ni/CeO_2-O. Ni/CeO_2-R shows the highest activity. It exhibits a CO_2 conversion of 84.2% at 225 °C with methane selectivity of 97.8%, corresponding to a high methane formation rate of 147.1 mmol g_{cat}^{-1} h^{-1}. This rate is 4.6 times and 12.8 times higher than that of Ni/CeO_2-C and Ni/CeO_2-O. Ni/CeO_2-R possesses the enriched frustrated Lewis pair (FLP) construction, as shown in Fig. 5.1, from the CeO_2 (110) nature and the steric hindrance between the oxygen vacancy (O_V) and hydroxyl species (OH). A synergy of the small size of the Ni catalyst and the FLP structure of the support leads to the high low-temperature activity for CO_2 methanation.

Wang et al. [16] applied a structural topological transformation of NiZrAl-layered double hydroxide (LDH) for the preparation of zirconium dioxide-supported Ni catalysts. The obtained catalysts show a unique interfacial structure ($Ni{-}O{-}Zr^{3+}{-}$Vö) formed between Ni nanoparticles and ZrO_{2-x} support ($0 < x < 1$). The catalyst with Zr loading of ~ 13.9% and Ni catalyst size of ~ 5.4 nm exhibits high CO_2 conversion (~ 72%) at 230 °C with a ~ 100% selectivity of methane at 30,000 mL g^{-1} h^{-1}. It shows a superior stability within a 110-h reaction test. The interfacial sites ($Ni{-}O{-}Zr^{3+}{-}$Vö) are active centers, with which CO_2 is activated at the interfacial oxygen vacancies ($Zr^{3+}{-}$Vö) and hydrogen is dissociated at metallic Ni sites and spillover to interfacial sites for hydrogenation of intermediate species (like formate and *CO) to methane. Compared with the pristine Ni catalyst, the catalyst with $Ni{-}O{-}Zr^{3+}{-}$Vö has a lower reaction activation energy.

Rontogianni et al. [17] studied the effect of the catalyst preparation method on the MCM-41-supported Ni catalyst for CO_2 methanation. They compared the Ni/MCM-41 catalysts prepared by one-pot synthesis and by the conventional impregnation. The one-pot synthesis means the nickel loading is carried out together with synthesis of MCM-41 support. Nickel nitrate was employed in this work. The Ni catalyst size of the one-pot synthesized catalyst is 19.8 nm when the Ni loading is 6 wt.%, while it is 10.7 nm by the impregnation. They found that the catalyst by the one-pot synthesis strongly favors the RWGS reaction, while the impregnation-prepared catalyst favors methanation during CO_2 hydrogenation. The small-sized and well-dispersed Ni catalyst formed in the impregnation was thought to be the main factor responsible for the good methanation behavior. In contrast, large Ni aggregates were found in the one-pot synthesized catalyst with methanation inactive terrace sites. The larger Ni particles appear to be the preferred sites for the RWGS reaction.

We recently developed a cold plasma decomposition of nickel precursor for the preparation of the supported Ni catalysts for CO_2 methanation. A plentiful of active species exist in cold plasma, leading to the fast nucleation of the nickel oxide nanocrystals. However, the operation temperature of the cold plasma is ~ 150 ºC, which helps to avoid the over growth of the nickel oxide nanocrystals on the support [18]. Followed by the hydrogen reduction thermally (without using cold plasma), the nickel catalyst of small size is obtained. Jia et al. [19] prepared a zirconia-supported nickel catalyst for CO_2 methanation via such plasma decomposition, followed by hydrogen reduction thermally at 500 °C. The obtained catalyst (Ni/ZrO_2-P) shows higher Ni dispersion (10.7%) with average catalyst size of 9.3 nm, compared with the catalyst thermally prepared (Ni/ZrO_2-C) with average catalyst size of 16.0 nm. The plasma decomposition results in an enhanced cooperation between Ni and interfacial active sites as well, leading to rapid dissociative adsorption of H_2 and hydrogen spillover. The sufficient H atoms thereby generated helps to create oxygen vacancies on ZrO_2 for CO_2 adsorption and promotes hydrogenation of adsorbed CO_2. An improved activity for CO_2 methanation is thus achieved. As exhibited in Fig. 5.2, both catalysts show almost 100% methane selectivity at reaction temperature below 270 °C. The activity of Ni/ZrO_2-P increases as the temperature rises until 350 °C, with the highest CO_2 conversion of 79.1% and methane yield of 76.5%. When the temperature reaches 400 °C, CO_2 conversion reduces to 77.5%. For Ni/ZrO_2-C, CO_2 conversion and methane yield only reach up to 74.2% and 70.8%, respectively, at 400 °C. From Fig. 5.2, when the temperature increases from 250 to 275 °C, the activity of Ni/ZrO_2-P increases more rapidly than Ni/ZrO_2-C at the temperature below 300 °C. Ni/ZrO_2-P presents a dramatically larger increase in CO_2 conversion from 15.4 to 50.2% than that of Ni/ZrO_2-C from 6.1 to 15.2%. The low-temperature activity of Ni/ZrO_2-P has been significantly improved.

The cold plasma decomposition has been also applied for the preparation of Ni/TiO_2 [20], Ni/$MgAl_2O_4$ [21] and Ni/CeO_2 [22] catalysts for CO_2 methanation. The same results as Ni/ZrO_2 have been achieved.

From the reported works in the supported Ni catalysts for CO_2 methanation, the catalyst of size below 10 nm normally shows higher activity. As shown in Fig. 5.3a, for Ni/SiO_2 catalysts, there exists an optimum catalyst size for CO_2 methanation based on the studies by Vogt et al. [23–25]. The highest TOF to methane presents with nickel metal nanoparticles of ~ 2.5 nm. If a supported single nickel atom catalyst or a supported nickel cluster catalyst is employed, the selective product of CO_2 hydrogenation would be CO or methanol, instead of methane, because of the big change in the electronic structure [9, 26, 27]. From the present understanding, the

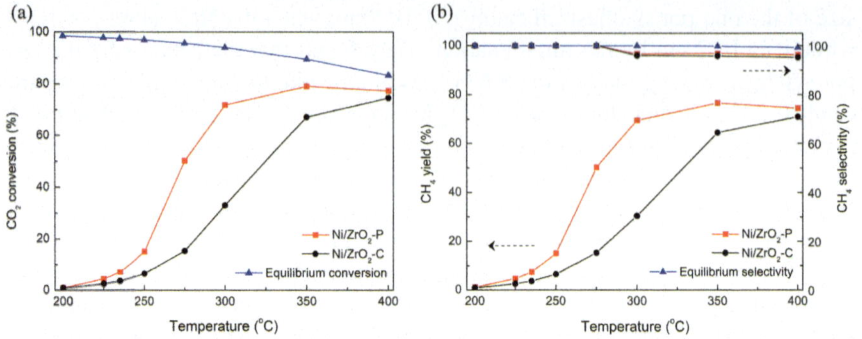

Fig. 5.2 **a** CO_2 conversion, **b** methane yield and selectivity of CO_2 methanation over Ni/ZrO_2-P and Ni/ZrO_2-C catalysts. Blue lines represent the thermodynamic equilibrium values of CO_2 conversion and methane selectivity. The reaction condition: $H_2/CO_2/N_2 = 32/8/10$; GHSV = 60,000 h^{-1}; atmospheric pressure. Adapted with permission from ref.19, Elsevier

single Ni atom catalyst or the nickel cluster catalyst can be formed on the oxide of intense interaction with nickel, like In_2O_3 and In_2O_3-ZrO_2 [9, 26, 27].

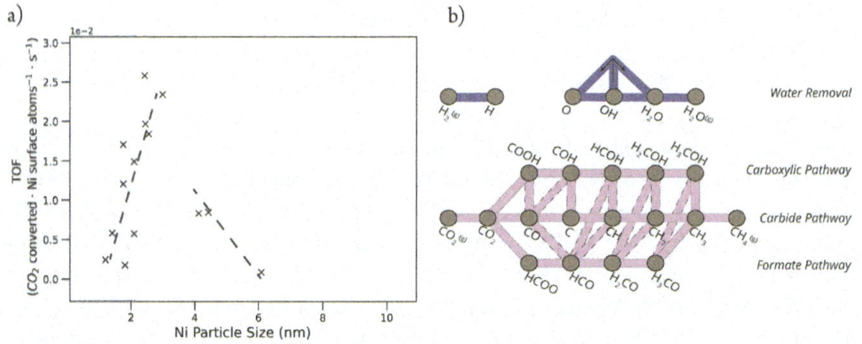

Fig. 5.3 **a** Data points from published turnover frequency (TOF) values for the catalytic CO_2 hydrogenation over Ni/SiO_2 catalysts [23–25]. **b** Schematic overview of relevant possible elementary reaction steps in the catalytic CO_2 methanation over nickel-based catalysts. The brown nodes correspond to the reactants, products, and reaction intermediates. The pink and purple edges between the nodes represent the elementary reaction steps that connect these compounds. The primary pathways are shown in pink: i.e., direct CO dissociation in the carbide pathway and hydrogen-assisted dissociation in the carboxylic and formate pathways. Interlinks are the elementary reactions that connect the primary pathways whose edges are boxed with a dashed line. Elementary reaction steps for the removal of water are shown in purple [28]. Adapted with permission from Ref. [28], ACS

5.3 The Reaction Mechanism

From the discussions above, one can clearly see that CO_2 methanation over the supported Ni catalysts is never a simple case. The reaction mechanism for it is still under debate. The morphology of Ni nanoparticle, the support (reducible or non-reducible), the Ni-support interaction, the catalyst preparation methodology and others can affect the methanation reactions in complex ways. The present understanding of the reaction mechanism over the supported nickel catalysts mostly relies on the density functional theoretical study and the in situ Fourier transform infrared (FTIR) analyses. Figure 5.3b presents three possible reaction pathways for CO_2 hydrogenation into methane with their interlinks shown in pink [28]. These pathways include direct CO_2 dissociation via the carbide pathway and two hydrogen-assisted CO_2 dissociation routes, referred to as the carboxylic and formate pathways. The elementary reaction steps for the formation of water are also shown in purple. The formation of the H_2O^* intermediate can occur via two pathways, i.e., by direct hydrogenation of OH^* and via proton shuffling between two OH^* species to produce O^* and H_2O^* [28]. The formate species in the formate pathway are the major intermediates formed during CO_2 methanation reaction. This pathway is also called as the CO_2 associative methanation, with which the chemisorbed CO_2^* species are firstly converted to bidentate formates ($HCOO^*$) and then to formic acid (HCOOH) and finally to methane. The CO pathway is also called the CO_2 dissociative methanation, with which the chemisorbed CO_2^* species is dissociated into CO^* and O^*. The formed CO^* species can further dissociate into carbon species (C^*), which will be further hydrogenated to methane by dissociated hydrogen, whereas the O^* species reacts with hydrogen to form H_2O.

The pathways shown in Fig. 5.3b can be extended to other oxide-supported Ni catalysts beyond Ni/SiO_2. From the DFT studies, Fig. 5.4 presents a modified reaction network of CO_2 hydrogenation to methane and methanol, based on a network created by Prof. J. G. Chen and his co-workers [29].

As mentioned above, CO_2 methanation over the supported nickel catalyst is complex. A real catalyst may take multi-pathways for the hydrogenation of CO_2.

During our study on Ni/ZrO_2 [19], the catalyst by the cold plasma decomposition of nickel nitrate takes the CO-hydrogenation route for CO_2 methanation. The high-coordinate Ni(111) facet of the plasma-decomposed catalyst facilitates the decomposition of CO_2 and formates into the adsorbed CO. The subsequent hydrogenation of adsorbed CO causes the methane formation. The catalyst by the thermal decomposition possesses more complex Ni crystal structure with more defects. It takes the pathway of direct formate hydrogenation.

With the use of sepiolite as the support, Han et al. [13] found the catalytic activity is almost linearly related to surface OH content of the clay mineral support. This means that the hydrogenation of CO_2 to methane is mainly affected by surface hydroxyl groups. The presence of magnesium in sepiolite significantly improves the activation of CO_2. The sepiolite-supported Ni catalyst takes both CO and formate routes for CO_2 methanation.

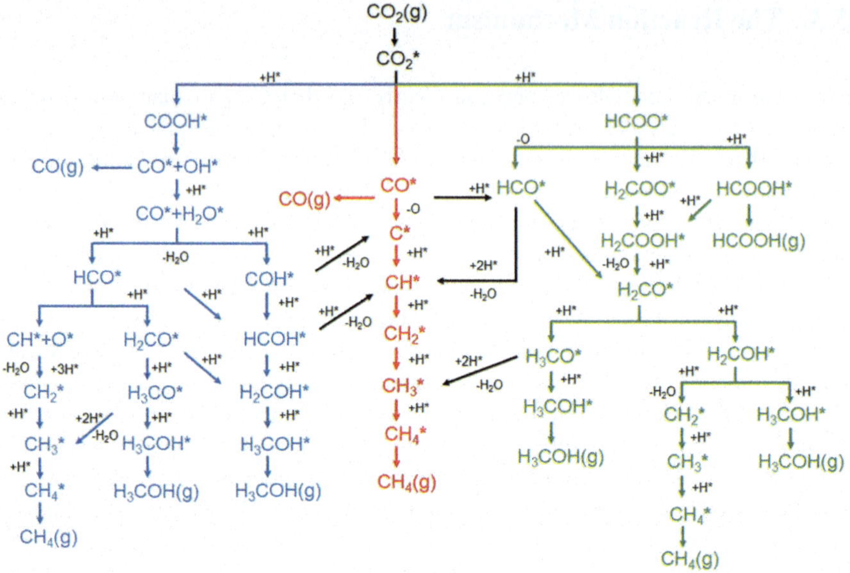

Fig. 5.4 Modified reaction network of CO_2 hydrogenation to methane and methanol over the supported nickel catalysts (* means the adsorbed species). Adapted with permission from Ref. [9], RSC

One can find many papers in the literature with the use of the FTIR analyses for the mechanism investigation. The FTIR analyses will be even more important in the future with the increasing research interests on CO_2 methanation. The DFT study of the supported Ni catalyst is still not easy to be carried out because of the unique characteristic of nickel.

5.4 Conclusions and Perspective

Intense research efforts have been made toward the understanding of the structural effect of Ni catalyst on CO_2 methanation. It remains as significant challenges for a structural controllable preparation of the supported nickel catalyst. For example, when one tries to study the size effect, she or he should keep the catalysts of different sizes with the same crystal structure. When one tries to compare the Ni catalysts with different oxide supports, she or he should make the Ni catalysts with the same catalyst size and the same crystal structure. However, none can be successful with them by far. Therefore, the present investigation obviously has many limitations. Moreover, the low-temperature activity (< 300 °C) of the supported Ni catalysts for CO_2 methanation needs to be further improved. Their carbon resistance performance at high temperatures (> 400 °C) also needs to be improved. These improvements require the understanding of the catalyst structure and the novel catalyst preparation methodologies. The strong interaction between the Ni catalyst and the support reduces the nickel size for improved activity and stabilizes the catalyst. However, it remains as a big challenge for how the supporting material affects the catalyst structure. More fundamental studies are requested to reach the goal of the preparation of nickel catalyst in the structure controllable way and to fully understand the structural effect on CO_2 methanation. We would thus suggest the following future studies.

First, the future theoretical studies should consider the multi-scale effects with the whole reaction network.

Second, the measurement of the chemical intermediates during CO_2 methanation is very important. The development of operando catalyst analysis technologies is needed, including those can be used for atomic-scale observations of catalyst structures during the real reaction.

Third, the novel methodologies for the preparation of model catalyst in a precise structural control of the catalyst and the support are required. By this way, one can study the phase structure effect at the same catalyst size, while she or he can also study the size effect in the phase structure of the support.

Finally, the study of the nucleation and growth of the nickel oxide or nickel nanocrystal on the porous supporting materials is immediately needed.

Acknowledgements We gratefully acknowledge the support from the National Key Research and Development Program of China (2022YFA1504801).

Competing Interests The authors declare no competing interests.

Contributions All authors contributed to this book chapter. C. S. drafted the manuscript. C. S. and C-j. L. conceived the content and revised the manuscript.

References

1. Ren J et al (2023) Methanation of CO/CO_2 for power to methane process: fundamentals, status, and perspectives. J Energy Chem 80:182–206
2. Li L et al (2022) Research progress and reaction mechanism of CO_2 methanation over Ni-based catalysts at low temperature: a review. Catalysts 12: 244
3. Han B et al (2022) Switchable tuning CO_2 hydrogenation selectivity by encapsulation of the Rh nanoparticles while exposing single atoms. Small 18: 2204490
4. Yang C et al (2023) Intrinsic mechanism for carbon dioxide methanation over Ru-based nanocatalysts. ACS Catal 13:11556–11565
5. Du Y et al (2022) Deep understanding into the effect of Fe on CO_2 methanation: a support-dependent phenomenon. Ind Eng Chem Res 61:10357–10364
6. Luo L et al (2020) Surface iron species in palladium-iron intermetallic nanocrystals that promote and stabilize CO_2 methanation. Angew Chem Int Ed 59:14434–14442
7. Liu Z et al (2022) Highly stable and selective layered Co-Al-O catalysts for low-temperature CO_2 methanation. Appl Catal B 310:121303
8. Fan T et al (2021) Cobalt catalysts enable selective hydrogenation of CO_2 toward diverse products: recent progress and perspective. J Phys Chem Lett 12:10486–10496
9. Zhang Z et al (2022) Advances in studies of the structural effects of supported Ni catalysts for CO_2 hydrogenation: from nanoparticle to single atom catalyst. J Mater Chem A 10:5792–5812

10. Zang Y et al (2023) CO_2 activation on Ni(111) and Ni(110) surfaces in the presence of hydrogen. J Phys Chem Lett 14:4381–4387
11. Gonzalez-Rangulan VVV et al (2023) CO_2 methanation over nickel catalysts: support effects investigated through specific activity and operando IR spectroscopy measurements. Catalysts 13: 448
12. Zheng H et al (2022) Unveiling the key factors in determining the activity and selectivity of CO_2 hydrogenation over Ni/CeO_2 catalysts. ACS Catal 12:15451–15462
13. Han F, Liu Q, Li D, Ouyang J (2023) An emerging and high-performance sepiolite-supported Ni catalyst for low-temperature CO_2 methanation: the critical role of hydroxyl groups. J Environ Chem Eng 11:110331
14. Hongmanorom P, Ashok J, Chirawatkul P, Kawi S (2021) Interfacial synergistic catalysis over Ni nanoparticles encapsulated in mesoporous ceria for CO_2 methanation. Appl Catal B 297:120454
15. Xie Y et al (2022) Frustrated Lewis pairs boosting low-temperature CO_2 methanation performance over Ni/CeO_2 nanocatalysts. ACS Catal 12:10587–10602
16. Wang H, Li Z, Cui G, Wei M (2023) Synergistic catalysis at the Ni/ZrO_{2-x} interface toward low-temperature CO_2 methanation. ACS Appl Mater Interfaces 15:19021–19031
17. Rontogianni A et al (2023) Efficient CO_2 hydrogenation over mono- and bi-metallic RuNi/MCM-41 catalysts: controlling CH_4 and CO products distribution through the preparation method and/or partial replacement of Ni by Ru. Chem Eng J 474:145644
18. Wang Z et al (2018) Catalyst preparation with plasmas: how does it work? ACS Catal 8:2093–2110
19. Jia X, Zhang X, Rui N, Hu X, Liu C-J (2019) Structural effect of Ni/ZrO_2 catalyst on CO_2 methanation with enhanced activity. Appl Catal B 244:159–169
20. Zhou R, Rui N, Fan Z, Liu C-J (2016) Effect of the structure of Ni/TiO_2 catalyst on CO_2 methanation. Int J Hydrogen Energy 41:22017–22025
21. Fan Z, Sun K, Rui N, Zhao B, Liu C-J (2015) Improved activity of Ni/$MgAl_2O_4$ for CO_2 methanation by the plasma decomposition. J Energy Chem 24:655–659
22. Rui N et al (2021) Highly active Ni/CeO_2 catalyst for CO_2 methanation: preparation and characterization. Appl Catal B 282
23. Vogt C et al (2019) Understanding carbon dioxide activation and carbon–carbon coupling over nickel. Nat Commun 10:5330
24. Vogt C et al (2018) Unravelling structure sensitivity in CO_2 hydrogenation over nickel. Nat Catal 1:127–134
25. Vogt C et al (2021) Dynamic restructuring of supported metal nanoparticles and its implications for structure insensitive catalysis. Nat Commun 12:7096
26. Jia X, Sun K, Wang J, Shen C, Liu C-J (2020) Selective hydrogenation of CO_2 to methanol over Ni/In_2O_3 catalyst. J Energy Chem 50:409–415
27. Zhu J et al (2021) Ni-In synergy in CO_2 hydrogenation to methanol. ACS Catal 11:11371–11384
28. Sterk EB et al (2022) Structure sensitivity of CO_2 conversion over nickel metal nanoparticles explained by micro-kinetics simulations. JACS Au 2:2714–2730
29. Kattel S, Liu P, Chen JG (2017) Tuning selectivity of CO_2 hydrogenation reactions at the metal/oxide interface. J Am Chem Soc 139:9739–9754

Chapter 6
CO_2 Reforming with Alkanes

Huanhao Chen, Xinrui Wang, Shihang Yu, and Xiaolei Fan

Abstract Direct CO_2 reforming with light alkanes is a promising approach for transforming CO_2 into value-added products, enabling carbon emission reduction and valorization (to reduce the reliance on fossil fuels). The development of highly active and stable catalytic systems is essential for the cost-effective and practical CO_2 reforming processes. In this chapter, we first summarized the state-of-the-art on the catalyst design for CO_2 reforming mainly regarding the support selection and structure modification to tackle the catalytic deactivation issues such as metal sintering or coke formation. Additionally, other advanced CO_2 reforming systems such as super-dry reforming and photocatalysis are discussed as well since they have the potential to activate the CO_2 conversion processes either with ultra-high CO_2 utilization efficiency or under milder conditions. Finally, recent developments on in situ/operando characterization techniques for mechanistic investigations are presented, which is beneficial for the rational design/optimization of catalysts to advance catalytic CO_2 reforming technologies.

6.1 Introduction

Carbon dioxide (CO_2) capture, utilization, and storage (CCUS) is considered as a promising technology for mitigating the carbon dioxide (CO_2) emission and its impact on the environment [1]. Especially, CO_2 utilization technologies using CO_2 as the feedstock has attracted huge attention, since they can valorize CO_2 into various

H. Chen
State Key Laboratory of Materials-Oriented Chemical Engineering, College of Chemical Engineering, Nanjing Tech University, Nanjing 211816, China

X. Wang · X. Fan (✉)
Department of Chemical Engineering, School of Engineering, The University of Manchester, Manchester M13 9PL, UK
e-mail: xiaolei.fan@manchester.ac.uk

S. Yu · X. Fan
Nottingham Ningbo China Beacons of Excellence Research and Innovation Institute, University of Nottingham Ningbo China, Ningbo, China

G. Zhang et al. (eds.), *Advances in CO_2 Utilization*, Green Chemistry and Sustainable Technology, https://doi.org/10.1007/978-981-99-8822-8_6

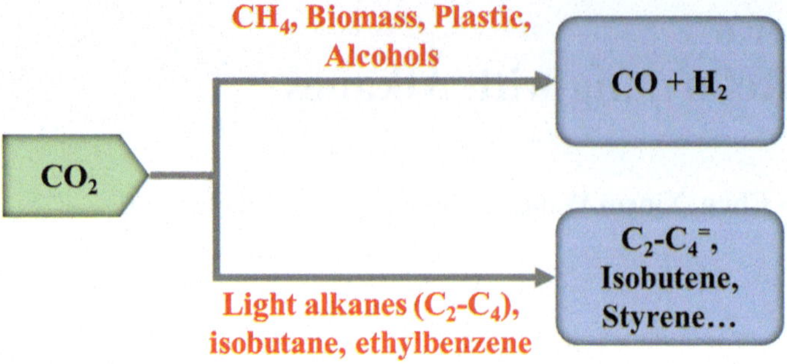

Fig. 6.1 Direct reforming routes of CO_2 with different reducing gases and hydrogen contained substances to produce syngas and other value-added chemicals

value-added products (such as chemicals, fuels, and concrete building materials) and reduce the dependence on fossil fuels (such as petroleum and natural gas). Among various CO_2 utilization pathways, direct reforming of CO_2 with a reducing gas (e.g., methane, ethane, propane, and butane) or hydrogen-containing substances (e.g., biomass, plastic, and alcohols) has been regarded as the promising route for valorizing CO_2 (Fig. 6.1). For example, catalytic dry reforming of methane (DRM) with CO_2 is considered as an ideal way to simultaneously convert green-house gases CO_2 and CH_4 into syngas with the H_2/CO ratio close to 1, which is suitable for the subsequent Fischer–Tropsch (FT) synthesis to produce fuels and chemicals [2–4]. CO_2-mediated oxidative dehydrogenation (CO_2-ODH) of light alkanes and ethylbenzene is another favorable route to produce different valuable platform chemicals including olefins (e.g., ethylene, propylene, and styrene) and others [5–7]. Dry reforming of biofuels (such as bio-ethanol and biodiesel) is considered as an effective way to close the carbon cycle, which can efficiently utilize the renewable biofuels [8–10]. In addition, CO_2 can also be used (to replace steam) in catalytic cracking of high molecular weight wastes (such as biomass and plastics, in the presence of heterogeneous catalysts, e.g., Ni-based catalysts) to produce syngas, which enables the circular economy [11–13]. Due to the rapid development of the CO_2-DRM and CO_2-ODH systems in the research community, in this chapter, we first focus on the recent advances in the design and development of catalysts and their activation methods in these fields.

6.2 Overview of CO_2 Reforming with Light Alkanes

Due to the human societal development, the consumption of fossil fuels and the associated carbon emissions to the atmosphere is ever-increasing, being the key contributor to the global warming. Meanwhile, due to the huge abundance of light alkanes (such as CH_4 and C_2–C_4) from natural and shale gas, efficient utilization of these gases has been remaining a great challenge as well. To tackle the two issues simultaneously, direct reforming of light alkanes with CO_2 is considered as an ideal route to convert CO_2 and alkanes into value-added chemicals (e.g., syngas and olefins). However, the DRM process (i.e., CO_2 reforming with CH_4) is the most widely investigated and has been demonstrated successfully in the pilot-scale in industry [2]. As a soft oxidant, CO_2 can be also utilized to transfer other light alkanes (i.e., C_2–C_4) into olefins [14]. However, in the reforming reactions, the challenge to achieve high catalytic activity is the effective CO_2 activation coupled with selective C–H activation, which requires effective catalyst design strategies and relevant activation methods [5]. Accordingly, current efforts on the investigation of CO_2 reforming with light alkanes mainly focus on the development of highly active and stable catalysts, the understanding of relevant reaction mechanisms, and the exploration of new catalytic systems (Fig. 6.2).

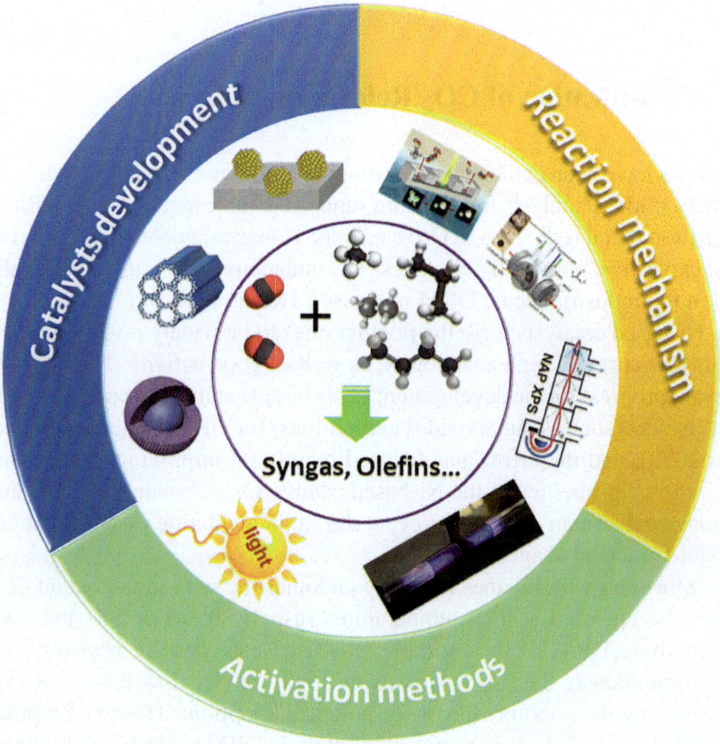

Fig. 6.2 Current studies on the CO_2 reforming reactions with light alkanes

6.3 Progress in Catalytic CO_2 Reforming with Light Alkanes

The cleavage of the first C–H bond and the activation of CO_2 are considered the rate-determining steps in the catalytic reforming reactions[15, 16]. Accordingly, different strategies are needed to develop highly active catalysts and activation means (for the overall catalytic systems) to achieve the facile activation of the steps above. In addition, catalyst stability is another priority in the CO_2 reforming reactions, for example, in thermally activated DRM, coke deposition and metal sintering are the two major factors accounting for catalyst deactivation. For the CO_2-ODH processes (e.g., CO_2 reforming with C_2–C_4 alkanes), the low olefin yield and poor catalyst stability are two major challenges for scaling up the processes. Herein, the typical classification of thermocatalytic CO_2 reforming catalysts (especially that for DRM) and relevant activation means will be discussed in detail.

6.3.1 Classification of CO_2 Reforming Catalysts

Many efforts have been dedicated to developing active and stable catalysts for DRM. The widely studied catalysts forDRM are supported noble metal (e.g., Ru, Rh, and Pt) and transition metal (e.g., Ni and Co) catalysts. However, noble metals are relatively rare and expensive, potentially being less advantageous (than transition metals) to be applied in the industrial-scale DRM processes. Hence, based on the state-of-the-art studies, Ni-based catalysts were the popular ones to be widely investigated in DRM due to its low cost and high availability, as well as good activity. Table 6.1 presents the relevant progress on the development of Ni-based catalysts supported on different catalyst carriers (such as metal oxides and zeolites) for DRM in the last decade (from 2012 to 2022), and the effort was focused mainly on improving the anti-sintering and anti-coking properties of the Ni-based catalysts, since metal sintering and coke formation are the main issues to prevent the implementation of Ni-based catalysts for possible practical usages.

CO_2 reforming with alkanes is highly endothermic with high content of carbon sources to be operated at high temperatures (usually between 500 and 1000 °C). Taking catalytic DRM as the example, Ni-based catalysts are probably the most promising candidates due to the relatively high activity, low cost, and extensive supply, showing the good prospects for practical adoptions. However, rapid deactivation of Ni-based catalysts is common in catalytic DRM processes at high temperatures, being the major issue needing to be addressed before the possible practical applications [32]. Catalyst deactivation is caused primarily by Ni sintering and carbon deposition (i.e., coking) at high temperatures [33]. Hence, various strategies to improve the anti-sintering and anti-coking properties of Ni-based catalysts were proposed, including bimetallic alloying [18, 34–36], designing unique microscopic catalyst structures (such as core–shell structures) [29], and encapsulation and

Table 6.1 Progress in the development of Ni-based catalysts for DRM (2012–2022)

Catalyst	Reaction temperature (°C)	CO_2/CH_4 conversion (%)	H_2/CO molar ratio	Stability (h)	References
$Ni_xCo_y/H\text{-}ZrO_2$	700	93/93	0.8	80	[17]
Ni–Fe/$MgAlO_x$	650	–/21	0.5	10	[18]
Ni/γ-Al_2O_3 (Ni^0Al_{MIL})	650	79/75	1.0	100	[19]
Ni/$MgFe_xAl_{2-x}O_4$	750	–/70	/	65	[20]
Ni-Mo/MgO	800	100/100	1.0	850	[21]
Ni-Pt/CeO_2	650	49/41	0.6	45	[22]
$Ni_xRu_yCeO_2$	600	–/91	1.0	65	[23]
Ni/Silicalite-1	800	82/81	0.8	150	[24]
Ni-Mo/ZSM-5	750	94/90	0.9	100	[25]
Ni@SiO_2-Silicalite-1	750	84/82	0.9	28	[26]
Ni/Ce-HS (hollow sphere) SiO_2	750	94/88	0.9	50	[27]
Ni@HZSM-5	550	31/35	/	50	[28]
NiCo@SiO_2 core–shell	800	89/87	/	1000	[29]
Ni-SBA-15	700	86/78	1.0	100	[30]
Ni/BN	700	74/64	0.9	20	[31]

confinement of metallic species (using framework materials such as zeolites) [25, 28]. Secondary metal doping is one of the popular strategies to overcome metal sintering and coking during catalytic DRM processes at high temperatures [37, 38]. Specifically, alloying Ni with a secondary metal (e.g., Pt, Pd, Ru, Co, and Fe) on metal oxide carriers could alter the physicochemical properties of the resulting bimetallic phases, and thus promoting the catalytic activity and selectivity toward the desired products, the resistance to coking and thermal sintering of the bimetallic phases [37, 39]. In the alloying strategy, several parameters (including the relative metal composition, degree of alloying, and microscopic morphology of the bimetallic phases) are key to enable the fine-tuning of the activity and stability of the bimetallic catalysts. For example, Li et al. [40]. studied the effect of the Ni/Fe ratio (supported on CeO_2 and CeO_2-ZrO_2) on the reaction behaviors of CO_2 reforming with *n*-butane (Fig. 6.3a), implying that higher Fe content (i.e., Ni:Fe molar ratio = 1:3) could form FeO_x species to promote the C–H bond cleavage and produce 1,3-butadiene through the oxidative dehydrogenation pathway, whilst a higher Ni content in the NiFe bimetallic catalyst (i.e., Ni_3Fe_1/CeO_2) mainly facilitated the C–C bond scission and promoted the dry reforming pathway for syngas production.

The confinement or encapsulation strategies (for regulating Ni species) using the well-controlled host structures were well explored and studied for catalyst design to improve CO_2 reforming reactions. As illustrated in Fig. 6.3b, these strategies include, but not limited to, the design of core–shell [41], core-mesopores [42], core-hollow

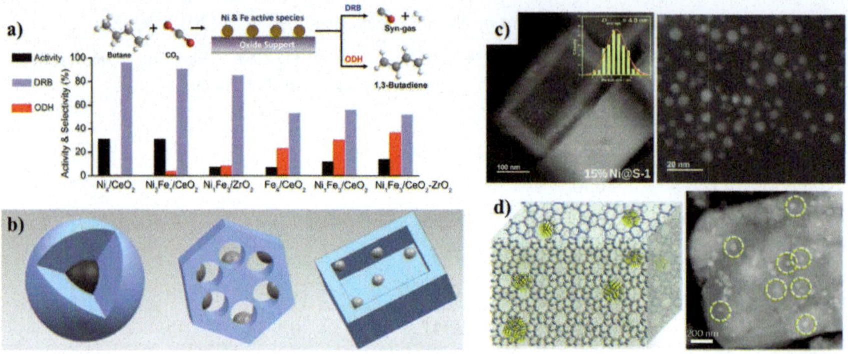

Fig. 6.3 Different strategies to improve the performance of Ni-based catalysts for CO_2 reforming processes: **a** effect of Ni/Fe ratio on CO_2 reforming with *n*-butane over bimetallic NiFe on CeO_2, ZrO_2 or CeO_2-ZrO_2 catalysts [40]; **b** exemplar confinement/encapsulation strategies [39]; **c** hollow silicalited-1 encapsulated Ni catalyst [24]; and **d** HZSM-5 encapsulated Ni catalyst [28]. Reproduced with permission from **a** Elsevier for Ref. [40], **b** John Wiley and Sons for Ref. [39], **c** Elsevier for Ref. [24], and **d** Springer Nature for Ref. [28]

matrix [32], york-shell [43], and sandwich structures [44]. For example, coating of Ni nanoparticles with a support shell could promote the metal–support interaction to prevent metal sintering and alleviate carbon deposition, being able to enhance the activity and stability of the resulting catalysts remarkably for DRM [39, 45]. Mesoporous silica and carbon with highly ordered porous structures could confine and stabilize metal nanoparticles to suppress the thermal sintering of metallic species significantly [42, 46]. Zeolites are class of framework materials being able to confine metal nanoparticles to improve the anti-sintering and anti-coking properties of the catalyst. Previously, encapsulation of Pt-Ni bimetallic nanoparticles inside hollow silicalite-1 zeolite was demonstrated to be able to reduce the size of metal nanoparticles and improve the metal dispersion effectively due to the confinement effect and strong interaction between the metal phase and support, which inhibited coke formation during DRM at high temperatures [32]. However, it is still challenging to encapsulate metal nanoparticles (e.g., Ni) within the zeolite framework, especially high metal loading while maintaining high metal dispersion, due to the poor compatibility between the metallic species and zeolites. It is highly beneficial to confine metal nanoparticles inside zeolite crystals with higher loading to enhance the mass-specific DRM reaction rate. To achieve this goal, Liu et al. [24] proposed a controlled 'dissolution–recrystallization' synthesis being able to embed Ni nanoparticles in the shell of hollow silicalited-1 zeolite with high loading up to 15 wt.%, as well as small particle sizes of ~ 4 nm (Fig. 6.3c). The developed catalyst presented outstanding turnover frequency (TOF) and reaction rate of 60 s^{-1} and 20 $mol_{CH4}\ g_{cat}^{-1}\ h^{-1}$, respectively, as well as good stability up to 150 h during DRM at 800 °C.

It is worth noting that the ideal DRM process is carried out with the stoichiometric feed ratio of CH_4 and CO_2 (i.e., equimolar), which is commonly not true in practice scenarios. Therefore, to answer the challenge of proceeding DRM with the non-ideal feed compositions, rational design of Ni-based catalysts with flexibility can be important. For example, the natural gas in the Southern China Sea is typically CO_2-rich containing ~ 70% of CO_2 and ~ 30% of CH_4. Accordingly, it is attractive to develop the catalysts and catalytic systems being able to reduce as much CO_2 as possible using the limited amount of CH_4. In view of this, Zhu et al. [28] encapsulated Ni nanoparticles within the matrix of HZSM-5 zeolite (Fig. 6.3d), which promoted H_2 spillover to facilitate CO_2 reduction. As a result, the developed catalyst was capable of reducing up to 2.9 mol of CO_2 per mol of CH_4, showing the promising prospect for being applied to upgrade CO_2-rich natural gases. In addition, Ni sintering and coke deposition were minimized effectively due to the steric effect of the rigid zeolite framework.

In catalytic DRM systems, carbon deposition, especially carbon nanotubes from catalytic decomposition of methane over Ni catalysts, could cause significant fouling in fixed-bed reactors, leading to high bed pressure drop and possible operational issues [47, 48]. Carbon deposition issues could be mitigated to some extents by improving the transport phenomena in the reactor, which was exemplified by the application of structured/monolithic catalysts and reactors to intensify the mass and heat transport in catalytic CO_2 reforming reactions, especially DRM [7, 33, 49–51]. Structured catalysts based on macroscopic cellular supports, such as silicon carbide (SiC), and monolithic supports, such as metallic honeycomb monolith and cordierite monolith, are particularly advantageous for enhancing the 'global' transport phenomena of CO_2 reforming reactions due to their outstanding properties, such as high-thermal conductivity and stability, high surface-to-volume ratio, high mechanical strength, as well as high permeability and low pressure drop. However, the pristine structured supports present a relatively low surface area and deficient surface functional groups, bringing challenges to stabilize the active metal sites on their surface. Therefore, functionalization of the structured supports with inorganic coatings (e.g., zeolites, carbon nanofibers, Al_2O_3, CeO_2, etc.) is necessary to facilitate metal dispersion [33, 52]. For example, Wei et al. [51] dispersed the silicalite-1 (S1) zeolite coating on the surface of SiC open-cell foams via hydrothermal synthesis and loaded Ni catalyst on it by impregnation (Fig. 6.4a and b). Compared to the bulk S1 supported Ni catalyst (i.e., pellets in packed bed configuration), the developed structured catalyst showed better catalytic activity and stability in catalytic DRM to produce syngas. Gao et al. [53] decorated SiC foams with a layer of carbon nanofibers (CNFs, with a loading amount of 6.6 wt.%), which enhanced the specific surface area of the SiC foam from 4 to 24 $m^2\ g^{-1}$ (Fig. 6.4c and d). Experimental results revealed that the CNFs layer served as the anchoring sites to promote the metal–support interaction and Ni dispersion, benefiting the activity, stability, and anti-coking ability of the resulting catalyst. Although the 'global' transport phenomena were enhanced by the structured supports, rational design of catalytic coating including the microstructure of inorganic coatings (e.g., pore structure) and the properties of the active metal sites (e.g., dispersion) is still needed. In Chen et al.'s study [33], a new yet simple strategy was proposed to fabricate S1 zeolite coating with the encapsulated Ni catalysts on SiC foams. In detail, encapsulation of metallic species during the synthesis

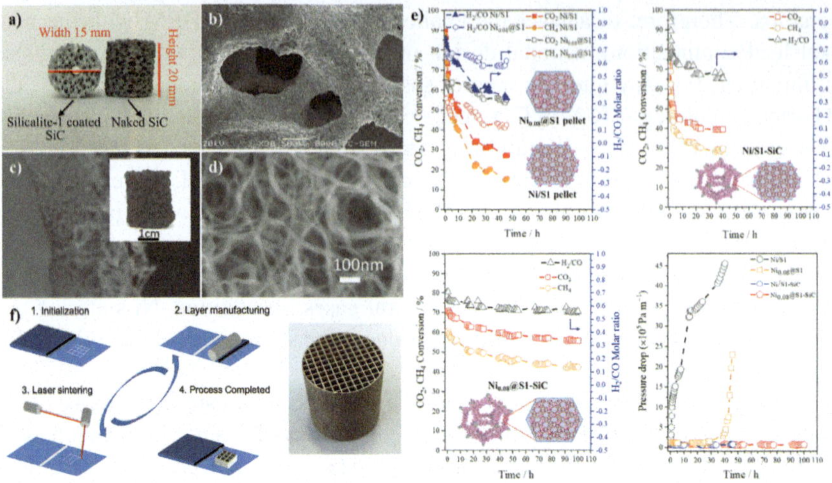

Fig. 6.4 Structured catalysts for CO_2 reforming processes: **a**, **b** Ni catalyst supported on silicalite-1 coated SiC foam [51]; **c**, **d** Ni catalyst supported on carbon nanofiber coated SiC foam [53]; **e** DRM performance over different bulk/structured catalysts [33]; and **f** 3D-printed stainless-steel honeycomb monolith with the supported Ni/CeO_2-ZrO_2 catalyst [49]. Reproduced with permission from Elsevier for **a**, **b** Ref. [51], Elsevier for **c**, **d** Ref. [53], John Wiley and Sons for **e** Ref. [33], and Elsevier **f** Ref. [49]

of the zeolite coating could reduce metal aggregation effectively, resulting in relatively small Ni particle size and improved metal dispersion. The combination of the structured foam support and zeolite encapsulated Ni coating, the structured catalyst showed high CO_2/CH_4 conversions of > 85% and H_2/CO molar ratio of > 0.85 at 700 °C (Fig. 6.4e), as well as significantly low bed pressure drop, outperforming other control catalysts such as packed beds. Structured catalysts are beneficial to other CO_2 reforming processes as well, e.g., oxidative dehydrogenation of alkanes (such as ethane, propane, and butane with CO_2 for producing olefins) [7, 50], in which heat and mass transfer were improved significantly. It is worth noting that 3D printing technology (i.e., additive manufacturing) is considered an emerging and promising method for fabricating structured/monolithic catalysts, with advantages of flexibility and versatility regarding design and material incorporation [54–56]. For example, Agueniou et al. [49] designed and 3D-printed a stainless-steel honeycomb monolith for supporting Ni/CeO_2-ZrO_2 catalysts (Fig. 6.4f), which showed enhanced heat transfer and activity for catalytic DRM.

6.3.2 Other Advanced Systems

Apart from the attempts on the catalytic design, development of advanced reaction systems is also an effective route to mitigate the catalytic deactivation and/or improve the CO_2 utilization efficiency. For example, to suppress the side reactions in CO_2 reforming reactions such as water–gas shift reaction (WSGR), a super-dry reforming strategy was proposed [57], in which three processes were coupled under isothermal condition, that is, (i) CH_4 reforming over the $Ni/MgAl_2O_4$ catalyst to produce syngas, (ii) Fe_3O_4 reduction by syngas to create CO_2 and H_2O with in situ CO_2 removal by CaO carbonation, (iii) inert gas sweeping to decompose $CaCO_3$ into CaO and CO_2 with CO_2 oxidation of Fe to produce high purity CO. The novel system is suitable for valorization of CO_2-rich natural gas, being able to deliver a high CO yield of 2.9 mol_{CO} mol_{CH4}^{-1}. Calcium-looping (CaL) coupled DRM is another attractive strategy to improve the energy efficiency of the CO_2 reforming systems [58, 59]. This can be achieved by designing bifunctional CaO-Ni sorbent-catalyst (Fig. 6.5) [59], and the integrated process could shift the reaction equilibrium by continuous CO_2 consumption, reducing the energy consumption of 22% compared to that of the conventional DRM process.

As CO_2 molecule is thermodynamically stable, and its activation is energy demanding, commonly requires high temperatures which is the main cause of carbon deposition and metal sintering in conventional thermocatalytic systems, activation of CO_2 reforming reactions under mild conditions using alternative means can be highly desirable, such as electrified (e.g., microwave, plasma, and magnetic induction heating) and photocatalysis [60–62]. For example, photocatalytic CO_2 reforming can potentially utilize the solar energy, being more sustainable than the conventional systems [60, 63], but the process efficiency of the photocatalytic systems is relatively low. Accordingly, photo-assisted thermal catalysis was proposed and developed for DRM under relatively mild conditions, i.e., low temperatures of < 500 °C [64–66]. The findings show that the light irradiation (i.e., ultraviolet, visible, and near infrared) could reduce the reaction activation energy required for thermocatalysis significantly, and the milder reaction conditions could alleviate catalyst deactivation and suppress WGSR to improve CO production. In this research field, the rational design of highly

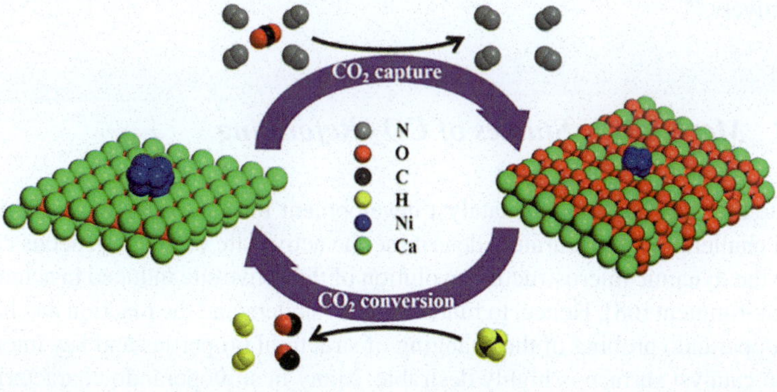

Fig. 6.5 Schematic representation of the bifunctional sorbent-catalyst for the integrated calcium-looping-DRM process [59]. Reproduced with permission from Ref. [59], open access with the CC BY-NC 4.0 license

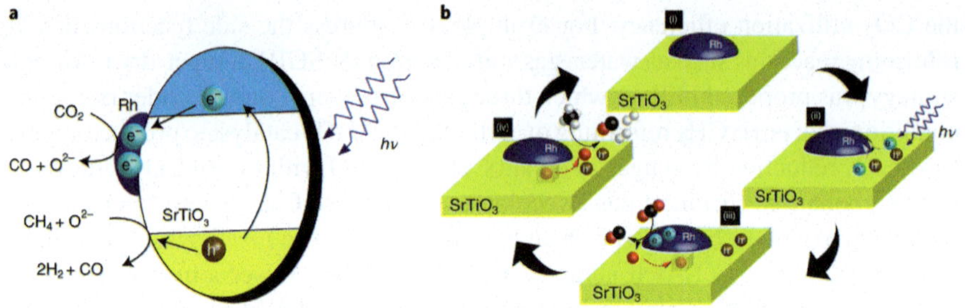

Fig. 6.6 Schematic of the mechanism of photothermal catalytic DRM over the Pt/TiO_2 catalyst: **a** band diagram, **b** schematic dynamics [63]. Reproduced with permission from Springer Nature for **a**, **b** Ref. [63]

active catalyst and the investigation of corresponding reaction mechanism are key to advance this emerging catalytic process [67]. However, most of reported works are conducted by the coupling of thermal heating with additional illumination, and hence, they are categorized as a photo-assisted catalysis rather than real photocatalysis [63]. Additionally, no clear evidence was reported so far to demonstrate that the energy for DRM reaction is induced by photoexcitation of electron–hole pairs. Therefore, to truly utilize the solar energy, it is highly desirable to develop an efficient photocatalyst for the DRM process and explore the corresponding reaction mechanism under light illumination alone condition. Motivated by this, Shoji et al. [63] reported a development of Rh catalyst supported on $SrTiO_3$ (i.e., Rh/STO) for photocatalytic DRM under UV light irradiation without additional heat input, achieving a considerably high conversions of CO_2/CH_4 (>50%) at relatively low temperatures (successfully overcome the thermodynamic limitation of DRM at 450 °C), which are impossible achieved by conventional thermal catalysis. Moreover, the mechanistic study of this work revealed that the oxidation of CH_4 over STO and the reduction of CO_2 over Rh were mainly induced by the photoexcited holes and electrons (Fig. 6.6), respectively.

6.3.3 *Mechanistic Studies of CO_2 Reforming*

Despite the great progress of catalyst development for CO_2 reforming reactions, it is still challenging to accurately determine the active site of heterogeneous catalyst due to the dynamic microstructure evolution of the active site induced in actual reaction environment [68]. Hence, to fundamentally understand the reaction mechanism, in situ/operando probing of the changing of structural properties and key intermediates on catalyst surface is highly desirable. Many in situ/operando characterization methods have been developed for heterogeneous catalytic systems, including transmission electron microscopy (TEM), diffuse reflectance infrared Fourier transform spectroscopy (DRIFTS), scanning transmission X-ray microscopy (STXM), X-ray diffraction (XRD), ambient pressure X-ray photoelectron spectroscopy (AP-XPS), and X-ray absorption spectroscopy (XAS), in which many of them were employed to study the CO_2 reforming mechanism. In the following text, the review will focus on the mechanistic investigation of the DRM reaction by employing various in situ characterization methods.

In situ DRIFTS (normally coupled with mass spectrometry, i.e., DRIFTS-MS) has been widely employed to study the mechanisms of DRM reaction over catalyst surface and provide abundant information on the chemical nature and surface dynamics of intermediate species [69–71]. Moreover, the transient in situ DRIFTS-MS has been demonstrated as promising tool to detect and identify the dynamics of surface reactive species adsorbed on catalyst surface in the timeframe of seconds rather than minutes [69, 72]. For example, using a in situ DRIFTS-MS method, Álvarez et al. [72] gained the dynamic evolution of the IR band area of relevant intermediates (i.e., carbonyls and carbonates) adsorbed on Ni/MgAl and Ni-Ru/MgAl catalysts (Fig. 6.7a), observing that the carbonyls are prone to form on the Ni-Ru/MgAl catalyst surface, which could be attributed to the CO dissociation, and their intensity reaches a peak value and declines till climbs slightly again. The results demonstrated that, during the DRM process, CO_2 could react with surface carbon species derived from CH_4 decomposition to produce CO via Boudouard reaction (i.e., $CO_2 + C \leftrightarrow 2CO$). Specifically, the alloying of Ni with Ru can promote the CO_2 dissociation but remarkably suppress the CO dissociation. In situ DRIFTS-MS has been also used to probe the surface DRM reaction mechanism in unconventional catalytic systems (e.g., plasma catalysis), revealing the dissociation of formats which adsorbed on metallic Pt surface to produce CO is main reaction route of the plasma catalytic DRM [73]. An in situ XRD was develped to be operated up to 700 °C, which was employed by Jiménez et al. [74] to reveal the active phases of Pd^0 and Ce^{4+} in the Pd-CeO_2 catalyst for DRM (Fig. 6.7b). However, it is still challenging to explore the dynamic changing of electronic structure and morphology of the active phase under real operating conditions. In situ STXM is a promising method which combined soft X-ray spectroscopy with 2D microscopy, which can probe and image the evolution of morphology and electronic structure of active sites. Specifically, Askari et al. first used in situ STXM characterization to examine the morphological and electronic structural changes of γ-Al_2O_3 supported NiCo bimetallic catalyst during DRM under industrially relevant conditions (Fig. 6.7c) [75]. Importantly, this study revealed that the loss of Ni from the outer surface of NiCo should be the main reason of the poor stability and catalyst deactivation. In situ XAS characterization is another powerful tool to gain evidence of the changes of both surface reactive intermediates and bulk metal structures. For example, AlSabban et al. [76] demonstrated that, during high temperature DRM over bimetallic CoNi nanoparticles, the metallic Co was significantly oxidized from the outer surface to the bulk, while NiCo and Ni phases mainly remained in reduced states, necessitating the importance of optimizing the composition of CoNi alloy. Very recently, Takami et al. [77] used in situ millisecond-order dispersive X-ray absorption spectroscopy (DXAS) to examine the transient information of local temperature of Rh nanoparticle during photothermal

DRM (Fig. 6.7d), and the findings revealed the chemical and morphological information of the metal sites at nanoscale. In addition to the methods discussed above, in situ AP-XPS were widely used to obtain the surface information including oxidation state of active metal sites during DRM reactions [78]. Previous studies also suggest that combining different in situ characterization techniques is highly attractive strategy to gain more comprehensively fundamental insights into the activity/deactivation of catalysts and propose the plausible reaction mechanisms [74, 78]. However, for thermocatalytic systems, such combined techniques operated at temperatures close to the realistic DRM conditions (e.g., at > 700 °C) are challenging. Also, probing of CO_2 reforming mechanisms with other alkanes in situ is also desirable to fundamentally understand the origin of activity and stability of relevant developed catalysts, which will bring tremendous assistance for advancing the practical adoption of these CO_2 utilization processes.

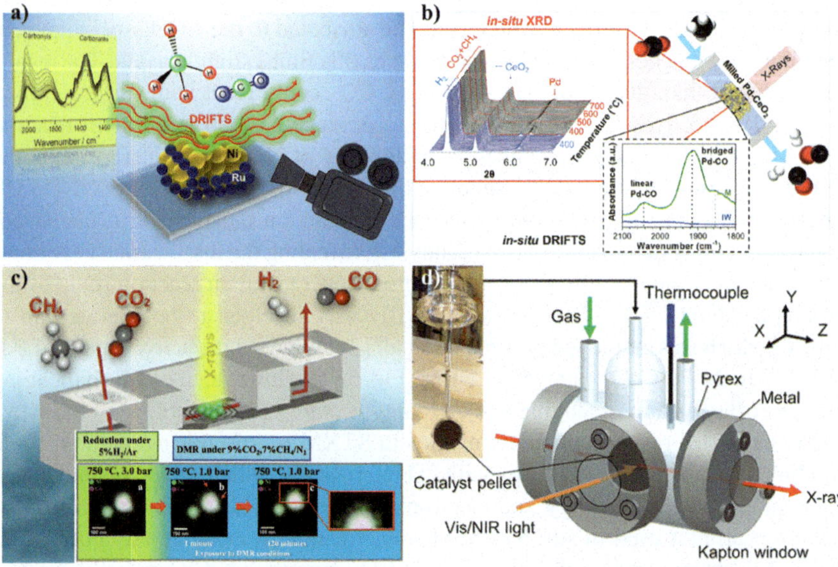

Fig. 6.7 In situ characterization methods for probing the DRM reaction mechanism: **a** DRIFTS [72], **b** XRD combined with DRIFTS [74], **c** STXM [75], and **d** XAS [77]. Reproduced with permission from Elsevier for **a** Ref. [72], **b** Ref. [74], open access with the CC BY 4.0 license, **c** Ref. [75], open access with the CC BY 4.0 license, and reproduced with permission from the American Chemical Society for **d** Ref. [77]

6.4 Conclusions and Perspectives

In the current sustainable development agenda, CO_2 as the feedstock for chemicals/fuels production needs to be realized with technical feasibility and economic viability, which encouraged many proposals and developments. CO_2 valorization via reforming reactions with light alkanes to chemicals/fuels (such as syngas and olefins) is one of the promising pathways for realizing CO_2 utilization in the future, and realization of this relies on rational catalyst/system design to prepare active, stable, and efficient ones for being used in practical settings. In this chapter, recent relevant advances regarding the catalytic design and mechanistic study for progressing CO_2 reforming reactions are reviewed. For the thermocatalytic systems, overcoming catalytic deactivation (mainly due to coking and metal sintering at high reaction temperatures) is considered as the priority in the catalyst design aspect. Previous studies have demonstrated the effectiveness and potential of active site and support engineering for addressing the deactivation issues, exemplified by secondary metal doping, encapsulation/confinement by framework materials and employment of macroscopic structured supports. For the catalytic support design particularly, the rational selection of support types, the construction of hierarchical microporous/mesoporous structure, and the modification of support chemical properties (e.g., zeolite de-alumination, graft with functional groups, etc.) will be the key for the deactivation elimination and CO_2 reforming activity elevation. Meanwhile, the attempt on the structured catalysts is also considered as another effective route to overcome the pressure drop due to carbon deposition and meet the demand of industrialization (e.g., reactor scale-up, high flowrate condition, etc.). Additionally, activation of CO_2 reforming reactions by non-conventional means is also one of the current hot topics in the academic community, and here, we commented on the recent developments in super-dry reforming and photocatalysis systems as examples, which can operate CO_2 reforming with high CO_2 utilization efficiency, and/or activate reforming reactions under milder condition (compared to the thermocatalytic counterparts) by using renewable energy sources, respectively. Yet, these developments are still in the infancy stage with the proof-of-concepts and relevant fundamental studies. It is worth noting that, in addition to the knowledge/experience for designing thermal catalysts, special consideration of the properties of the new systems is highly necessary for their catalyst design to optimize the reaction efficiency. Fundamental and mechanistic understanding of the catalytic systems above is very necessary, which is the foundation for developing them toward the real-world applications, which is indeed very challenging and requires the development of relevant in situ/operando characterization techniques. Relevant techniques such as in situ XAS, XRD, and DRIFTS are available for characterizing some heterogeneous catalytic systems, which could provide some correlations between the surface/structure properties (e.g., types and dynamics of active sites, key intermediates, etc.) and bulk-phase catalytic performance, being very beneficial to facilitate the rational design of catalysts for the targeted conversions. Such advanced techniques are needed for catalytic CO_2 reforming reactions, ideally under the conditions close to the realistic ones.

For future development, cost-effective catalysts with simple synthesis, as well as the engineering design for the catalysis based on them, could be attractive for real applications. Also considering the possible varied composition of the feed gases, a generic and flexible catalyst platform could be useful, i.e., it could be iterated and derived for preparing the improved catalysts to accommodate the changes in feed composition with the maintained activity and stability. All the investigations discussed here in this chapter present significant opportunities for the academia and industry to develop CO_2 reforming systems further which can contribute to the realization of CO_2 utilization processes.

References

1. Rogelj J et al (2016) Paris Agreement climate proposals need a boost to keep warming well below 2 °C. Nature 534:631–639
2. Li M, Sun Z, Hu YH (2021) Catalysts for CO_2 reforming of CH_4: a review. J Mater Chem A 9:12495–12520. https://doi.org/10.1039/D1TA00440A
3. Sharifianjazi F et al (2022) A review on recent advances in dry reforming of methane over Ni-and Co-based nanocatalysts. Int J Hydrogen Energy 47:42213–42233
4. Shen D et al (2022) Synergistic Pt-CeO_2 interface boosting low temperature dry reforming of methane. Appl Catal B 318:121809. https://doi.org/10.1016/j.apcatb.2022.121809
5. Gambo Y et al (2021) CO_2-mediated oxidative dehydrogenation of light alkanes to olefins: advances and perspectives in catalyst design and process improvement. Appl Catal A 623:118273. https://doi.org/10.1016/j.apcata.2021.118273
6. Mishanin II, Bogdan TV, Koklin AE, Bogdan VI (2022) Design of highly selective heterogeneous catalyst for CO_2-mediated ethane oxidative dehydrogenation based on nonoxidative catalysis in stainless-steel reactor. Chem Eng J 446:137184
7. Yan P, Chen Y, Cheng Y (2022) Industrially potential MoVNbTeOx@ FoamSiC structured catalyst for oxidative dehydrogenation of ethane. Chem Eng J 427:131813
8. Yu J, Odriozola JA, Reina TR (2019) Dry reforming of ethanol and glycerol: mini-review. Catalysts 9:1015
9. Li F et al (2022) Sandwich-type Co core@ shell nanocomposite (SiO_2@ Co@ CeO_2): coke resistant catalyst toward CO_2 reforming with ethanol. Appl Catal A 638:118605
10. Abdullah N, Ainirazali N, Setiabudi HD (2022) Recent development in catalyst and reactor design for CO_2 reforming of alcohols to syngas: a review. Chem Eng Res Des 178:438–453
11. Kwon D et al (2022) Strategic management of harmful chemicals produced from pyrolysis of plastic cup waste using CO_2 as a reaction medium. Chem Eng J 437:135524
12. Jung J-M et al (2022) Control of the fate of toxic pollutants from catalytic pyrolysis of polyurethane by oxidation using CO_2. Chem Eng J 442:136358
13. Bhattacharjee S et al (2023) Photoelectrochemical CO_2-to-fuel conversion with simultaneous plastic reforming. Nat Synth 2:182–192. https://doi.org/10.1038/s44160-022-00196-0
14. Li G, Liu C, Cui X, Yang Y, Shi F (2021) Oxidative dehydrogenation of light alkanes with carbon dioxide. Green Chem 23:689–707. https://doi.org/10.1039/D0GC03705B
15. Schwach P, Pan X, Bao X (2017) Direct conversion of methane to value-added chemicals over heterogeneous catalysts: challenges and prospects. Chem Rev 117:8497–8520. https://doi.org/10.1021/acs.chemrev.6b00715
16. Latimer AA et al (2017) Understanding trends in C-H bond activation in heterogeneous catalysis. Nat Mater 16:225–229. https://doi.org/10.1038/nmat4760
17. Sheng K et al (2019) Ni_xCo_y nanocatalyst supported by ZrO_2 hollow sphere for dry reforming of methane: synergetic catalysis by Ni and Co in alloy. ACS Appl Mater Interfaces 11:24078–24087. https://doi.org/10.1021/acsami.9b05822

18. Kim SM et al (2017) Cooperativity and dynamics increase the performance of NiFe dry reforming catalysts. J Am Chem Soc 139:1937–1949. https://doi.org/10.1021/jacs.6b11487
19. Karam L, Reboul J, Casale S, Massiani P, El Hassan N (2020) Porous nickel-alumina derived from metal-organic framework (MIL-53): a new approach to achieve active and stable catalysts in methane dry reforming. ChemCatChem 12:373–385. https://doi.org/10.1002/cctc.201901278
20. Theofanidis SA et al (2018) Fe-containing magnesium aluminate support for stability and carbon control during methane reforming. ACS Catal 8:5983–5995. https://doi.org/10.1021/acscatal.8b01039
21. Song Y et al (2020) Dry reforming of methane by stable Ni–Mo nanocatalysts on single-crystalline MgO. Science 367:777–781. https://doi.org/10.1126/science.aav2412
22. Chen H et al (2021) Dry reforming of methane on bimetallic Pt–Ni@CeO_2 catalyst: a in situ DRIFTS-MS mechanistic study. Catal Sci Technol 11:5260–5272. https://doi.org/10.1039/D1CY00382H
23. Tang Y et al (2019) Synergy of single-atom Ni1 and Ru1 sites on CeO_2 for dry reforming of CH_4. J Am Chem Soc 141:7283–7293. https://doi.org/10.1021/jacs.8b10910
24. Liu Y et al (2022) Embedding high loading and uniform Ni nanoparticles into silicalite-1 zeolite for dry reforming of methane. Appl Catal B 307:121202. https://doi.org/10.1016/j.apcatb.2022.121202
25. Zhang X et al (2021) High-performance binary Mo–Ni catalysts for efficient carbon removal during carbon dioxide reforming of methane. ACS Catal 11:12087–12095. https://doi.org/10.1021/acscatal.1c02124
26. Xu S et al (2022) Developing silicalite-1 encapsulated Ni nanoparticles as sintering-/coking-resistant catalysts for dry reforming of methane. Chem Eng J 446:137439. https://doi.org/10.1016/j.cej.2022.137439
27. Kosari M et al (2022) Strong coke-resistivity of spherical hollow Ni/SiO_2 catalysts with shell-confined high-content Ni nanoparticles for methane dry reforming with CO_2. Appl Catal B 310:121360. https://doi.org/10.1016/j.apcatb.2022.121360
28. Zhu Q et al (2022) Enhanced CO_2 utilization in dry reforming of methane achieved through nickel-mediated hydrogen spillover in zeolite crystals. Nat Catal 5:1030–1037. https://doi.org/10.1038/s41929-022-00870-8
29. Zhao Y, Li H, Li H (2018) NiCo@SiO_2 core-shell catalyst with high activity and long lifetime for CO_2 conversion through DRM reaction. Nano Energy 45:101–108. https://doi.org/10.1016/j.nanoen.2017.12.023
30. Zhang Q et al (2017) A sintering and carbon-resistant Ni-SBA-15 catalyst prepared by solid-state grinding method for dry reforming of methane. J CO_2 Utiliz 17:10–19. https://doi.org/10.1016/j.jcou.2016.11.002
31. Zhang X et al (2022) Coking- and sintering-resistant Ni nanocatalysts confined by active BN edges for methane dry reforming. ACS Appl Mater Interfaces 14:25439–25447. https://doi.org/10.1021/acsami.2c04149
32. Dai C et al (2015) Hollow zeolite encapsulated Ni–Pt bimetals for sintering and coking resistant dry reforming of methane. J Mater Chem A 3:16461–16468. https://doi.org/10.1039/C5TA03565A
33. Chen H et al (2021) Structured silicalite-1 encapsulated Ni catalyst supported on SiC foam for dry reforming of methane. AIChE J 67:e17126. https://doi.org/10.1002/aic.17126
34. Xie Z, Winter LR, Chen JG (2021) Matter 4:408–440
35. Sasson Bitters J et al (2022) Utilizing bimetallic catalysts to mitigate coke formation in dry reforming of methane. J Energy Chem 68:124–142. https://doi.org/10.1016/j.jechem.2021.11.041
36. Yan B et al (2016) Dry reforming of ethane and butane with CO_2 over PtNi/CeO_2 bimetallic catalysts. ACS Catal 6:7283–7292. https://doi.org/10.1021/acscatal.6b02176
37. Yentekakis IV, Panagiotopoulou P, Artemakis G (2021) A review of recent efforts to promote dry reforming of methane (DRM) to syngas production via bimetallic catalyst formulations. Appl Catal B 296:120210. https://doi.org/10.1016/j.apcatb.2021.120210

38. Myint M, Yan B, Wan J, Zhao S, Chen JG (2016) Reforming and oxidative dehydrogenation of ethane with CO_2 as a soft oxidant over bimetallic catalysts. J Catal 343:168–177. https://doi.org/10.1016/j.jcat.2016.02.004
39. Bian Z, Das S, Wai MH, Hongmanorom P, Kawi S (2017) A review on bimetallic nickel-based catalysts for CO_2 reforming of methane. ChemPhysChem 18:3117–3134. https://doi.org/10.1002/cphc.201700529
40. Li X et al (2018) Oxidative dehydrogenation and dry reforming of n-butane with CO_2 over NiFe bimetallic catalysts. Appl Catal B 231:213–223. https://doi.org/10.1016/j.apcatb.2018.02.063
41. Wang F, Xu L, Shi W (2016) Syngas production from CO_2 reforming with methane over core-shell Ni@SiO_2 catalysts. J CO_2 Utiliz 16:318–327. https://doi.org/10.1016/j.jcou.2016.09.001
42. Ma Q et al (2016) Ordered mesoporous alumina-supported bimetallic Pd–Ni catalysts for methane dry reforming reaction. Catal Sci Technol 6:6542–6550. https://doi.org/10.1039/C6CY00841K
43. Lu Y et al (2021) Enhanced catalytic performance of Nix-V@HSS catalysts for the DRM reaction: the study of interfacial effects on Ni-VO_x structure with a unique yolk-shell structure. J Catal 396:65–80. https://doi.org/10.1016/j.jcat.2021.02.005
44. Bian Z, Kawi S (2018) Sandwich-Like Silica@Ni@Silica multicore-shell catalyst for the low-temperature dry reforming of methane: confinement effect against carbon formation. ChemCatChem 10:320–328. https://doi.org/10.1002/cctc.201701024
45. Li Z, Mo L, Kathiraser Y, Kawi S (2014) Yolk–satellite–shell structured Ni–Yolk@Ni@SiO_2 nanocomposite: superb catalyst toward methane CO_2 reforming reaction. ACS Catal 4:1526–1536. https://doi.org/10.1021/cs401027p
46. Huang X et al (2017) Ordered mesoporous CoO-NiO-Al_2O_3 bimetallic catalysts with dual confinement effects for CO_2 reforming of CH_4. Catal Today 281:241–249. https://doi.org/10.1016/j.cattod.2016.02.064
47. Rostrup-Nielsen JR (1997) Industrial relevance of coking. Catal Today 37:225–232. https://doi.org/10.1016/S0920-5861(97)00016-3
48. Bradford MCJ, Vannice MA (1999) Industrial relevance of coking. Catal Rev 41:1–42. https://doi.org/10.1081/CR-100101948
49. Agueniou F et al (2021) 3D-printing of metallic honeycomb monoliths as a doorway to a new generation of catalytic devices: the Ni-based catalysts in methane dry reforming showcase. Catal Commun 148:106181
50. Chen Y et al (2022) MoVNbTeOx M1@ CeO_2@ Cordierite structured catalysts for ODHE process. Chem Eng Sci 253:117597
51. Wei Q et al (2017) Ni/Silicalite-1 coating being coated on SiC foam: a tailor-made monolith catalyst for syngas production using a combined methane reforming process. Chem Eng J 327:465–473. https://doi.org/10.1016/j.cej.2017.06.109
52. Zhang R et al (2020) Structured Ni@NaA zeolite supported on silicon carbide foam catalysts for catalytic carbon dioxide methanation. AIChE J 66:e17007. https://doi.org/10.1002/aic.17007
53. Gao X et al (2017) Carbon nanofibers decorated SiC foam monoliths as the support of anti-sintering Ni catalyst for methane dry reforming. Int J Hydrogen Energy 42:16547–16556. https://doi.org/10.1016/j.ijhydene.2017.05.164
54. Lawson S, Li X, Thakkar H, Rownaghi AA, Rezaei F (2021) Recent advances in 3D printing of structured materials for adsorption and catalysis applications. Chem Rev 121:6246–6291. https://doi.org/10.1021/acs.chemrev.1c00060
55. Wei Q et al (2020) Metal 3D printing technology for functional integration of catalytic system. Nat Commun 11:4098. https://doi.org/10.1038/s41467-020-17941-8
56. Wei Y et al (2022) One-pot three-dimensional printing robust self-supporting MnO_x/Cu-SSZ-13 zeolite monolithic catalysts for NH_3-SCR. CCS Chem 4:1708–1719. https://doi.org/10.31635/ccschem.021.202100942
57. Buelens LC, Galvita VV, Poelman H, Detavernier C, Marin GB (2016) Super-dry reforming of methane intensifies CO_2 utilization via Le Chatelier's principle. Science 354:449–452

58. Löfberg A, Kane T, Guerrero-Caballero J, Jalowiecki-Duhamel L (2017) Chemical looping dry reforming of methane: toward shale-gas and biogas valorization. Chem Eng Process 122:523–529. https://doi.org/10.1016/j.cep.2017.05.003
59. Tian S, Yan F, Zhang Z, Jiang J (2019) Calcium-looping reforming of methane realizes in situ CO_2 utilization with improved energy efficiency. Sci Adv 5:eaav5077. https://doi.org/10.1126/sciadv.aav5077
60. Zhou L et al (2020) Light-driven methane dry reforming with single atomic site antenna-reactor plasmonic photocatalysts. Nat Energy 5:61–70
61. Yang Y et al (2022) Light-induced redox looping of a rhodium/Ce_xWO_3 photocatalyst for highly active and robust dry reforming of methane. Angew Chem 134:e202200567
62. Wang X, Guo W, Xu S, Chen H, Fan X (2023) Stainless steel membrane distributor-type dielectric barrier discharge plasma reactor for co-conversion of CH_4/CO_2. AIChE J 69:e18059. https://doi.org/10.1002/aic.18059
63. Shoji S et al (2020) Photocatalytic uphill conversion of natural gas beyond the limitation of thermal reaction systems. Nat Catal 3:148–153. https://doi.org/10.1038/s41929-019-0419-z
64. Takami D, Ito Y, Kawaharasaki S, Yamamoto A, Yoshida H (2019) Low temperature dry reforming of methane over plasmonic Ni photocatalysts under visible light irradiation. Sustain Energy Fuels 3:2968–2971. https://doi.org/10.1039/C9SE00206E
65. Wu S, Li Y, Hu Q, Wu J, Zhang Q (2021) Photothermocatalytic dry reforming of methane for efficient CO_2 reduction and solar energy storage. ACS Sustain Chem Eng 9:11635–11651. https://doi.org/10.1021/acssuschemeng.1c03692
66. Takami D, Yamamoto A, Yoshida H (2020) Dry reforming of methane over alumina-supported rhodium catalysts at low temperatures under visible and near-infrared light. Catal Sci Technol 10:5811–5814. https://doi.org/10.1039/D0CY00858C
67. Zhang Z-Y et al (2022) Promoted solar-driven dry reforming of methane with Pt/mesoporous-TiO_2 photo-thermal synergistic catalyst: performance and mechanism study. Energy Convers Manag 258:115496. https://doi.org/10.1016/j.enconman.2022.115496
68. Feng K et al (2021) In-situ/operando techniques to identify active sites for thermochemical conversion of CO_2 over heterogeneous catalysts. J Energy Chem 62:153–171. https://doi.org/10.1016/j.jechem.2021.03.054
69. Chen H et al (2021) Dry reforming of methane on bimetallic Pt–Ni@ CeO_2 catalyst: a in situ DRIFTS-MS mechanistic study. Catal Sci Technol 11:5260–5272
70. Chansai S, Burch R, Hardacre C, Oh H, Epling WS (2013) An investigation of the role of surface nitrate species in the oxidation of propene on a Pt-based diesel oxidation catalyst. Catal Sci Technol 3:2349–2356
71. Németh Ms, Somodi F, Horváth A (2019) Interaction between CO and a coke-resistant NiIn/SiO_2 methane dry reforming catalyst: a DRIFTS and CO pulse study. J Phys Chem C 123:27509–27518
72. Álvarez MA, Bobadilla LF, Garcilaso V, Centeno MA, Odriozola JA (2018) CO_2 reforming of methane over Ni-Ru supported catalysts: on the nature of active sites by operando DRIFTS study. J CO_2 Utiliz 24:509–515. https://doi.org/10.1016/j.jcou.2018.01.027
73. Vakili R et al (2020) Plasma-assisted catalytic dry reforming of methane (DRM) over metal-organic frameworks (MOFs)-based catalysts. Appl Catal B 260:118195
74. Jiménez JD et al (2022) Identification of highly selective surface pathways for methane dry reforming using mechanochemical synthesis of Pd–CeO_2. ACS Catal 12:12809–12822
75. Beheshti Askari A et al (2020) In situ X-ray microscopy reveals particle dynamics in a NiCo dry methane reforming catalyst under operating conditions. ACS Catal 10:6223–6230

76. AlSabban B et al (2017) In-operando elucidation of bimetallic CoNi nanoparticles during high-temperature CH_4/CO_2 reaction. Appl Catal B 213:177–189. https://doi.org/10.1016/j.apcatb.2017.04.076
77. Takami D, Yamamoto A, Kato K, Shishido T, Yoshida H (2022) Transient temperature response of supported Rh nanoparticles in photothermal dry reforming of methane—an operando dispersive X-ray absorption spectroscopy study. J Phys Chem C 126:15736–15743
78. Xie Z et al (2018) Dry reforming of methane over CeO_2-supported Pt-Co catalysts with enhanced activity. Appl Catal B 236:280–293. https://doi.org/10.1016/j.apcatb.2018.05.035

Chapter 7
CO_2 Electrocatalytic Conversion: Outlooks, Pitfalls and Scientific Gaps

Gabriele Centi and Siglinda Perathoner

Abstract CO_2RR electrocatalysis is a key technology/process for transformative (industrial) chemical production but requires accelerating the development of novel materials and approaches by turning the current design strategies. This contribution does not review this topic but instead highlights pitfalls and scientific gaps in the current approaches. The discussion is focused on electrode/reactor engineering and related aspects, including how to control the nanoffenvironment at the electrocatalyst surface. Their better understanding, especially the development of reliable modelling able to account for the complexity of the phenomena present in electrocatalysis, is the condition to accelerate the progress. It is also remarked that the mechanistic and design strategies used have an intrinsic methodological issue. Approaching catalysis from the perspective of selectively transferring energy for the activation barriers rather than only considering the nature of the active sites and the energetic pathways of transformation (the current approach) is discussed. The concept is related to the role of localised phonons, polarons and interaction with excitons in determining the CO_2 electrocatalytic conversion pathways.

7.1 Introduction

The electrocatalytic conversion of CO_2 (CO_2RR) is of increasing research attention. As an index, a search in SciFinder indicates an average of 50 per year reviews in English using as keywords "CO_2 and electro", about ten times higher than a decade ago. At the same, this poses the question of whether it could be necessary to have additional reviews on the topic. Looking more in detail at recent reviews highlighting the mechanistic and design aspects of electrocatalysts [1–37], it appears that despite

G. Centi (✉) · S. Perathoner
Department of ChiBioFarAm, University of Messina, Messina, Italy
e-mail: centi@unime.it

European Research Institute of Catalysis (ERIC) aisbl, Brussels, Belgium

S. Perathoner
e-mail: perathon@unime.it

G. Zhang et al. (eds.), *Advances in CO_2 Utilization*, Green Chemistry and Sustainable Technology, https://doi.org/10.1007/978-981-99-8822-8_7

the richness of indications and the undoubtful scientific advances, there is a lack of a generalised approach as the bases for the industrialisation of the process overcoming current limits (stability, Faradaic selectivity and specific current densities for reliable electrode sizes, above at least 10 cm^2).

Industrial targets, even if dependent on the products, are current densities above 200 mA/cm^2 (better above 0.5 A/cm^2) and Faradaic selectivities above 80–90% at high current densities. Other relevant aspects are the type of byproducts and how they impact downstream separation. At the same time, stability should be at least 1000 h under variable loading. Other requisites are a nearly constant selectivity in an enough wide potential (at least 0.5 V) and current density range. Avoiding critical raw materials in developing the electrodes is also necessary. Purity and concentration of CO_2 are other key aspects of industrial scalability. The electrolyte choice or, preferably, its elimination are also critical issues, as commented later. Often, top-performance electrocatalysts were tested in unreliable conditions, and a drastic performance loss is observed when studies under conditions relevant to industrialisation. Thus, even if for some reactions (particularly the production of CO or ethylene), high current densities and Faradaic selectivities and efficiencies have been reported in the literature, as described in the cited reviews, the effective possible industrialisation is still far.

Overcoming this gap would require developing suitable strategies for CO_2RR. The focus in the literature has been mainly on catalyst design strategies. They include aspects such as.

(i) surface morphology changes, for example, the control of the nanostructure, doping, defect engineering and formation of heterostructures, and modification by surface organic groups or molecular functionalisation;
(ii) development of specific active sites, such as single-atom catalysts (SAC) or dual-atom catalytic sites, facets engineering of nanoparticles, formation of bimetallic or core–shell nanoparticles and generation of high-entropy alloys;
(iii) nano-architecture of the electrocatalysts, such as the development of dendritic nanostructures and rough surfaces, spatial confinement and hierarchical-porosity, generation of grain boundaries and mesostructural effects.

Typically, the design criteria are supported by theoretical modelling of the reaction mechanism and the nature of the active sites. However, often these insights are valid for the specific materials investigated, and there are several contradictions when generalised to the whole set of the electrocatalysts proposed to derive general trends in the electrocatalyst design. There are few comparative studies in this direction. One example regards the reaction of CO_2 electrocatalytic methanation [38]. Even limiting the analysis to a subset of relatively homogeneous electrocatalysts for which a detailed mechanistic approach has been proposed, there is an evident lack of general relations between reaction mechanisms, the nature of the active sites and catalytic performances.

This issue derives largely from the experimental observation of many authors that the surface microenvironment, type of electrode and operation conditions, cell design and other aspects not related to the electrocatalysts determine the performances in CO_2RR (selectivity, type of products, specific current density and side reaction of H_2 formation). These effects often predominate over the characteristics of the electrocatalyst itself [39]. Even unselective catalysts may become selective under proper conditions [40]. This and other shreds of evidence question the relevance of focusing on electrocatalysts only.

While attention is often given in the literature on aspects such as verification of the absence of products deriving by contamination (recommending, for example, tests with labelled CO_2), the need from an industrial perspective is rather on rigorous protocols of testing (including the use of reliable electrocatalytic cells and electrode sizes) which allow reproducibility of the results and understanding of their scalability. This question is far beyond the issue of contamination. Several literature results are largely not reproducible under conditions and reactors relevant from an industrial perspective.

On the other hand, CO_2RR is a key technology to address the carbon circularity and CO_2 reduction targets in energy-intensive industries [41]. It is part of the necessary processes to establish a new sustainable industrial chemical production not based on the use of fossil fuels as energy and C-source [42–44]. There is a need to accelerate the development of industrial processes based on this technology to meet climate change targets.

The availability of renewable electrical energy to transform chemical production in this direction is often questioned. However, the same is valid for all emerging directions related to energy transition, from the production of green H_2 to electrical mobility. Chemical production is at the basis of an extensive value chain impacting all modern life. Investing in the decarbonisation of chemical production through electrification and closure of the carbon cycle would have a cascade effect on all the subsequent value chain. Thus, the impact should not be measured as a direct amount of CO_2 recycled or fossil fuels avoided but as a global impact on the transformative economy. From this perspective, the low volume of chemicals with respect to fuels has minor relevance.

This comment remarks that the choice of the target products of CO_2RR also has to be viewed from this perspective, in addition to considering the integration of the technology in the value chain. Producing CO from CO_2 rather than other chemicals that involve a more complex transformation (ethylene production is a 12-electron reduction compared to a two-electron reduction in CO formation; in addition, a C–C coupling reaction is necessary) is easier and can be achieved with higher selectivity and energy efficiency. However, CO has to be further processes, which requires additional costs and energy. Suppose the final target is ethylene to have a straight comparison. In that case, H_2 must also be generated by electrolysis. Then, CO and H_2 must be pressurised and catalytically transformed to ethylene, typically in a multistep process [45, 46]. Direct electrocatalytic production of ethylene, eventually in a two-step integrated electrocatalytic process (via intermediate CO further electrocatalytically converted to ethylene) [47], would be competitive over the hybrid electrocatalytic-heterogenous catalysis approach. In addition, the electrocatalytic path is well-suitable for small-scale, decentralised production. It is more resilient than the hybrid process, avoids the cost and impact of transport, and better integrates with local resources. Thus, technological change has to be also considered from the perspective of changing the modality of production, which impacts the economics and environment in a far more complex way than accounted for by typical techno-economic estimations [48].

While the hybrid approach is currently at a higher Technology Readiness Level (TRL), it is necessary to accelerate the progress in direct electrocatalytic routes, similar to the case of producing H_2 by electrolysis and then using it to convert CO_2 by heterogeneous catalysis [49, 50]. The potential and selection of CO_2 electrocatalytic routes and technologies have to be assessed given the transformative scenario rather than using the traditional assessment criteria. This point is the challenge facing discussion on the future of electrocatalytic routes of CO_2 conversion, but instead, often not properly recognised. However, it is out of the scope to discuss these aspects further, being analysed in other contributions [42–44, 49, 50].

In addition, this contribution does not aim to be a further review of CO_2RR, which is already covered by many other reviews. Instead, it aims to highlight some less-covered aspects but relevant from the general perspective of fostering obtaining reliable results from the perspective of industrialisation of the technology.

7.2 The Role of Electrode and Cell Design

As commented in the introduction, design strategies in the literature focus mainly on the electrocatalyst itself. These strategies can be lumped into two main categories: those influencing the electronic nature of the active sites, and thus the energetic path of conversion, and those determining the density of the active sites, which influence current density, but also the selectivity (for the possibility of pairing reactions, for example). In other words, the mechanism and reaction pathways are often assumed to be independent of the electrocatalytic reactor design, engineering and the choice of operative conditions. Many experimental results proved that this assumption is not valid [39]. While an intertwined relationship exists between electrocatalysts and electrode/reactor design, the latter often dominates the overall behaviour. The engineering of the electrode/reactor determines factors such as local pH around the active centres (and changes during the reaction), mass transport conditions and three-phase boundaries, e.g. the region at the interface between solid, liquid and gas [51–56].

The theoretical mechanistic studies do not account for these aspects. In addition, a crucial missing element is that the potential at the electrocatalyst surface is not uniform, as typically assumed. A not uniform charge and potential distribution are rather present, depending on the nanostructure and electrode design. This fact determines gradient profiles in the electrodes beyond not uniform double layers, which generates not uniform electrolyte ions distributions, mass transfer, bubble generation and sticking, which sum to a non-uniform distribution in electrodes (particularly the porous ones) of the electrochemical active surface area (ESCA). Often what is interpreted as an electrocatalytic feature (for example, the exposed crystal facets in nanoparticles) is a modification of the potential distribution and related microenvironment. Thus, the effect observed depends on the latter aspect rather than to change in the electrocatalyst's electronic nature or geometry. However, these aspects are typically not even considered.

More aspects are discussed later. However, these brief comments prove that many of the mechanistic conclusions and strategies for CO_2RR are based on an oversimplification of the effective processes occurring. From this fact, they have limitations in developing effective strategies for improving CO_2RR.

7.2.1 The Use of Representative Electrode Sizes and Reactor Configurations

The presence of significant effects related to charge distribution and fluidodynamics of electrode–electrolyte interface remarks that testing with a minimum electrode size would be necessary to have reliable information on scalability and performance. In addition, the analytical results' reliability also requires using electrodes with a minimum size. As an indicative value for CO_2RR testing, the electrode size should be at least 10 cm^2 using flow electrocatalytic reactors, as commented later. Several high-performance electrocatalysts reported in the literature do not meet these criteria, with the size of the electrodes often below 1 cm^2. Specific studies regarding the reproducibility and scalability of the results in the literature are missing. The practice is that valuable literature results are often not reproducible under conditions and reactors relevant to industrial scalability. Many aspects could determine irreproducibility and may raise endless questions out of the scope of this contribution. It is thus not relevant to analyse which specific results fail from this perspective but only advise the readers of care in using literature references and of the need to use reliable testing protocols.

Using suitable electrocatalytic reactors is also critical for the proper scalability and reliability of the results and mechanistic conclusions. The electrocatalytic tests are often made in an H-type cell, e.g. a cell with two liquid compartments separated by a membrane. The CO_2RR electrocatalyst acts as the working electrode in the cathodic chamber. The low solubility of CO_2 in aqueous media, the high availability of protons, the high electrolyte volume and the presence of often mass transfer limitations determine conditions at the surface of the electrode very different from those present under electrodes relevant for industrial exploitability that operate at high current density. Figure 7.1 illustrates this concept by presenting schematically the local concentration and pH gradients in conventional laboratory-scale reactors (H-type), gas-diffusion electrode (GDE) flow reactors and industrial conditions. The reactor, electrodes and operative conditions can drastically change the availability of reactants at the surface and selectivity.

The optimal electrocatalysts should be different, being the microenvironment different. Thus, ranking in the H-cell may not give reliable results for industrial

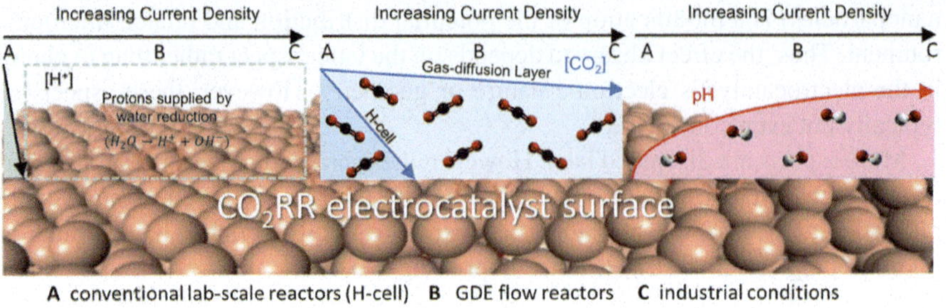

Fig. 7.1 Effect of current density on the availability of protons, concentration of CO_2 and pH at the surface of a CO_2RR electrocatalyst. Reproduced with permission from ref. [39] (Copyright Elsevier, 2023), reporting an elaboration of the indications given by Burdyny and Smith [57]

conditions. The behaviour in electrocatalysis is often dominated by the concentration of species at the catalyst surface and interface with the electrolyte rather than the electrocatalyst features. The dynamic of mass transport phenomena at a charged (porous) electrode and the dependence on the electrode and reactor design determine selectivity, onset potential and other aspects, which are instead typically ascribed to the intrinsic properties of the electrocatalyst. Understanding these comments better would require to discuss preliminary in more detail the type of electrodes and reactors used in CO_2RR testing.

7.2.2 Electrode

The simpler electrode is a planar metallic foil, which is unsuitable for CO_2RR. Many studies have proven the need to have size-controlled nanoparticles, which have thus to be supported over a conductive porous substrate such as carbon paper or mesh. However, to overcome limitations related to CO_2 solubility in the electrolyte, the use of gas-diffusion electrodes (GDEs) has become increasingly relevant. In this configuration, gas CO_2 is fed directly on one side of the electrode (typically treated with a hydrophobic substance such as Teflon to limit water back diffusion). It reaches the electrocatalyst localised on the other side of the electrode. The electrocatalyst could be in contact with the electrolyte or instead with the membrane, acting both as a separator of the cathodic and anodic zones and as a solid electrolyte. This configuration is indicated with different names (electroless, gas-phase and zero-gap electrodes). The latter is becoming the most popular [58–61]. Figure 7.2 illustrates schematically these electrodes.

Planar or porous electrodes are the most used CO_2RR electrodes due to their easy manufacture [62]. They operated immersed in a CO_2-saturated electrolyte, as outlined in Fig. 7.2. Their performances strongly depend on the electrochemical cell design because their behaviour depends on CO_2 and ionic diffusion. For this reason, they cannot be used for high-current density tests. The porous electrode

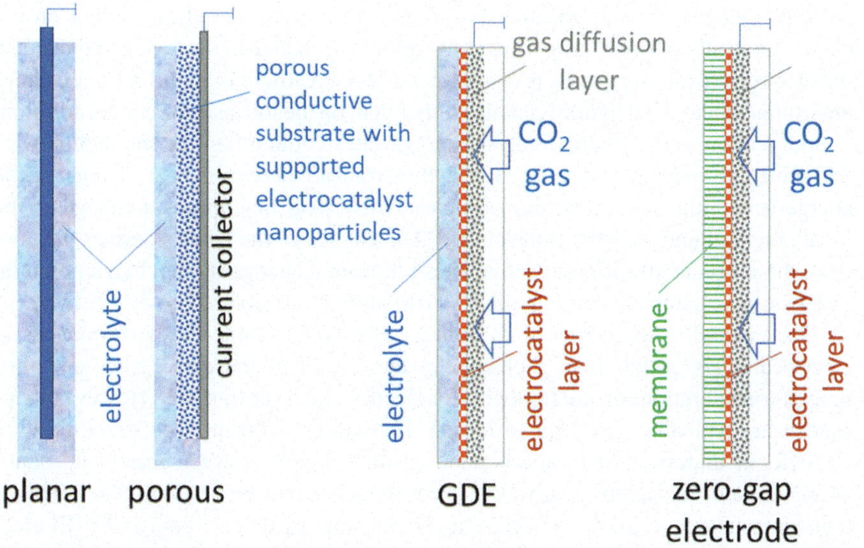

Fig. 7.2 Simplified drawing of planar, porous, gas-diffusion electrode (GDE) and zero-gap electrodes

design with respect to the planar increases the active electrode area, reducing overall cell voltage at equivalent current density. While designing 3D-type porous electrodes, e.g. with higher thickness of the porous conductive substrate, could be a valuable way to increase the productivity per geometrical area of the electrode, the main issues to overcome are (i) the intra-electrode mass transport, especially at high current densities, and (ii) the inhomogeneities in electron distribution. While 3D-structured electrodes are desirable [63], and new 3D-printing methods open new perspectives [64], the main issues to overcome is understanding the complex multiphase flow in the pores, the interface interactions and multiscale kinetics at the catalysts. Still, effective 3D electrodes for CO_2RR must be developed, even the recent interesting advances [65–67].

The ohmic resistance of the 3D electrode depends on the nature and geometry of the substrate and the interfacial conductivity. A hierarchical morphology is necessary, which could be realised by functionalising metallic foams. Crucial aspects are the wettability and macroporous porosity because they affect the saturation/capillary-pressure relationship and, thus, the multiphase fluidodynamic inside the 3D electrodes [68].

While it is often considered that the electrode kinetics is dependent only on the electrocatalyst, the overall kinetics instead depend significantly on aspects such as the nature of the conductive substrate and interface with the electrocatalyst, the type of multiphase flows within the electrode and local environments. In heterogeneous catalysis, it is possible to measure the true kinetics and differentiate from the overall kinetics that heat and mass transfer limitations may influence. This approach is much more challenging in electrocatalysis. What is often considered depending on the electrocatalyst's intrinsic properties (for example, a change in the facets or defects in the electrocatalyst properties) is an indirect effect, e.g. there is an influence on nano-environment and local hydrodynamics rather than on the intrinsic kinetic. In addition, a modification in the electrocatalyst nanoparticles could affect the interaction with the substrate and the resistance of electron transfer determining a change in the charge accumulation on the nanoparticles. This effect determines a change in the local potential and electron transfer rate. The change in the kinetic derives thus not from the presence of different active sites but from a change in the charge/potential distribution causing a change in the CO_2RR surface reaction rate and selectivity.

This example shows how often indirect effects determine the performances in electrocatalysis, rather than a change in the nature of the active sites and thus related considerations on the strategies for electrocatalyst design [69]. However, these aspects are typically not even considered. There are significant gaps, particularly in CO_2RR, in understanding aspects such as the relation between interfacial nano-architecture of the electrocatalyst/electrode catalysts and electrode kinetics. This is a major issue determining the lack of a general strategy in developing CO_2RR electrodes suitable for the industrialisation of the processes, as commented in the introduction. Electrocatalysis is far more complex than heterogeneous catalysis. Nevertheless, still, this complexity is not well understood. There is a tendency to consider electrocatalysis similar to heterogeneous catalysis and use similar approaches (for example, in modelling and mechanistic studies) while not recognising the intrinsic differences [38, 70].

There are many additional issues. Due to the potential applied, an easier surface reconstruction and change may occur during electrocatalysis. Metal-oxides or hydroxide nanostructures may generate during CO_2RR, depending on the size and location of the nanoparticles [71–73]. They change the nature of the active sites, but at the same time, many other properties, from the wettability and nano environment to the change localisation, because they are poorly conductive.

GDEs typically show better CO_2RR behaviour with respect to planar and porous electrodes because the transfer rate of CO_2 to the electrode is enhanced [62, 74]. CO_2 diffuses directly through a gas-diffusion layer (GDL) to the electrocatalyst layer. The latter may be in contact with the electrolyte or the membrane separating cathodic and anodic zones, as outlined in Fig. 7.2. Depending on various aspects, as commented later, the electrocatalyst layer can be located directly at the gas–liquid interface or fully wetted by the electrolyte. The last situation is apparently closer to porous configuration. However, in the GDE, the electrocatalyst, even if fully wetted, is still very close to the gas–electrolyte interface. The transfer rate is at least one order of magnitude larger.

Wettability and hydrophilic-hydrophobic characteristics of the electrocatalyst and GDE, catalyst loading and porosity are among the key aspects determining the exact location of the electrolyte-CO_2 gas interface inside the GDL and with respect to the location of the electrocatalyst [74–76]. The operative conditions and current density also influence this interface. Common GDLs are metal mesh/foam or a conductive porous macrostructured conductive carbon substrate (e.g. carbon cloth or felt) treated typically by polytetrafluoroethylene (PTFE) to induce hydrophobicity. Structured GDLs (formed, for example, by macroporous/microporous layers) are common. The optimisation of GDE regarding porosity and electrolyte fluid dynamics is critical, even typically not made at a laboratory scale. Changes in these aspects (associated with changes in the electrocatalysts) may influence the observed CO_2RR performances often more than the intrinsic changes in the electrocatalyst properties [62].

The optimised engineering of GDE should be made in parallel to that of the cell and operative conditions, being these aspects strictly correlated. For example, Yang et al. [77] reported that the applied potential plays a central role in transitioning from a non-flooding to a flooding situation during CO_2RR, as summarised in Fig. 7.3. This is also in relation to the size of the electrodes, being flooding easier at high current densities with respect to very small (below 1 cm^2) electrodes. This issue is one of the aspects of why results using very small electrode sizes could not be reproduced with larger sizes or different cell configurations. However, these aspects are considered to a very limited extent in the literature and are part of the missing aspects determining reproducibility.

The GDE electrode internal gradients and, consequently, the selectivity and product distribution in CO_2RR depend on the change in the fluidodynamics when turning from non-flooded to flooded conditions. Möller et al. [78] modelled the selectivity changes inside a GDE in a cathodic current range of 50–700 mA cm^{-2}. Closer to the GDL side in contact with gas CO_2, the formation of C2+ products (e.g.

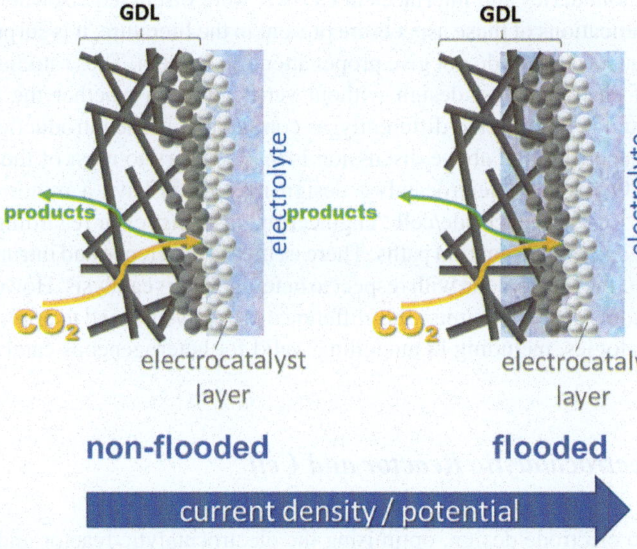

Fig. 7.3 Simplified illustration of the transition from no-flooding to flooding conditions for GDE electrodes on increasing the potential and current densities. Revised from the indications given in Ref. [77]

those with two or more carbon atoms) is favoured, while C1 products form closer to the electrolyte side. A change in C2/C1 formation ratio occurs within the GDE due to the variation in the gradients. While many literature studies, supported by extensive theoretical modelling, remark on the need for specific characteristics of the electrocatalyst/active centres to produce C2 rather than C1 products in CO_2RR, these and other experimental and theoretical evidence [74, 79–81] remark that the concentration profiles at the electrocatalyst (protons, CO_2, other ions present in the electrolytes) determine the path and selectivity in the transformation. The selectivity of CO_2RR is strongly influenced by the structure of the electrode and cell characteristics, which influence the effective gradients along the catalyst layer. In the zero-gap configurations (Fig. 7.2), where the electrolyte is absent, the gradients are different, although specific literature studies are lacking.

There are also gradients in the local potential and current density along the electrode, which also determine changes in the reaction paths and selectivity. Kas et al. [81] showed the presence of a spatial distribution of the current density and changes in local pH within the electrode. Large local variations in the current density (>150 mA cm^{-2}) could be present. Nesbitt et al. [56] indicate the presence of different phase boundaries along the GDE profile and how high current–density CO_2RR performances could be obtained only by realising a high CO_2 diffusion. A change in the performances and selectivity may not depend on the electrocatalyst but only on how the electrode is realised, the presence of flooding aspects, the distribution of the electrocatalyst along the GDE, a variation in wettability and hydrodynamics, and other related aspects. Further aspects and an analysis of state of the art on the role of three-phase boundaries and interfaces in CO_2RR were discussed elsewhere [39].

Even if indications of these aspects are present in the literature, it is surprising that most mechanistic studies do not give proper attention to them. These studies indicate strategies of electrocatalyst design without verification on whether the variations observed could be interpreted differently, as commented in the introduction.

The message from the above discussion is that contrary to most of the literature studies in CO_2RR, the electrocatalyst design may often have a minor relevance with respect to other electrode/cell "engineering" aspects, even regarding reaction mechanisms and transformation paths. There is, thus, a difference and intrinsic higher complexity of the phenomena with respect to heterogeneous catalysis. However, most literature studies do not recognise this difference properly and tend to apply concepts and methodologies, including in modelling, valid for heterogeneous catalysis.

7.2.3 *Electrocatalytic Reactor and Cell*

In parallel to electrode design, optimising the electrocatalytic reactor and cell also has a marked influence on the performances and possibility of scaling up the results. As commented above, reactor and cell engineering are not an additional aspect to consider after defining the electrocatalyst and its performance. The ranking and selection of the electrocatalyst depend significantly on the choice of the reactor/cell, and the results may not be translatable to a different cell. The Faradaic efficiency and onset potential (associated with the electrocatalyst characteristics and typically assumed to be independent of the choice of the reactor/cell) depend instead on the reactor configuration and design. The reasons are the same discussed regarding electrode design, e.g. a behaviour highly dominated by the microenvironment rather than the nature of the electrocatalyst. Methodologies not accounting for these aspects, including theoretical methods, do not accurately describe the phenomena dominating the behaviour. From here, the lack of a generalised approach in CO_2RR electrocatalysis is commented in the introduction.

Figure 7.4 reports a simplified overview of the main CO_2RR electrocatalytic reactors/cells. The H-type cell is the most used laboratory-scale CO_2RR reactor/cell. It is characterised by two containers containing the anode and cathode electrodes immersed in the electrolyte. The two containers are joined and separated using (optionally) a membrane to limit cross-over. In a review of some years ago, over 98% of literature studies on CO_2RR analysed (over 1100 papers) used H-type cells. Likely this percentage has decreased recently, but these reactors remain the most used in laboratory-scale and mechanistic studies. Weekes et al. [82] recommended flow-type rather than H-type CO_2RR electrolysers because the latter do not provide reliable data on the dynamic environment of an electrolyser [83, 84]. Even in an initial screening or for mechanistic studies, they may not provide the correct information [57]. Thus, even if the issue is documented, H-type cells continue to be used, and related results are utilised to propose design strategies for CO_2RR.

The zero-gap flow cell is preferable to perform tests at high current density and obtain reliable results for scaling the systems. Alinejad et al. [61] evidenced how the same catalyst behaves differently in H-type and zero-gap cells, in agreement with previous comments. An issue in zero-gap cells is the possible salt precipitation which affects stability, although strategies to mitigate the effect have been proposed [85].

The salt precipitation derives from the generation of hydroxide ions during the reaction, the use of high alkaline electrolytes, the formation of bicarbonate/carbonate ions by reaction with CO_2

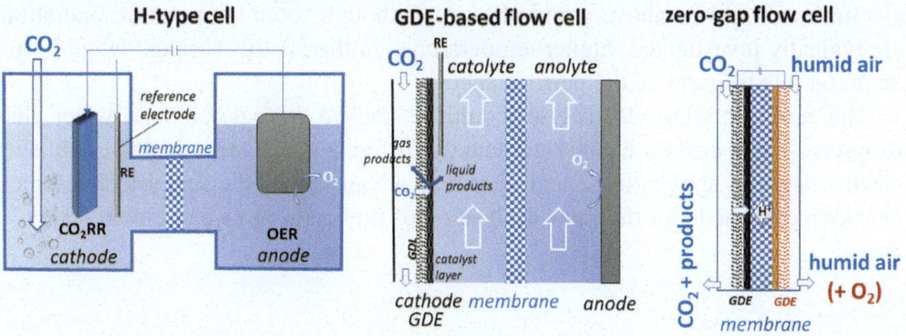

Fig. 7.4 Simplified illustration of CO_2RR reactors/cells: H-type cell; GDE in a membrane-based flow cell, and advanced compact design for zero-gap flow cells with GDE on both anodic and cathodic sections. *Note*: *OER* oxygen reduction reaction, *RE* reference electrode

$$OH^- + CO_2 \rightarrow HCO_3^- \quad (7.1)$$

$$HCO_3^- + OH^- \rightarrow CO_3^{2-} + H_2O \quad (7.2)$$

and precipitation of potassium (bi)carbonate if KOH is used as the electrolyte. This process is illustrated in Fig. 7.5 for reducing CO_2 to CO in a zero-gap cell using silver as the electrocatalyst [85].

The precipitation occurs at high electrolyte pH and high reaction rates. Although high pHs (up to 15) are used in several literature experiments to limit the side reaction of H_2 formation (HER—hydrogen evolution reaction) for many reasons (high conductivity and thus a high current density, reduced potential to apply, higher selectivity, particularly in ethylene production from CO_2) [86–88], these conditions have to be discouraged from an industrial perspective.

The overall CO_2RR energy efficiency is strongly negatively affected by the reaction of CO_2 with OH^- to form carbonate because it causes CO_2 losses leading to a net negative energy balance when considering the regeneration of the electrolyte [89]. There is an apparent high energy efficiency because a high pH minimises cell voltage due to chemical potential and the dependence of the thermodynamic electrode potentials on the pH. However, the OH^- ions in the electrolyte are consumed by exergonic $CO_3{}^{2-}$ formation. The energy required to regenerate CO_2 and 2 OH^- from aqueous $CO_3{}^{2-}$ is larger than the energy stored by CO_2 electrolysis. A negative energy balance of around ~100–130 kJ•mol^{-1} is present. Therefore, CO_2RR must be studied with electrolytes with a pH close to neutrality.

A comparative analysis of the three types of discussed electrocatalytic reactors is presented in Table 7.1. The focus is on aspects critical for the scalability and industrialisation of CO_2RR processes.

The zero-gap flow cell is the most promising configuration for obtaining high current density under an industrial scale. Typical current densities achieved in CO_2RR are 100–500 mA cm^{-2} range with a cell voltage of 2–5 V and energy efficiency of about 20–45%. Power, e.g. the energy required to run the system, is in the range of 0.1–10 W (per cm^2 of electrode surface area). As commented before, the size of the electrodes should be above about 10 cm^2. Although room temperature operations are typically investigated, higher temperatures, in the 50–70 °C range, would often be preferable to improve the performances.

The zero-gap GDE electrolyser could be stacked as fuel cells and water electrolysers. Their costs are not significantly different from these better-established electrochemical applications, and they take advantage of the many technological developments made for the industrialisation of fuel cells and water electrolysers.

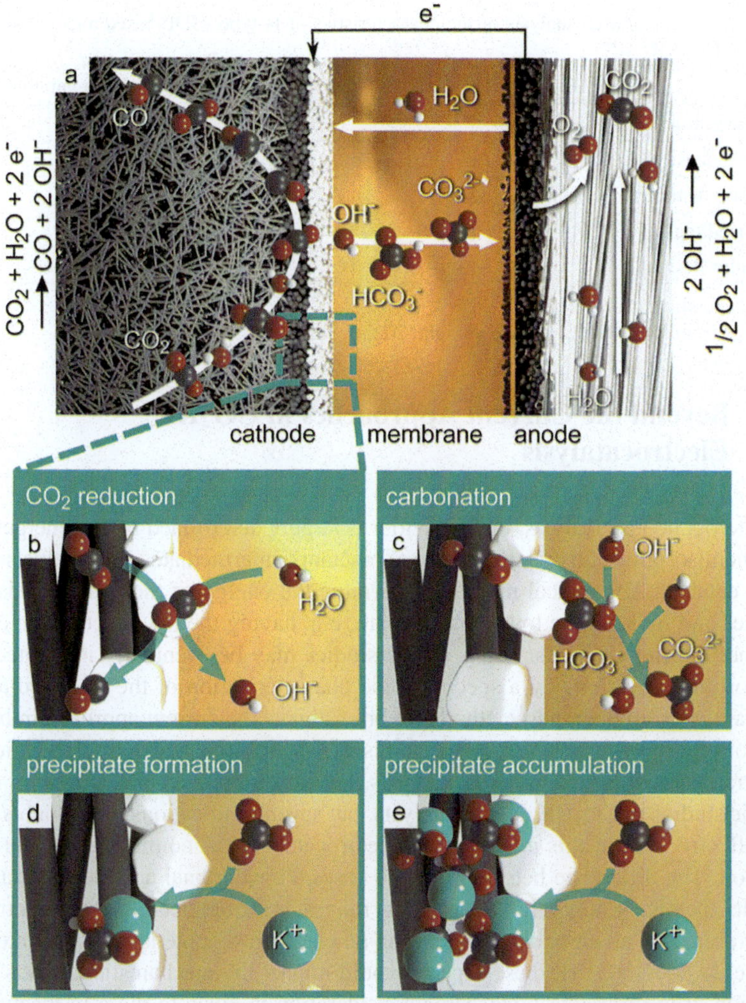

Fig. 7.5 Salt precipitation process in zero-gap CO_2 electrolysis. **a** Desired reactions and transport during CO_2 reduction. **b** Reduction of CO_2 producing CO and OH^-. **c** Carbonation reaction: CO_2 reacts with the produced hydroxide ions. **d** Precipitation of potassium (bi)carbonate. **e** Precipitates accumulating in the gas-diffusion electrode blocking pores. In the figure, oxygen is in red, carbon in black, hydrogen in white and potassium in petro colour. Reproduced with permission from Ref. [85]. Royal Chemical Society copyright, 2023

Table 7.1 Comparative analysis of the characteristics of H-type, GDE-based and zero-gap flow cells for CO_2RR

	H-type cell	GDE-based flow cell	Zero-gap flow cell
Pressure drop	↑	○	↑
Current density	↓	○	↑
Uniform distribution of current	↓	○	↑
Gas handling	↓	↑	↑
Complex design	↑	○	↓

Note ↑: good; ○: average; ↓: bad
Adapted from Ref. [39]. Elsevier copyright, 2023

7.3 Beyond the Current Approaches in CO_2RR Electrocatalysis

In heterogeneous catalysis, the behaviour is largely determined by the energetics of the chemisorption and transformation of reactants, intermediates and products on the solid catalyst. Theoretical methods determine the energy of the possible adsorbed species and indicate the lower energy path, e.g. having the lower activation energy for the limiting elementary step. These studies may be eventually integrated with supporting indications by a spectroscopic characterisation of the intermediates or indications from operando methods. In brief, this is the current approach in heterogeneous catalysis to establish the mechanism of the reaction and derive indications on the strategies to design improved catalysts. The same approach is also used for mechanistic studies in CO_2RR electrocatalysts, but several indications raise doubts about the effectiveness of this approach for electrocatalysis, as commented in the introduction. The difference between the two cases (conventional and electrocatalysts), besides the diverse temperature operation ranges, is the application of a potential and the presence of an electrolyte generating an electrical double layer at the interface. While it is obvious that these aspects would impact the mechanistic/kinetics of the reaction, it is surprising how scarce consideration is given to them in mechanistic CO_2RR studies.

As commented before, the electrolyte (concentration, species, buffer capacity and pH value) influences the local reaction conditions impacting the pathways of CO_2 conversion. These and other aspects (electrode and reactor design, as indicated before) determine the mass transfer and availability of protons at the surface of the electrocatalysts, determining the rate of the side HER (hydrogen evolution reaction) and the transformation paths of CO_2, influencing both the nature and concentration of C-based intermediates. These aspects strongly correlate to electrode and reactor design because they influence the interface's local fluidodynamics. The previous section has given some examples. Many proofs in the literature exist of these aspects. However, the mechanistic studies on CO_2RR are largely based on approaches which not consider these aspects properly, for example, in density functional theory (DFT) theoretical methods. Note that we do not intend to deny the contribution of theoretical methods to understanding catalysis but only remark that the current approaches cannot catch the complexity of electrocatalysis. Thus, new theoretical methods and approaches would be necessary. In contrast, indications using the current method do not provide enough reliable indications to develop a generalised design strategy for CO_2RR, which is necessary to progress towards the industrialisation of the technology, as commented in the introduction.

In electrocatalysis, differently from heterogeneous catalysis, applying a potential determines the generation of surface charges which can induce direct inner- or outer-sphere electron and/or energy transfer, thus not necessarily involving chemisorbed species. The electrical field at the surface determines a change in the structure of the electrolyte at the interface (the Stern layer), which strongly influences the rate of these processes (in classical Marcus theory of electrocatalysis, the rate of electron transfers depends on solvent reorganisation). In nanostructured electrocatalysts, the surface localised charges are not homogeneously distributed but depend on the nanostructure, even if few studies exist. A change in the nanostructure (for example, the nano shape of supported active metal nanoparticles) determines a change in the local electrical fields and charge distribution, which determines changes in the rate of electron/energy transfer. The change in the reaction rate of CO_2RR may be thus due to other reasons than a change in the exposed facts, as typically assumed [38].

A charged surface affects the mass transfer and chemisorption of species (CO_2, but also ions in the electrolyte and protons). Charged intermediate species (such as $^{\bullet}CO_2^-$ anion radical formed by one electron transfer to CO_2) may diffuse back from the catalyst when a cathodic (negative) potential is applied. Diffusion and the local environment/pH strongly determine the path of transformation, with the presence of significant intra-electrode gradients, as earlier commented. These are some of the issues present in electrocatalysis, not present in heterogeneous catalysis, making the mechanistic studies complex and the application of simplified approaches unreliable [38].

Electrocatalysis requires a specific approach different from heterogeneous catalysis, while still, scarce recognition exists on this issue [90]. In addition to the presence of an applied potential and thus the presence of surface charges, the new open area is a non-uniform charge distribution at the electrocatalyst, depending strongly on the nanostructure of the electrocatalyst and electrode. The near-surface localised electrical field is a crucial aspect to consider, which induces a non-uniform double layer and near-surface electrolyte reorganisation, causing modifications in the surface adlayer and electrocatalyst reactivity and/or selectivity [91–94].

7.3.1 A Disruptive Model for Approaching Electrocatalysis

The above discussion remarked that the current approaches are not an acceptable description of complex phenomena occurring. In the modelling, it is necessary to consider the effect of the applied potential explicitly, the role of the electrolyte, the possible in-situ reconstructions during the electrocatalytic process, the low surface coverage or the presence of mass transfer limitations, the presence of gradients in concentration inside the electrode, a non-uniform distribution of the potentials and changes, the presence of an electrical field, etc.

It may also be advanced that even the fundaments of catalysis for the design of electrocatalysts should be reconsidered. As commented before, DFT and analogous methods calculate the activation energies of individual reaction steps using a model surface. How to provide the energy necessary to overcome the activation energies of transition states is not considered because thermal energy provides sufficient energy to overcome the activation barrier at the operational temperatures of heterogeneous catalysts. This fact is one reasons for a threshold temperature in heterogeneous catalytic reactions.

Due to the applied potential, the electrocatalytic reaction occurs at much lower temperatures than the equivalent reaction by heterogeneous catalysis. The conversion rate is usually negligible without applying external energy (the electrical potential). The energy that drives the process is associated with the potential that generates localised charges. It should be thus asked how the energy to the incoming or adsorbed molecule is transferred to overcome the activation barriers.

We suggested [38, 95, 96] that the energy transfer involves surface phonons being the mechanism by which a solid can transfer energy to an adsorbed molecule [97]. A surface phonon is the quantum of a lattice vibration mode associated with a solid surface [98]. The energy transfer occurs by coupling the (localised) surface phonons with the molecule bond vibrations. Usually, surface phonons are extended vibrations, but when defects or nanostructures are present, they could be localised and have the energy to transfer to adsorbed molecules. It is thus analogous to the concept of active sites but translated into solid-state physics. Extended models based on Langevin dynamics can describe the phenomena [99]. This approach was never used to describe electrocatalytic processes and design strategies for their development.

Localised phonons, at a point and extended defects [100], step edges, twin boundaries, etc., are necessary to transfer energy to the chemisorbed molecules selectively. There is a likely dependence on the nanostructure, even if this aspect was never investigated. It is evident, however, how the classical aspects related to the concept of the active site (in heterogeneous catalysis) could be translated in terms of factors breaking down the "propagating" modes of surface phonons localising them. In this way, there is a selective and effective energy transfer to the adsorbed molecule [101].

Selectivity in energy transferring to the adsorbed molecule is the key to controlling the transformation pathway. Classical electrochemical theories or the current theoretical approaches do not account for this aspect. The design strategy is to tune the nature of the electrocatalyst to create localised surface phonons which match the energy required to overcome a specific activation barrier and drive the transformation along a specific path.

Furthermore, surface phonons interact with strong localised electrical fields and generate polaritons which have been suggested to play a relevant role in the transformative pathways of molecules whose high-frequency vibrational modes are strongly coupled with them (resonant catalysis) [102, 103].

When excitons (charges) are present, the coupling between excitons-phonons markedly influences the local electrical field and transfer rate to molecules [104]. Trapping of excitons by the lattice induces a strong electron–phonon coupling [105], which could induce a multi-electron transfer in (electro)catalysis. In electrocatalysis, a simultaneous multi-electron transfer is important to provide a low-activation and overpotential path as in enzymatic mechanisms [70, 106].

Scattered indications of these aspects are present in the literature. A theory to interpret the mechanisms of electrocatalysis in terms of localised phonons and modes to control them is absent. However, the above discussion offers thought for food for a disruptive interpretation of why current modelling approaches are ineffective in passing from conversion heterogeneous to electrocatalysis. They remark on the need for CO_2RR theoretical methods to address the mechanism to transfer the energy to chemisorbed molecules selectively. The same elements used to describe the active sites and related mechanisms could be used for an alternative mechanistic explanation.

7.4 Outlook/Future Perspectives

Electrocatalysis, specifically for the CO_2RR area, is a key technology/process for transformative (industrial) chemical production that meets climate change targets, sustainability and resilience. Even with the progress in the last years, it is necessary to accelerate the implementation of these technologies on an industrial scale. We advanced the idea that this acceleration requires developing novel materials and approaches by turning the current design strategies focused on the electrocatalysts, giving instead attention to electrode/reactor engineering and related aspects, including how to control the nano environment at the electrocatalyst surface.

We have shown that these aspects dominate the behaviour, including aspects typically related to electrocatalyst only, such as selectivity and onset potential. Their better understanding, especially the development of modelling able to account for the complexity of the phenomena present in electrocatalysis, is the condition to accelerate the progress. Current approaches are oversimplified. Consequently, the indications of the strategies for developing CO_2RR (industrial) processes are unreliable.

A key aspect remarked is that the current approaches in electrocatalysis derive from applying methods largely derived from heterogeneous catalysis. The mechanistic and design strategies used have an intrinsic methodological issue, which is not only related to a lack of analysis of what happens (in mechanistic terms) when a potential is applied. The existing approaches are not an acceptable simplification of the complex phenomena in electrocatalysis. We indicated some key aspects to consider that, even if challenging, are a necessary direction to explore.

We also introduced the concept of approaching catalysis from the perspective of selectively transferring energy for the activation barriers rather than only considering the nature of the active sites and the energetic pathways of transformation (the current approach). The concept is related to the role of localised phonons, polarons and interaction with excitons in determining the CO_2 electrocatalytic conversion pathways. These species represent a solid-state alternative mechanistic interpretation of the active sites. Their role and how to selectively design the electrocatalysts and their nanostructure to make selective the energy transfer and thus control the path of transformation is a novel area of investigation which can be a breakthrough in understanding electrocatalysis and designing the next-generation electrocatalysis, overcoming current limitations.

In conclusion, CO_2 electrocatalytic conversion is a key technology for a sustainable future and carbon circularity. The outlooks are positive, but it is necessary to accelerate the progress in the area, particularly overcoming the current limits, due to an approach which is translated from heterogeneous catalysis rather than by properly understanding the critical elements characterising electrocatalysis. From here, the many pitfalls in literature and the lack of reliability in a generalised approach to CO_2RR electrocatalysis are commented.

There are thus many scientific gaps to overcome, some discussed in the paper, especially to develop strategies to design the electrocatalysts, electrodes and reactors/ cells from an industrial perspective. They differ from the usually discussed aspects, such as scalability, purify of the feed, costs and technology maturity. These aspects are important but not the aim of the discussion in this contribution. Similarly, the aim was not to review state of the art and indicate the best electrocatalysts as a function of the target molecules. Many reviews are available on these aspects. We highlighted some less-covered aspects but relevant from the general perspective of analysing the effective impact of published results and derived indications on their reliability of the industrialisation of the technology. The aim is to evidence the general issues and missing aspects rather than a point-to-point analysis of the literature data, which may have a limited value.

Acknowledgements This work was supported by the EU ERC Synergy SCOPE project (810182). GC also thanks the Alexander von Humboldt-Stiftung/ Foundation (Humboldt Research Award).

References

1. Bai J, Liu D, Yang J, Chen Y (2019) Nanocatalysts for electrocatalytic oxidation of ethanol. Chemsuschem 12:2117–2132. https://doi.org/10.1002/cssc.201803063
2. Chen W, Wang Y, Li Y, Li C (2022) Electrocatalytic CO_2 reduction over bimetallic bi-based catalysts: a review. CCS Chem 5:544–567. https://doi.org/10.31635/ccschem.022.202202357
3. Chen ZW, Chen LX, Wen Z, Jiang Q (2019) Understanding electrocatalysis by using density functional theory. Phys Chem Chem Phys 21:23782–23802. https://doi.org/10.1039/c9cp04430b
4. Choi JY, Choi W, Park JW, Lim CK, Song H (2020) Strategies for designing nanoparticles for electro- and photocatalytic CO_2 reduction. Chem Asian J 15:253–265. https://doi.org/10.1002/asia.201901533
5. Choi W, Won DH, Hwang YJ (2020) Catalyst design strategies for stable electrochemical CO_2 reduction reaction. J Mater Chem A 8:15341–15357. https://doi.org/10.1039/D0TA02633F

6. Dongare S, Singh N, Bhunia H, Bajpai PK, Das AK (2021) Electrochemical reduction of carbon dioxide to ethanol: a review. ChemistrySelect 6:11603–11629. https://doi.org/10.1002/slct.202102829
7. Etim UJ, Semiat R, Zhong Z (2021) CO_2 valorisation reactions over Cu-based catalysts: characterisation and the nature of active sites. Am J Chem Eng 9:53–78. https://doi.org/10.11648/j.ajche.20210903.12
8. Fan L et al (2020) Strategies in catalysts and electrolyser design for electrochemical CO_2 reduction toward C2+ products. Sci Adv 6:eaay3111. https://doi.org/10.1126/sciadv.aay3111
9. Fujita E, Grills DC, Manbeck GF, Polyansky DE (2022) Understanding the role of inter- and intramolecular promoters in electro- and photochemical CO_2 reduction using Mn, Re, and Ru catalysts. Acc Chem Res 55:616–628. https://doi.org/10.1021/acs.accounts.1c00616
10. Goud D, Gupta R, Maligal-Ganesh R, Peter SC (2020) Review of catalyst design and mechanistic studies for the production of olefins from anthropogenic CO_2. ACS Catal 10:14258–14282. https://doi.org/10.1021/acscatal.0c03799
11. Govindarajan N, Xu A, Chan K (2022) How pH affects electrochemical processes: three mechanisms underlie the impact of pH on the activity of electrochemical reactions. Science 375:379–380. https://doi.org/10.1126/science.abj2421
12. Handoko AD, Steinmann SN, Seh ZW (2019) Theory-guided materials design: two-dimensional MXenes in electro- and photocatalysis. Nanoscale Horiz 4:809–827. https://doi.org/10.1039/c9nh00100j
13. Hua Y, Wang J, Min T, Gao Z (2022) Electrochemical CO_2 conversion towards syngas: recent catalysts and improving strategies for ratio-tunable syngas. J Power Sources 535:231453. https://doi.org/10.1016/j.jpowsour.2022.231453
14. Huynh HL, Yu Z (2020) CO_2 methanation on hydrotalcite-derived catalysts and structured reactors: a review. Energy Technol 8:1901475. https://doi.org/10.1002/ente.201901475
15. Iqbal MZ, Imteyaz S, Ghanty C, Sarkar S (2022) A review on electrochemical conversion of CO_2 to CO: Ag-based electrocatalyst and cell configuration for industrial application. J Ind Eng Chem 113:15–31. https://doi.org/10.1016/j.jiec.2022.05.041
16. Li C-F, Guo R-T, Wu T, Pan W-G (2022) Progress and perspectives on 1D nanostructured catalysts applied in photo(electro)catalytic reduction of CO_2. Nanoscale 14:16033–16064. https://doi.org/10.1039/d2nr04063h
17. Li X, Yu J, Jaroniec M, Chen X (2019) Cocatalysts for selective photoreduction of CO_2 into solar fuels. Chem Rev 119:3962–4179. https://doi.org/10.1021/acs.chemrev.8b00400
18. Lim KRG et al (2020) Rational design of two-dimensional transition metal carbide/nitride (MXene) hybrids and nanocomposites for catalytic energy storage and conversion. ACS Nano 14:10834–10864. https://doi.org/10.1021/acsnano.0c05482
19. Liu Y-L et al (2022) Two-dimensional metal-organic framework nanosheets: synthesis and applications in electrocatalysis and photocatalysis. Chemsuschem 15:e202102603. https://doi.org/10.1002/cssc.202102603
20. Loffler T, Ludwig A, Rossmeisl J, Schuhmann W (2021) What makes high-entropy alloys exceptional electrocatalysts? Angew Chem Int Ed 60:26894–26903. https://doi.org/10.1002/anie.202109212
21. Pan S et al (2022) Halide perovskite materials for photo(electro)chemical applications: dimensionality, heterojunction, and performance. Adv Energy Mater 12:2004002. https://doi.org/10.1002/aenm.202004002
22. Ramírez-Valencia LD, Bailón-García E, Carrasco-Marín F, Pérez-Cadenas AF (2021) From CO_2 to value-added products: a review about carbon-based materials for electro-chemical CO_2 conversion. Catalysts 11:351. https://doi.org/10.3390/catal11030351
23. Zhang M, Zhang K, Ai X, Liang X, Zhang Q, Chen H, Zou X (2022) Theory-guided electrocatalyst engineering: from mechanism analysis to structural design. Chin J Catal 43:2987–3018. https://doi.org/10.1016/S1872-2067(22)64103-2
24. Ross MB et al (2019) Designing materials for electrochemical carbon dioxide recycling. Nat Catal 2:648–658. https://doi.org/10.1038/s41929-019-0306-7

25. Wan C-P, Yi J-D, Cao R, Huang Y-B (2022) Conductive metal/covalent organic frameworks for CO_2 electro-reduction. Chin J Struct Chem 41:2205001–2205014. https://doi.org/10.14102/j.cnki.0254-5861.2022-0075
26. Wang Y, He D, Chen H, Wang D (2019) Catalysts in electro-, photo- and photoelectrocatalytic CO_2 reduction reactions. J Photochem Photobiol C 40:117–149. https://doi.org/10.1016/j.jphotochemrev.2019.02.002
27. Wang Y et al (2021) Engineering two-dimensional metal oxides and chalcogenides for enhanced electro- and photocatalysis. Sci Bull 66:1228–1252. https://doi.org/10.1016/j.scib.2021.02.007
28. Wu B, Chen J, Qian L (2022) Recent advances in heterogeneous electroreduction of CO_2 on copper-based catalysts. Catalysts 12:860. https://doi.org/10.3390/catal12080860
29. Wu Q-J, Liang J, Huang Y-B, Cao R (2022) Thermo-, electro-, and photocatalytic CO_2 conversion to value-added products over porous metal/covalent organic frameworks. Acc Chem Res 55:2978–2997. https://doi.org/10.1021/acs.accounts.2c00326
30. Xue Y, Guo Y, Cui H, Zhou Z (2021) Catalyst design for electrochemical reduction of CO_2 to multicarbon products. Small Methods 5:2100736. https://doi.org/10.1002/smtd.202100736
31. Yang W, Dastafkan K, Jia C, Zhao C (2018) Design of electrocatalysts and electrochemical cells for carbon dioxide reduction reactions. Adv Mater Technol 3:1700377. https://doi.org/10.1002/admt.201700377
32. Younus HA et al (2023) Molecular catalysts for CO_2 electroreduction: progress and prospects with pincer type complexes. Coord Chem Rev 493:215318. https://doi.org/10.1016/j.ccr.2023.215318
33. Zhang J et al (2022) Single-atom catalysts for thermal- and electrocatalytic hydrogenation reactions. J Mater Chem A 10:5743–5757. https://doi.org/10.1039/d1ta07910g
34. Zhang W, Jin Z, Chen Z (2022) Rational-designed principles for electrochemical and photoelectrochemical upgrading of CO_2 to value-added chemicals. Adv Sci 9:2105204. https://doi.org/10.1002/advs.202105204
35. Zhu Y et al (2021) Emerging dynamic structure of electrocatalysts unveiled by in situ X-ray diffraction/absorption spectroscopy. Energy Environ Sci 14:1928–1958. https://doi.org/10.1039/d0ee03903a
36. Wang S, Wang L, Wang D, Li Y (2023) Recent advances of single-atom catalysts in CO_2 conversion. Energy Environ Sci 16:2759–2803. https://doi.org/10.1039/D3EE00037K
37. Sun Q, Jia C, Zhao Y, Zhao C (2022) Single atom-based catalysts for electrochemical CO_2 reduction. Chin J Catal 43:1547–1597. https://doi.org/10.1016/S1872-2067(21)64000-7
38. Kang H et al (2023) Understanding the complexity in bridging thermal and electrocatalytic methanation of CO_2. Chem Soc Rev 52:3627–3662. https://doi.org/10.1039/D2CS00214K
39. Ampelli C et al (2023) Electrode and cell design for CO_2 reduction: a viewpoint. Catal Today 421:114217. https://doi.org/10.1016/j.cattod.2023.114217
40. Marepally BC et al (2017) Enhanced formation of >C1 products in electroreduction of CO_2 by adding a CO_2 adsorption component to a gas-diffusion layer-type catalytic electrode. Chemsuschem 10:4442–4446. https://doi.org/10.1002/cssc.201701506
41. Perathoner S, Van Geem KM, Marin GB, Centi G (2021) Reuse of CO_2 in energy intensive process industries. Chem Commun 57:10967–10982. https://doi.org/10.1039/D1CC03154F
42. Centi G, Perathoner S (2023) The chemical engineering aspects of CO_2 capture, combined with its utilisation. Curr Opin Chem Eng 39:100879. https://doi.org/10.1016/j.coche.2022.100879
43. Centi G, Perathoner S (2022) Status and gaps toward fossil-free sustainable chemical production. Green Chem 24:7305–7331. https://doi.org/10.1039/D2GC01572B
44. Papanikolaou G, Centi G, Perathoner S, Lanzafame P (2022) Catalysis for e-chemistry: need and gaps for a future de-fossilized chemical production, with focus on the role of complex (direct) syntheses by electrocatalysis. ACS Catal 12:2861–2876. https://doi.org/10.1021/acscatal.2c00099
45. Centi G, Iaquaniello G, Perathoner S (2011) Can we afford to waste carbon dioxide? Carbon dioxide as a valuable source of carbon for the production of light olefins. Chemsuschem 4:1265–1273. https://doi.org/10.1002/cssc.201100313

46. Ronda-Lloret M, Rothenberg G, Shiju NR (2019) A critical look at direct catalytic hydrogenation of carbon dioxide to olefins. Chemsuschem 12:3896–3914. https://doi.org/10.1002/cssc.201900915
47. Romero Cuellar NS et al (2020) Two-step electrochemical reduction of CO_2 towards multicarbon products at high current densities. J CO_2 Utiliz 36:263–275. https://doi.org/10.1016/j.jcou.2019.10.016
48. Centi G, Iaquaniello G, Perathoner S (2019) Chemical engineering role in the use of renewable energy and alternative carbon sources in chemical production. BMC Chem Eng 1:5. https://doi.org/10.1186/s42480-019-0006-8
49. Centi G, Perathoner S, Genovese C, Arrigo R (2023) Advanced (photo) electrocatalytic approaches to substitute the use of fossil fuels in chemical production. Chem Commun 59:3005–3023. https://doi.org/10.1039/D2CC05132J
50. Papanikolaou G, Centi G, Perathoner S, Lanzafame P (2022) Transforming catalysis to produce e-fuels: prospects and gaps. Chin J Catal 43:1194–1203. https://doi.org/10.1016/S1872-2067(21)64016-0
51. Birdja YY et al (2019) Advances and challenges in understanding the electrocatalytic conversion of carbon dioxide to fuels. Nat Energy 4:732–745. https://doi.org/10.1038/s41560-019-0450-y
52. Linnemann J, Kanokkanchana K, Tschulik K (2021) Design strategies for electrocatalysts from an electrochemist's perspective. ACS Catal 11:5318–5346. https://doi.org/10.1021/acscatal.0c04118
53. Senocrate A, Battaglia C (2021) Electrochemical CO_2 reduction at room temperature: status and perspectives. J Energy Stor 36:102373. https://doi.org/10.1016/j.est.2021.102373
54. Wakerley D et al (2019) Bio-inspired hydrophobicity promotes CO_2 reduction on a Cu surface. Nat Mater 18:1222–1227. https://doi.org/10.1038/s41563-019-0445-x
55. Melchionna M, Fornasiero P, Prato M, Bonchio M (2021) Electrocatalytic CO_2 reduction: role of the cross-talk at nano-carbon interfaces. Energy Environ Sci 14:5816–5833. https://doi.org/10.1039/D1EE00228G
56. Nesbitt NT et al (2020) Liquid-solid boundaries dominate activity of CO_2 reduction on gas-diffusion electrodes. ACS Catal 10:14093–14106. https://doi.org/10.1021/acscatal.0c03319
57. Burdyny T, Smith WA (2019) CO_2 reduction on gas-diffusion electrodes and why catalytic performance must be assessed at commercially-relevant conditions. Energy Environ Sci 12:1442–1453. https://doi.org/10.1039/C8EE03134G
58. Sassenburg M, Kelly M, Subramanian S, Smith WA, Burdyny T (2023) Zero-gap electrochemical CO_2 reduction cells: challenges and operational strategies for prevention of salt precipitation. ACS Energy Lett 8:321–331. https://doi.org/10.1021/acsenergylett.2c01885
59. Subramanian S et al (2023) Geometric catalyst utilisation in zero-gap CO_2 electrolysers. ACS Energy Lett 8:222–229. https://doi.org/10.1021/acsenergylett.2c02194
60. Lee WH et al (2021) Highly selective and stackable electrode design for gaseous CO_2 electroreduction to ethylene in a zero-gap configuration. Nano Energy 84:105859. https://doi.org/10.1016/j.nanoen.2021.105859
61. Alinejad S et al (2022) Electrochemical reduction of CO_2 on Au electrocatalysts in a zero-gap, half-cell gas diffusion electrode setup: a systematic performance evaluation and comparison to an H-cell setup. ChemElectroChem 9:e202200341. https://doi.org/10.1002/celc.202200341
62. Garg S et al (2020) Advances and challenges in electrochemical CO_2 reduction processes: an engineering and design perspective looking beyond new catalyst materials. J Mater Chem A 8:1511–1544. https://doi.org/10.1039/C9TA13298H
63. Sun H et al (2019) Hierarchical 3D electrodes for electrochemical energy storage. Nat Rev Mater 4:45–60. https://doi.org/10.1038/s41578-018-0069-9
64. Chu T, Park S, Fu K (2021) 3D printing-enabled advanced electrode architecture design. Carbon Energy 3:424–439. https://doi.org/10.1002/cey2.114
65. Vedharathinam V et al (2019) Using a 3D porous flow-through electrode geometry for high-rate electrochemical reduction of CO_2 to CO in ionic liquid. ACS Catal 9:10605–10611. https://doi.org/10.1021/acscatal.9b03201

66. Hui S et al (2020) Three-dimensional cathodes for electrochemical reduction of CO_2: from macro- to nano-engineering. Nanomaterials 10:1884. https://doi.org/10.3390/nano10091884
67. Vaněčková E et al (2021) Electrochemical reduction of carbon dioxide on 3D printed electrodes. ChemElectroChem 8:2137–2149. https://doi.org/10.1002/celc.202100261
68. Weber AZ et al (2014) A critical review of modeling transport phenomena in polymer-electrolyte fuel cells. J Electrochem Soc 161:F1254. https://doi.org/10.1149/2.0751412jes
69. Li F, MacFarlane DR, Zhang J (2018) Recent advances in the nanoengineering of electrocatalysts for CO_2 reduction. Nanoscale 10:6235–6260. https://doi.org/10.1039/C7NR09620H
70. Perathoner S, Centi G (2019) Catalysis for solar-driven chemistry: the role of electrocatalysis. Catal Today 330:157–170. https://doi.org/10.1016/j.cattod.2018.03.005
71. Velasco-Vélez J-J et al (2019) The role of the copper oxidation state in the electrocatalytic reduction of CO_2 into valuable hydrocarbons. ACS Sustain Chem Eng 7:1485–1492. https://doi.org/10.1021/acssuschemeng.8b05106
72. Genovese C et al (2018) Operando spectroscopy study of the carbon dioxide electro-reduction by iron species on nitrogen-doped carbon. Nat Commun 9:935. https://doi.org/10.1038/s41467-018-03138-7
73. Arrigo R et al (2022) Dynamics at polarised carbon dioxide-iron oxyhydroxide interfaces unveil the origin of multicarbon product formation. ACS Catal 12:411–430. https://doi.org/10.1021/acscatal.1c04296
74. Weng L-C, Bell AT, Weber AZ (2018) Modeling gas-diffusion electrodes for CO_2 reduction. Phys Chem Chem Phys 20:16973–16984. https://doi.org/10.1039/C8CP01319E
75. Liang Y et al (2022) Wettability control in electrocatalyst: a mini review. J Energy Chem 70:643–655. https://doi.org/10.1016/j.jechem.2021.09.005
76. Rabiee H et al (2022) Regulating the reaction zone of electrochemical CO_2 reduction on gas-diffusion electrodes by distinctive hydrophilic-hydrophobic catalyst layers. Appl Catal B 310:121362. https://doi.org/10.1016/j.apcatb.2022.121362
77. Yang K, Kas R, Smith WA, Burdyny T (2021) Role of the carbon-based gas diffusion layer on flooding in a gas diffusion electrode cell for electrochemical CO_2 reduction. ACS Energy Lett 6:33–40. https://doi.org/10.1021/acsenergylett.0c02184
78. Möller T et al (2021) The product selectivity zones in gas diffusion electrodes during the electrocatalytic reduction of CO_2. Energy Environ Sci 14:5995–6006. https://doi.org/10.1039/D1EE01696B
79. Suter S, Haussener S (2019) Optimizing mesostructured silver catalysts for selective carbon dioxide conversion into fuels. Energy Environ Sci 12:1668–1678. https://doi.org/10.1039/C9EE00656G
80. Weng L-C, Bell AT, Weber AZ (2020) A systematic analysis of Cu-based membrane-electrode assemblies for CO_2 reduction through multiphysics simulation. Energy Environ Sci 13:3592–3606. https://doi.org/10.1039/D0EE01604G
81. Kas R et al (2021) Along the channel gradients impact on the spatioactivity of gas diffusion electrodes at high conversions during CO_2 electroreduction. ACS Sustain Chem Eng 9:1286–1296. https://doi.org/10.1021/acssuschemeng.0c07694
82. Weekes DM, Salvatore DA, Reyes A, Huang A, Berlinguette CP (2018) Electrolytic CO_2 reduction in a flow cell. Acc Chem Res 51:910–918. https://doi.org/10.1021/acs.accounts.8b00010
83. Li YC et al (2016) Electrolysis of CO_2 to syngas in bipolar membrane-based electrochemical cells. ACS Energy Lett 1:1149–1153. https://doi.org/10.1021/acsenergylett.6b00475
84. Endrődi B et al (2017) Continuous-flow electroreduction of carbon dioxide. Prog Energy Combust Sci 62:133–154. https://doi.org/10.1016/j.pecs.2017.05.005
85. Disch J, Bohn L, Metzler L, Vierrath S (2023) Strategies for the mitigation of salt precipitation in zero-gap CO_2 electrolysers producing CO. J Mater Chem A 11:7344–7357. https://doi.org/10.1039/D2TA09966G
86. Ma M et al (2020) Insights into the carbon balance for CO_2 electroreduction on Cu using gas diffusion electrode reactor designs. Energy Environ Sci 13:977–985. https://doi.org/10.1039/D0EE00047G

87. De Gregorio GL et al (2020) Facet-dependent selectivity of Cu catalysts in electrochemical CO_2 reduction at commercially viable current densities. ACS Catal 10:4854–4862. https://doi.org/10.1021/acscatal.0c00297
88. Dinh CT et al (2018) CO_2 electroreduction to ethylene via hydroxide-mediated copper catalysis at an abrupt interface. Science 360:783–787. https://doi.org/10.1126/science.aas9100
89. Rabinowitz JA, Kanan MW (2020) The future of low-temperature carbon dioxide electrolysis depends on solving one basic problem. Nat Commun 11:5231. https://doi.org/10.1038/s41467-020-19135-8
90. Tang C, Zheng Y, Jaroniec M, Qiao S-Z (2021) Electrocatalytic refinery for sustainable production of fuels and chemicals. Angew Chem Int Ed 60:19572–19590. https://doi.org/10.1002/anie.202101522
91. Zhao X, Levell ZH, Yu S, Liu Y (2022) Atomistic understanding of two-dimensional electrocatalysts from first principles. Chem Rev 122:10675–10709. https://doi.org/10.1021/acs.chemrev.1c00981
92. Ringe S, Hörmann NG, Oberhofer H, Reuter K (2022) Implicit solvation methods for catalysis at electrified interfaces. Chem Rev 122:10777–10820. https://doi.org/10.1021/acs.chemrev.1c00675
93. Dattila F, Seemakurthi RR, Zhou Y, López N (2022) Modeling operando electrochemical CO_2 reduction. Chem Rev 122:11085–11130. https://doi.org/10.1021/acs.chemrev.1c00690
94. Sundararaman R, Vigil-Fowler D, Schwarz K (2022) Improving the accuracy of atomistic simulations of the electrochemical interface. Chem Rev 122:10651–10674. https://doi.org/10.1021/acs.chemrev.1c00800
95. Centi G, Perathoner S (2022) Catalysis for an electrified chemical production. Catalysis Today 423:113935. https://doi.org/10.1016/j.cattod.2022.10.017
96. Centi G, Perathoner S (2022) Redesign chemical processes to substitute the use of fossil fuels: a viewpoint of the implications on catalysis. Catal Today 387:216–223. https://doi.org/10.1016/j.cattod.2021.03.007
97. Vayssilov GN (1995) Theoretical models for description of the gas-solid surface vibrational interactions. Adv Coll Interface Sci 57:123–159. https://doi.org/10.1016/0001-8686(95)00237-K
98. Srivastava G, Tütüncü H (2009) Theoretical modelling of surface phonons. Open Physics 7:209–219. https://doi.org/10.2478/s11534-009-0032-x
99. Tamm A et al (2018) Langevin dynamics with spatial correlations as a model for electron-phonon coupling. Phys Rev Lett 120:185501. https://doi.org/10.1103/PhysRevLett.120.185501
100. Scott EA et al (2018) Phonon scattering effects from point and extended defects on thermal conductivity studied via ion irradiation of crystals with self-impurities. Phys Rev Mater 2:095001. https://doi.org/10.1103/PhysRevMaterials.2.095001
101. Caldwell JD et al (2015) Low-loss, infrared and terahertz nanophotonics using surface phonon polaritons. Nanophotonics 4:44–68. https://doi.org/10.1515/nanoph-2014-0003
102. Campos-Gonzalez-Angulo JA, Ribeiro RF, Yuen-Zhou J (2019) Resonant catalysis of thermally activated chemical reactions with vibrational polaritons. Nat Commun 10:4685. https://doi.org/10.1038/s41467-019-12636-1
103. Thomas A et al (2019) Tilting a ground-state reactivity landscape by vibrational strong coupling. Science 363:615–619. https://doi.org/10.1126/science.aau7742
104. Antonius G, Louie SG (2022) Theory of exciton-phonon coupling. Phys Rev B 105:085111. https://doi.org/10.1103/PhysRevB.105.085111
105. McCall KM, Stoumpos CC, Kostina SS, Kanatzidis MG, Wessels BW (2017) Strong electron-phonon coupling and self-trapped excitons in the defect halide perovskites A3M2I9 (A = Cs, Rb; M = Bi, Sb). Chem Mater 29:4129–4145. https://doi.org/10.1021/acs.chemmater.7b01184
106. Mallamace D, Papanikolaou G, Perathoner S, Centi G, Lanzafame P (2021) Comparing molecular mechanisms in solar NH_3 production and relations with CO_2 reduction. Int J Mol Sci 22:1–139

Chapter 8
Bioconversion of CO_2 into Valuables

Neha Sharma, Sovik Das, and Deepak Pant

Abstract The world today is experiencing adverse effects of climate change due to the escalated levels of greenhouse gases such as CO_2 in the atmosphere. Consequently, there is an urgent need for minimizing the CO_2 level to tackle climate change. Utilization of CO_2 for synthesizing valuable commodities, such as biofuel and biochemical, is an effective strategy toward mitigating climate change as well as the energy crisis. Moreover, the conversion of CO_2 into valuable products is also a promising step for achieving carbon neutrality and exemplifying a circular economy. In this regard, several methods, such as physical, chemical, and biological, are available for the utilization and conversion of CO_2. However, biological methods for the conversion of CO_2 into biofuels and biochemicals are more sustainable as they are environmentally friendly and cost-effective compared to the chemical methods. Thus, in this chapter, different CO_2 bioconversion processes with their efficacy and their limitations have been discussed. Besides this, future developments in terms of CO_2 bioconversion have also been presented for alleviating the limitations, so that the production of biocommodities from CO_2 can be achieved at a large scale.

8.1 Introduction

The pathways for CO_2 utilization can be broadly classified into two categories, namely direct use and indirect use. In direct use, no chemical alteration is done with CO_2 molecules, while in indirect use, CO_2 molecules are converted into valuable products, such as fuel, chemicals, and consumables [1]. Furthermore, indirect pathways for the utilization of CO_2 are classified into chemical and biological. In the biological method, plants or microorganisms consume the CO_2 and convert it into

N. Sharma · S. Das
Department of Civil Engineering, Indian Institute of Technology Delhi, New Delhi 110016, India
e-mail: dassovik@iitd.ac.in

D. Pant (✉)
Separation and Conversion Technology, Flemish Institute for Technological Research (VITO), Boeretang 200, 2400 Mol, Belgium
e-mail: deepak.pant@vito.be

G. Zhang et al. (eds.), *Advances in CO_2 Utilization*, Green Chemistry and Sustainable Technology, https://doi.org/10.1007/978-981-99-8822-8_8

valuable bioproducts [1, 2]. On the other hand, in the chemical method of CO_2 utilization, the bonds inside the CO_2 molecule are broken and transformed into fuels and chemicals. In this veneration, different chemical technologies for CO_2 conversion are available like reforming, hydrogenation, carboxylation, mineralization, electrochemical, photochemical, and plasma catalysis [1]. These technologies are useful for the valorization of CO_2 into numerous end products, such as syngas, methanol, urea, and methane. However, most of these technologies require higher input of temperature and pressure for operating conditions, which make them costly. Also, photochemical technology of CO_2 conversion needs expensive and toxic metals, which is less environmentally friendly. Furthermore, the requirement of electric current for the electrochemical technology also limits its application [3].

Additionally, the cost and stability of the catalyst used in the chemical processes for CO_2 conversion is also a matter of concern. Due to these drawbacks of chemical processes, the biological process for CO_2 conversion has grabbed a lot of attention. The biological conversion using microorganisms and enzymes can potentially substitute the chemical process, which is very energy intensive [3]. With regard to the biological processes, microorganisms that are used to utilize the CO_2 to turn it into bioproducts can be broadly classified as photosynthetic and non-photosynthetic [2, 4]. The energy requirement in the bioconversion processes of CO_2 is lesser as these require only light energy compared to the chemical processes, which demand higher input of the energy [4]. In biological CO_2 conversion methods, microorganisms consume CO_2 as carbon source and the conversion of CO_2 occurs at the temperature range between 20 and 35 °C and at mild pressure, which makes them less energy-intensive [2, 4]. Moreover, the fixation of CO_2 using microorganisms is a promising option because of the minimum cost of feedstock and operation to produce bioproducts [5]. For these reasons, biological technology is advantageous over chemical technology for CO_2 utilization and is preferred over chemical technology (Fig. 8.1).

8.2 Different Types of CO_2 Bioconversion Methods

8.2.1 Plant and Tree-Based Method

Plants and trees are considered as the natural carbon sink for ages as they intake CO_2 from the atmosphere and turn it into organic carbon through photosynthesis, which is very useful for the growth of plants and trees. Elevated levels of CO_2 in the atmosphere benefit the plants to elevate the photosynthesis rate by partially opening the stomata which leads to a reduction in water loss and in turn, promotes the growth and yield of the plant and is known as the carbon fertilizing effect. Depending upon the pathways used by the plants for CO_2 fixation, plants are broadly classified into three categories, namely C3, C4, and crassulacean acid metabolism. In C3 plants, the photorespiration process occurs as the rubisco enzyme works as oxygenase due to the higher concentration of O_2 over CO_2 and fixes the O_2, and generates CO_2.

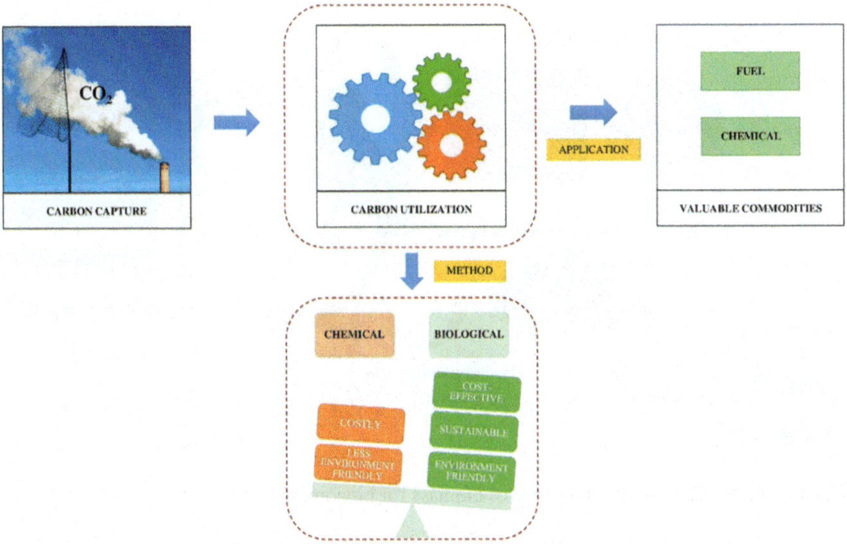

Fig. 8.1 Carbon capture and utilization for circular economy

On the contrary, C4 and crassulacean acid metabolism plants have a carbon concentration mechanism that helps to concentrate the CO_2 on the rubisco enzyme and, hence, fix the CO_2 through the photosynthesis process [6]. However, at the elevated level of CO_2, the rubisco enzyme in C3 plants works as the carboxylase and fixes the CO_2 by overpowering the photosynthesis process over the photorespiration [7]. Furthermore, biomass derived from plants and trees also referred as lignocellulosic biomass (LCB), contains cellulose, hemicellulose, and lignin, which can be used as a feedstock for producing fuels, and chemicals are becoming popular due to its abundant availability and biorenewability [8]. The LCB is generally classified as crop residues, energy crops, and forestry biomass, and this biomass is non-edible, so do not have any competition with food crops and is counted as the feedstock for the next generation. Further, this LCB can be converted into biofuels and biochemicals through numerous technologies, such as anaerobic digestion, fermentation, pyrolysis, gasification, carbonization, and liquefaction [9].

Different types of biofuel, e.g., bioethanol, biobutanol, biomethane, and biopolymers, have been produced using LCB (Fig. 8.2). Mariano et al. [10] investigated the production of bioethanol using the coconut residue and reported the production of 27.19 g/L of bioethanol after 48 h of fermentation. Similarly, 95 L biobutanol per ton of the pre-treated residue of the sugarcane was generated using fermentation as reported by Pratto et al. [11]. Additionally, Bhatia et al. [12] also used the *Miscanthus* biomass hydrolysate as the substrate for *Ralstonia eutropha* 5119 to produce bioplastic and reported 2.1 g/L of polyhydroxyalkanoate production. Moreover, other types of LCB were also utilized for deriving the biofuel, and different yields were obtained (Table 8.1). Undoubtedly, LCB is the better alternative to transit

Table 8.1 (continued)

Process	Feedstock	Inoculum	Conversion method	Commodity (yield)	Reference
	Synthetic wastewater and *Leersia hexandra*	–	–	Bioelectricity (64.2 mW/m^2)	[26]
Gas fermentation	$CO_2 + H_2$	Anaerobic sludge	–	Acetic acid (3.0 g/L) Butyric acid (0.7 g/L)	[27]
	CO_2		–	Acetic acid (0.84 g/L)	[28]
Enzymatic	CO_2	–	–	Formic acid (4.76 mM)	[29]

** LCB* lignocellulosic biomass, *MES* microbial electrosynthesis, *CW* constructed wetland, *MFC* microbial fuel cell, *PHB* polyhydroxy butyrate, *DW* dry weight, *MCC* microbial carbon capture

For instance, high lipid concentrating algal and cyanobacterial species are preferred for biodiesel production, while high carbohydrate concentrating algal and cyanobacterial species are preferred for bioethanol, biomethane, and biohydrogen production [32]. The selection of proper algal and cyanobacterial species and its tolerance for high CO_2 concentrations, nutrient limitation, toxicity, and pH, as well as technologies employed for the cultivation, harvesting, and drying of biomass, are crucial for efficacious biofuel production. Several physical, mechanical, biological, and chemical methods have been reported in the literature that have been used to extract lipid and oil from the algal and cyanobacterial biomass for producing biodiesel [33]. Biodiesel derived from *Chlorella vulgaris* BDUG 91,771 species was successfully used in the Kirloskar ignition engine, which also complies with the European and American standards [31]. Similarly, algal and cyanobacterial biomass also contain plentiful amount of carbohydrates and sugars that can be converted into bioethanol using saccharification and microbial fermentation. In this regard, Trivedi et al. [34] utilized green seaweed *Ulva* to produce bioethanol and reported 0.45 g/g of bioethanol yield using the process of microbial fermentation. Biohydrogen, considered as clean fuel, can also be produced from the algal biomass, but for that pretreatment of algal biomass is prerequisite due to the conversion of complex carbohydrate into simple sugar so that fermentative bacteria can easily degrade simple sugar during dark fermentation to produce biohydrogen.

In contrast, microalgae can also synthesize biohydrogen by dissociating water molecules into hydrogen and oxygen through direct photolysis. Similarly, biohydrogen production can also be achieved by indirect photolysis in which hydrogen is produced by splitting of water molecules using hydrogenases enzymes secreted by cyanobacteria species, namely *Anabaena, Phormidium* , *Synechocystis,* and *Nostoc* [35]. Moreover, algal and cyanobacterial biomass can also be converted into biomethane by employing anaerobic digestion or fermentation in which complex polymers, such as lipids, carbohydrates, and protein, are broken down into simple monomers that can be further converted into biomethane (Fig. 8.2). Biooil, biochar, and syngas can also be produced by employing algal biomass as feedstock for hydrothermal liquefaction, pyrolysis, and gasification processes, respectively [36]. Other than this, pigments, nutraceuticals, fertilizers, and protein, with potential applications in cosmetics, healthcare, agriculture, and the food industry, respectively, can also be derived from algal and cyanobacterial biomass (Fig. 8.2) [30, 31, 37]. Some of the algal species, i.e., *Spirulina, Chlorella,* and cyanobacterial species i.e., *Chlorogloea fritschii, Nostoc muscorum,* accumulate the polyhydroxyalkanoate or polyhydroxybutyrate, which can be extracted for the production of bioplastic; hence, assist in reducing the environmental impact resulting from the use of conventional plastic [38, 39]. Other types of algae and cyanobacterial species were also utilized for deriving biofuel or biochemical and considerable yield was obtained (Table 8.1).

Furthermore, bioelectricity can also be generated by employing algae and cyanobacteria in microbial carbon capture cell (MCC), a bioelectrochemical system comprising of an anodic and cathodic chamber separated by a proton exchange membrane (PEM). This PEM facilitates the migration of protons from the anodic to the cathodic chamber. Furthermore, the anodic chamber of the MCC contains exoelectrogens that are anaerobic in nature and consume organic matter as substrate present in the wastewater, subsequently converting the organic matter into CO_2 and electrons (Fig. 8.3). This CO_2 is further recirculated to the cathodic chamber housing algae or cyanobacteria, leading to electricity and biomass production, CO_2 sequestration and wastewater treatment [40, 41]. In this regard, Zhou et al. [41] reported the power generation of 2485.35 mW/m^3 and chemical oxygen demand removal of 84.8% with the current density of 7.9 A/m^3 using algal culture *C. vulgaris* grown in the cathodic chamber of MCC.

Further, Pandit et al. [42] cultured cyanobacterial species *Anabaena* in the cathodic chamber and observed 57.8 mW/m^2 of power generation with CO_2-air mixture as feedstock as compared to 27.9 mW/m^2 of power generation when sparged only with air. Besides this, other studies also reported the production of bioelectricity using MCC (Table 8.1). In addition to power generation, algal and cyanobacterial biomass is also produced via MCC, which can be utilized for biofuel and other biomaterial production (Fig. 8.3) [40]. Further, in recent years, efforts have also been directed to enhance the growth rate of photosynthetic microbes by modifying their genes, which in turn influences the metabolic pathways and results in optimized growth of algal and cyanobacterial species leading to better yield [43]. Although these photosynthetic microbes have proven to be a promising alternative to produce third-generation biofuel, bioelectricity, and other valuable bioproducts; however, its commercialization on a larger scale is still limited due to the high cost and energy requirement for cultivation, harvesting, and conversion methods. Moreover, the identification of the most suitable algal and cyanobacterial species for the given conditions that can produce higher biomass is still challenging [30]. Consequently, more efforts

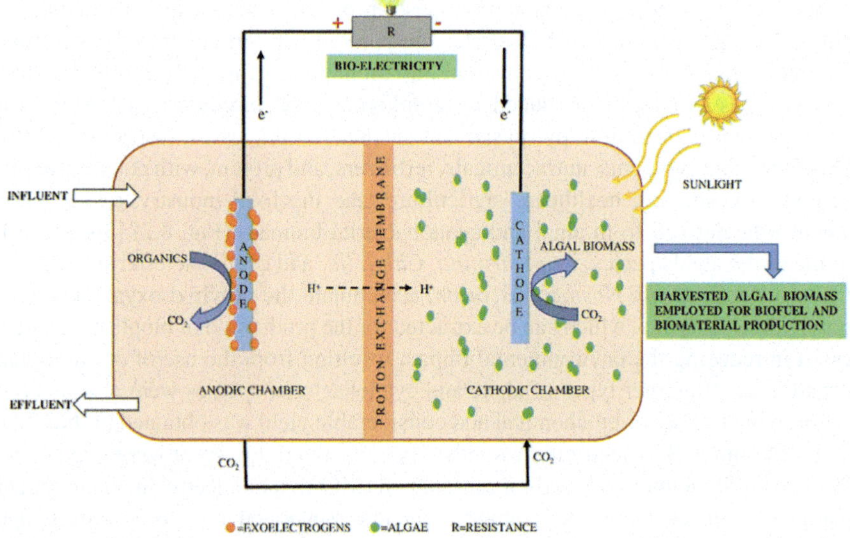

Fig. 8.3 Microbial carbon capture cell

should be focused on minimizing the cost associated with algal and cyanobacterial-based biofuel production and exploring the most appropriate algal and cyanobacterial species, so that this technology can be made more viable in the near future and subsequently commercialized at a large scale.

8.2.3 Microbial Electrosynthesis

Microbial electrosynthesis (MES) is a bioelectrochemical process in which CO_2 is converted into multi-carbon organic compounds, also referred as the electro-commodities employing anaerobic electrotrophic microbes as biocatalysts [44]. A typical MES consists of a biotic cathodic chamber under anaerobic conditions and an abiotic anodic chamber separated by PEM, which facilitates the migration of protons from the anodic to the cathodic chamber (Fig. 8.4). In the anodic chamber of MES, splitting of water results in the formation of protons, electrons, and oxygen. These protons are transferred to the cathodic chamber via PEM, which combine with electrons supplied by a potentiostat and CO_2 in the presence of electrotrophic microbes to produce electro-commodities [40]. In MES, electron uptake capacity and metabolic properties of microbes govern the productivity and energy conversion efficiency of MES process. Microbes that are employed in MES to convert CO_2 into electro-commodities are chemolithoautotrophic and can be attached to the cathode as biofilm or can be planktonic or both simultaneously. Generally, direct and mediated electron transfer between microbes and cathode surface takes place for microbes present in the form of biofilm and planktonic cells, respectively [44, 45]. Mainly acetogens and methanogens as biocatalysts in the form of both pure and mixed microbial cultures have been extensively explored in MES for converting CO_2 into electro-commodities [46].

Different pure microbial strains, namely *Clostridium* and *Sporomusa*, and mixed microbial culture have been reported in the literature to synthesize organic compounds like acetate, formate, methane, butanol, propanol, etc. [44]. In particular, pure microbial culture leads to the selective production of valuables, while in mixed culture, the formation of side products takes place due to the presence of different microbes in mixed culture by initiating non-targeted reactions. However, the application of mixed culture microbes as biocatalyst in MES is mostly preferred as it can be easily implicated at the field-scale. Besides electrotrophic microbes, the type of cathode material also affects the performance of MES since it acts as electron donor for the electrotrophic microbes [47]. Hence, biocompatible cathode possessing a high surface area and high conductivity is desirable for application in MES [45]. Also, different operational factors, i.e., catholyte pH, operating temperature, source of inoculum, and applied potential affect the performance of MES, which need to be optimized for desired yield of valuables [48, 49]. Till now, MES process has been successfully utilized for CO_2 reduction and simultaneously synthesizing different electro-commodities, which can be divided into various categories, such as C1, C2, C3, and C4 depending upon the carbon atoms present in the product.

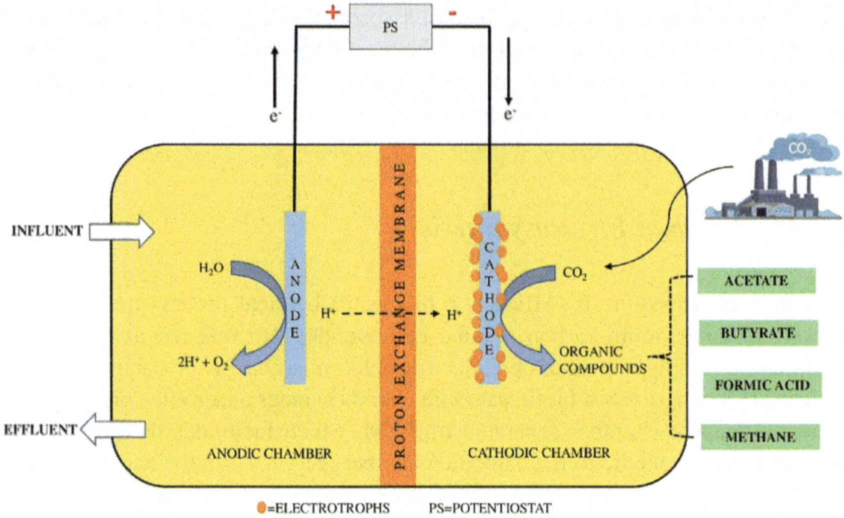

Fig. 8.4 Microbial electrosynthesis

The C1 compounds like methane, methanol, and formic acid have been successfully synthesized by MES employing CO_2 as substrate. Among these C1 compounds, methane is one of the widely reported electro-commodity produced through MES. In this regard, Enzmann and Holtmann [50] reported methane yield of 11.7 mmol/day using *Methanococcus maripaludis* with CO_2 as a substrate in MES. Also, the production of 63.2 mmol/m^2.day of formate via MES using thermophilic microbe *Moorella thermoautotrophica* was reported by Yu et al. [51]. In contrast to methane, acetate is the other major commodity chemical that has been widely synthesized from CO_2 using MES. LaBelle and May [52] reported 0.78 g/L h of acetate production using acetogenic microbe in mixed culture inoculated MES with CO_2 as the substrate. Moreover, longer-chain organic compounds like isopropanol with a yield of 1.17 g/m^2.day was also produced via MES by Arends et al. [53].

Apart from these, different compounds, such as butyrate, isobutyrate, butanol, and succinate have also been reported to be produced via MES from CO_2. Other studies also reported the synthesis of biocommodities using MES with considerable yield (Table 8.1). Although the process of MES is a futuristic solution to control the concentration of CO_2 in the atmosphere by transforming the CO_2 into valuable bioproducts, scale-up of this process to a larger scale is still limited. Low CO_2 absorption rate and transfer efficiency of electrons from cathode to the microbes are the major bottlenecks of this process as it limits the yield of valuables. Therefore, to alleviate these drawbacks, proper selection of highly efficient chemolithoautotrophic microbes that can fix CO_2 faster and efficient cathode material are required for better performance of MES. Furthermore, while targeting a particular product, formation of side products lowers the purity of the targeted product and downstream processing for extraction becomes indispensable, which adds to the cost. However, efforts have been made to curtail the downstream processing cost by optimizing the reactor design in which separation can take place cost-effectively.

To address this, Das et al. [54] designed a three-chamber circular MES and reported acetic acid yield of 104.25 ± 1.75 g/m^2.day in the extraction chamber compared to 81.35 ± 1.15 g/m^2.day of acetic acid yield in the cathodic chamber. In addition, fluctuation in the concentration of CO_2 is also a limiting factor causing low cell density in the system and ultimately leads to lower yield of the product. For this, continuous supply of CO_2 is needed to maintain the optimum growth of microbes leading to higher product formation. Additionally, the high cost associated with the construction and operation of MES is also one of the major constraints to its field-scale implementation. Incorporation of renewable energy with MES system to fulfill the energy demand can be a probable alternative to curtail the operating cost and render the process economically feasible. Also, this high cost can be compensated by synthesizing higher chain organic compounds with high market value, which would also make this process economically feasible. Moreover, further efforts are required for deriving low-cost and efficient electrode material to reduce construction cost and escalate the electrocatalytic activity and microbial adherence. These further efforts will boost the performance of MES by overcoming the limitations of this process, so that it can be implemented at a large scale to maintain a circular economy.

8.2.4 Constructed Wetland Coupled Microbial Fuel Cell

Natural wetlands play a prominent role in carbon sequestration, despite covering only six to eight percent of Earth's surface area [55]. These wetlands sequester carbon through photosynthesis and simultaneously store it in vegetation both above and below the ground, in the form of litter, peats, and sediments that build up over the years [56]. However, the destruction of natural wetlands due to farming, road construction, and other developmental projects has posed a serious threat to the ecosystem due to the disruption of the associated carbon cycle. Consequently, constructed wetlands (CWs) have been envisaged to rectify this carbon balance, which emulates a similar mechanism by mimicking the natural wetlands with concomitantly treating polluted water. These CWs are engineered systems consisting of shallow ponds; packed with substrates like gravel and sand, macrophytes, and liners to control the direction of the wastewater flow (Fig. 8.5a) [57]. Besides carbon sequestration, CW possesses enormous potential to treat wastewater as macrophytes uptake the nutrients and pollutants present in wastewater [24]. However, the application of CW is not limited to carbon sequestration and wastewater treatment only; as CW produces abundant biomass, which can also be utilized for the production of valuable commodities, such as biofuel and biochar [58, 59].

As per the study by Liu et al. [58], if entire amount of wastewater generated in China is treated using CW, it can produce 8.2×10^7 GJ/year of bioenergy, which is sufficient to fulfill the demand of two million households of China. Further, biomass harvested from CW can be utilized for biofuel, i.e., bioethanol and biogas production,

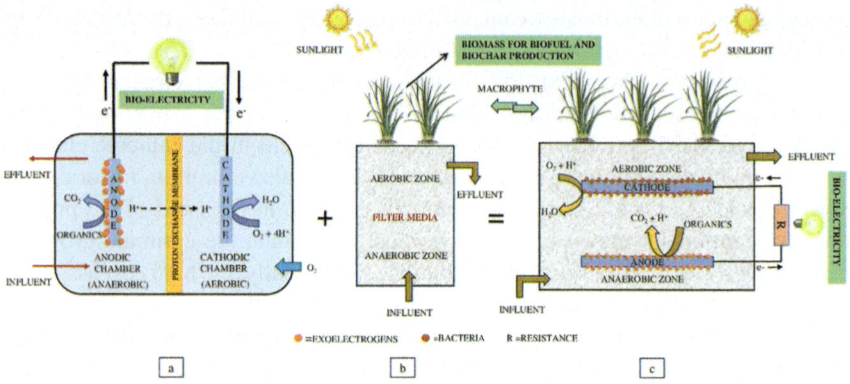

Fig. 8.5 **a** Microbial fuel cell, **b** constructed wetland, **c** constructed wetland-microbial fuel cell

as the biomass of CW plants contains the same chemical composition and calorific value like other biofuel feedstocks, such as switch grass and corn residues [60, 61]. Notably, biofuel production depends upon the type of macrophyte employed; hence, macrophytes suitable for biofuel production and which result in higher yield are desirable to be used as vegetation in CWs. In this regard, Lin et al. [24] reported the bioethanol concentration of 34.89, 32.46, and 30.14 g/L produced from *Phragmites australis*, *Thalia dealbata*, and *Juncus effusus* L, respectively, when used as vegetation in CW. Similarly, Gizińska-Górna et al. [62] also conducted a study using four macrophyte species, namely *P. australis*, *Salix viminalis*, *Helianthus tuberosus*, and *Miscanthus giganteus* in CW to produce biomethane. In this investigation, methane with the yield of 2888, 1309, 1151, and 1532 m^3/ha was produced from the biomass of *P. australis*, *S. viminalis*, *H. tuberosus*, and *M. giganteus*, respectively, via anaerobic digestion. Besides this, other plants were also used in CW for deriving the biofuel and considerable yield was obtained (Table 8.1).

In addition to the macrophyte species, their richness or diversity in CW also influences biofuel production as biomass production increases with species richness. In this regard, Du et al. [63] examined the bioenergy production from CWs employing four, three, two and single species of macrophyte in a single CW separately and found that CW employing four species exhibited maximum bioenergy production of 19.01 MJ/m^2 year followed by 12.65, 11.65, and 8.65 MJ/m^2 year for three, two, and single species, respectively, which signifies that species richness in CWs influences biofuel production. Besides biofuel, biomass harvested from CW was also used to derive biochar for nitrogen removal in CW and reported removal efficiencies of 85.62% and 81.16% for total nitrogen and NO_3-N, respectively [59]. Prominently, CW is a cost-effective technology that possesses the benefits of CO_2 sequestration and renewable energy production. However, prevalence of anaerobic conditions in the bottom layer of CW supports the growth of methanogenic bacteria, which results in a significant amount of methane generation contributing to global warming [64]. Hence, the minimization of this greenhouse gas (GHG) emission from the CWs needs to be critically looked into to reduce the global warming potential of this innovative technology.

To control the CH_4 emissions from CWs, microbial fuel cell (MFC) has been integrated with CW as both aerobic and anaerobic conditions exist in CW that can be utilized for MFC, which favor the integration of these technologies. Briefly, MFC comprises of biotic anode and abiotic cathode, separated by PEM, thus treating wastewater and simultaneously generating bioenergy. In anodic chamber, which is anaerobic, electrogens oxidize the organic matter present in wastewater, resulting in the release of electrons and protons. The electrons generated are transferred to the cathode via an external circuit to generate bioelectricity, and protons are transferred to cathode via PEM (Fig. 8.5b). At the cathode, which is aerobic, protons and electrons combine with oxygen (oxygen reduction) and form water in the presence of catalysts [65]. Notably, the integration of CW with MFC leads to the competition between methanogens and electrogens for the consumption of organic matter, and subsequently, the entire organic matter is not available for methanogens resulting in less CH_4 generation as compared to CWs [64]. Besides alleviating CH_4 emissions, this integrated CW-MFC technology also generates bioelectricity simultaneously treating wastewater (Fig. 8.5c). Using this approach, Xu et al. [66] reported maximum power density of 3714.08 mW/m^2 employing *P. australis* in CW-MFC with $82.32 \pm 12.85\%$ chemical oxygen demand removal from synthetic wastewater.

Furthermore, bioelectricity generation in CW-MFC is also influenced by macrophyte species as different macrophytes have their morphology and physiology, which affects the diversity of the microbes. In this veneration, a study compared the power generation from CW-MFC employing three different macrophyte species and a higher power generation of 229 mW/m^3 was noted for *Cyperus prolifer*, which was two folds higher than CW-MFC planted with *P. australis* and *Wachendorfia thyrsiflora* [67]. Moreover, other studies also reported the generation of bioelectricity using CW-MFC indicating the suitabality of CW-MFC for bioconversion of CO_2 (Table 8.1). In addition, other factors, such as electrode material, spacing between the electrodes, distance between rhizosphere and anode, and organic loading rate, also govern the performance of CW-MFC, which are needed to be taken into consideration for achieving higher bioresource recovery [68]. Although CW-MFC presents a sustainable solution for abating the CH_4 emissions from CWs and simultaneously generating bioelectricity, this integrated technology is limited to laboratory-scale due to the lower power output, which limits its field-scale application. Since this integrated technology is still in its infancy, further research can be focused on understanding the in-depth mechanism of CW-MFC to maximize the power output. Also, efforts are needed to optimize different parameters, i.e., electrode material, electrode spacing, macrophyte species, electrode and macrophyte positioning, and organic loading rate to increase the power generation for implementation of CW-MFC at the field scale.

8.2.5 Gas Fermentation

Gas fermentation is a biological process in which microbes utilize gases, i.e., carbon monoxide (CO), CO_2, and hydrogen (H_2), as substrates to produce different biocommodities like butanol, ethanol, acetate, butyrate, lactate, and hexanoate. Different waste gas streams from industries, such as steel mills and refineries, synthetic gas or syngas generated from gasification of biomass or solid waste, rich in CO, CO_2, and H_2 can be utilized as feedstock in gas fermentation to derive valuable biocommodities via gas fermentation [69, 70]. Notably, microbes used in gas fermentation are homoacetogens, which employ Wood–Ljungdahl pathway to synthesize acetyl coenzyme A using H_2 as an energy source. Eventually, this acetyl coenzyme A fixes carbon, which results in the formation of different valuable biocommodities [71]. Till now, different homoacetogens, namely *Clostridium ragasdalei*, *Clostridium autoethanogenum*, *Clostridium carboxidivorans*, and *Clostridium ijungdahlii*, were widely explored for the process of gas fermentation to synthesize different biocommodities, such as ethanol, butanol, isopropanol, butyrate, and 3-hydroxybutyrate [72, 73]. These homoacetogens can be employed as pure culture or can be co-cultured to synthesize biocommodities using gas fermentation. In this regard, Mohammadi et al. [74] reported the generation of 6.5 g/L of ethanol employing *C.ljungdahlii*, which consumed syngas as a substrate for gas fermentation.

Another study reported a butanol and hexanol production of 3.5 ± 0.69 mmol/L.day and 2.0 ± 0.46 mmol/L.day, respectively, via gas fermentation employing co-culture of *C. autoethanogenum* and *Clostridium kluyveri* utilizing syngas as substrate [75]. In contrast, mixed culture from various sources, such as animal feces and sewage digester, can also be used for gas fermentation. For instance, Tang et al. [76] used mixed culture from rabbit feces for the fermentation of a mixture CO and CO_2 and reported ethanol concentration of 1.41 g/L [76]. Notably, in gas fermentation, formation of the product is governed by composition of the gaseous substrate. Jack et al. [77] investigated the effect of different compositions of syngas on the yield of product and reported that by increasing the H_2/CO ratio from 0.5 to 2, acetate production increased from 22.64 ± 0.55 to 35.21 ± 0.84 mM employing *C. ljungdahlii* in gas fermentation [77]. Additionally, different factors, such as mass transfer of the substrate from the gaseous phase to the liquid phase, fed gas pressure, pH, and operating temperature, also affect the performance of the gas fermentation process [78]. Undoubtedly, gas fermentation is a potential alternative for transforming waste gases into valuable commodities and mitigating GHG. However, this process has some bottlenecks, which are major hurdles toward scaling up this process. In recent years, companies such as Lanza Tech have made significant progress in upscaling of this process. Their plant in Belgium at Arcelor Mittal steel mill has started production of ethanol recently using the off-gases from the steel mill [79].

Further, waste gases from industries contain inhibitory gases like acetylene, ethylene, hydrogen cyanide, and ammonia, which affect the enzymatic activity of the microbes; hence, results in lower yield of valuables produced through gas fermentation. To circumnavigate this bottleneck, installation of scrubbing system at the point of emission becomes mandatory for removing inhibitory compounds that would enhance the microbial activity for attaining better process efficiency but would also add to the cost of the process. Furthermore, inferior gas–liquid mass transfer of substrate and lower yield are major limitations for scaling up this process. Different types of reactors, such as continuous stirred tank, airlift, and bubble column, have been widely explored for improving gas–liquid mass transfer by imparting more agitation. However, agitation used in these types of reactors causes shear stress on microbes and adds to the cost due to higher energy consumption. To overcome this, other types of reactors, such as trickling bed and hollow fiber membrane-based reactor, have also been used to enhance gas–liquid mass transfer in gas fermentation [80]. Nevertheless, more efforts are still needed for efficient and economical utilization of the gas fermentation process, which can pave the way for its scaling up. Further advancement can be focused on designing an economical reactor in which better gas–liquid mass transfer of substrate can take place. Moreover, genetic modification of microbial strain can also be envisaged for optimizing the metabolic pathway and enhancing microbial tolerance toward toxic gases, so that better yield can be attained.

8.2.6 Enzymatic Method

Enzymatic method for bioconversion of CO_2 has gained significant attention as enzymes can perform complex reactions and results in higher yield of valuables with high specificity. In enzymatic conversion method, enzymes, namely oxidoreductase and lyase, which are extracted from specific organisms, utilize CO_2 as a substrate that results in the formation of valuable biofuels and biochemicals. In particular, oxidoreductase enzymes catalyze electron transfer from one molecule to another, which can be electron acceptors or donors. This transfer of electrons is accomplished by the co-factors, such as nicotinamide adenine dinucleotide phosphate and nicotinamide adenine dinucleotide for better functioning of these enzymes [81]. Different oxidoreductase enzymes, such as formate dehydrogenase, CO_2 reductase, carbon monoxide dehydrogenase, and remodeled nitrogenase, have been used for CO_2 fixation to synthesize formate, carbon monoxide, and methane [82]. On the contrary, lyases enzymes facilitate the formation of valuable biocommodities by catalyzing the cleavage of bonds present in CO_2 and generates a new bond [81].

In this regard, lyases enzymes, namely carbonic hydrase and pyruvate decarboxylase, have been reported for synthesizing bicarbonate and biochemicals including pyruvate [82]. However, instead of using single enzyme; these enzymes can be used in cascade form, where multi-enzymes are employed for synthesizing the specific product. In this regard, Marpani et al. [83] synthesized methanol using multi-enzymes, namely formate dehydrogenase, formaldehyde dehydrogenase, and alcohol dehydrogenase with nicotinamide adenine dinucleotide based co-factor. In addition, other studies also reported the synthesis of different biocommodities using various enzymes (Table 8.1). Enzymatic method for bioconversion of CO_2 is a sustainable method for fixing CO_2 and transforming it into valuable biocommodities. However, this method has limitations, such as instability and sensitivity of enzymes, no self-generation of enzymes, and high cost of co-factors, which restrict its implementation at a larger scale. Notably, research was also focused on integrating enzymatic method with photocatalytic or electrocatalytic methods, regeneration of co-factors, co-factor independent reaction of enzymes, and immobilization of enzymes for better stability, thus eliminating the limitations of this technique [84, 85]. However, further research should be carried out on exploring new methods to create new enzymes with specific properties for efficient bioconversion of CO_2, so that this method can be scaled up at a pragmatic scale.

8.3 Life Cycle Assessment and Techno-Economic Analysis

Life cycle assessment (LCA) is a tool used for evaluating the environmental impacts associated with the process or product throughout its life cycle, from the extraction of raw material to its final disposal. On the contrary, techno-economic analysis (TEA) examines the economic feasibility of the process or product, which facilitates a step toward commercialization of the process or product at a large scale. Both of these investigations are of utmost importance for any process or product as it should be both environmentally friendly and economically viable for its implementation at a large scale. Wu et al. [86] conducted the LCA and TEA for the production of bioethanol and biodiesel using microalgal biomass. In this study, it was found that GHG emissions for producing bioethanol and biodiesel were 0.112 kg CO_2-eq/MJ and 0.039 g CO_2-eq/MJ, respectively. Also, breakeven prices for bioethanol and biodiesel were US$ 2.61/kg and US$ 0.49/kg, respectively, with 29.21% of internal rate of return [86].

Furthermore, Huang et al. [87] compared the environmental suitability of biodiesel production from microalgal biomass using different methods and studied their environmental impacts. In this study, four methods, namely lipid extraction via chloroform and methanol after disruption of cell wall using pulsed electric field treatment (LE-PEF), lipid extraction via hexane after microwave assisted in-situ transesterification (IST-MAT), in-situ transesterification of wet algal biomass employing ethyl acetate (IST-EAC), and lipid extraction using methanol after mild hydrothermal treatment (LE-MHT) were employed to derive biodiesel from algal biomass. The GHG emissions of – 41.12 g $CO_{2\text{-eq}}$/MJ, 4.87 g $CO_{2\text{-eq}}$/MJ, 467.96 g $CO_{2\text{-eq}}$/MJ, and 4044.83 g $CO_{2\text{-eq}}$/MJ were reported for LE-MHT, IST-EAC, IST-MAT, and LE-PEF, respectively. These findings showed that only LE-MHT is environmentally sustainable due to the negative GHG emissions. In addition, eco-toxicity study disclosed that LE-MHT is less toxic, while due to the usage of chloroform, LE-PEF was highly toxic among all methods [87].

Another study conducted by Ramirez and Rainey [88] to produce liquid crude oil via pyrolysis, thermal liquefaction, and gasification of sugarcane bagasse evaluated the unit margin referring to profit or loss per volume of fuel production. For pyrolysis and thermal liquefaction, the unit margin was US $ – 0.21/L and US $ – 0.24/L, respectively. Since revenue for the production of liquid crude oil using the above-mentioned three methods was less than the operating cost, the unit margin was negative. On the other hand, the unit margin for gasification process was US $ – 0.75/L, more than three folds lower than the unit margin for pyrolysis and liquefaction due to the higher operating cost and lower production [88]. Furthermore, MES requires electricity for transforming the CO_2 into valuables that can add to global warming if the electricity is produced from non-renewable sources. In this regard, Okoroafor et al. [89] presented a study, which evaluated the contribution to global warming termed as global warming potential due to electricity consumption for three different processes, namely MES, abiotic electrochemical reduction, and hydrogenation of CO_2 to synthesize formic acid using CO_2. This study reported that the global warming potential for MES was 2120 tons CO_2 equivalent, which was four and ten times lesser than abiotic electrochemical reduction and hydrogenation of CO_2, respectively, indicating that MES is beneficial over the other routes for synthesizing formic acid.

Other study conducted by Christodoulou et al. [90] concluded that MES utilizes more carbon for one year than it produced during the synthesis of methanol, ethanol, and formic acid. Also, TEA of this study showed the economic feasibility for ethanol and formic acid production with synthesis costs of 0.88 £/kg and 0.30 £/kg, respectively; which were lesser than the market price of 1.09 £/kg and 0.38 £/kg, for the above-mentioned chemicals, respectively [90]. Furthermore, GWP and ozone layer depletion of integrated electrochemical constructed wetland (ECW) were reported to be 69.1% and 62.1% lower than CW, respectively. However, use of environmentally unsustainable materials, such as copper and stainless steel for electrode in ECW, results in toxicity to humans [91]. These aforementioned studies presented the environmental impacts and economic feasibility of different methods for the bioconversion of CO_2 into valuables. Nevertheless, more comprehensive LCA and TEA studies are further required for different bioconversion methods of CO_2 to understand the environmental impacts and cost economics, so that the best feasible method with minimum associated environmental impacts can be implemented.

8.4 Future Perspectives

Bioconversion of CO_2 into valuable commodities, such as biofuels and biochemicals, is a sustainable approach to boost bioeconomy and fuel security, especially when natural resources of fossil fuels are on the decline. However, due to the lower yield and efficiency, the majority of these methods are limited to the laboratory scale only and further interventions are needed for their field-scale applications. According to the database of CCU projects created by CO_2 value Europe, of the 250 projects listed currently, 45 are using one of the biological approaches described in this chapter [92]. Optimization of different operating parameters, i.e., pH, temperature, pressure, etc., of the process affecting its performance is required for attaining a higher yield of bioconversions that can increase the viability of these processes at a commercial scale. Another major constraint for the bioconversion of CO_2 is the lower efficiency of microbes that can be resolved by genetic modification as well as in-depth understanding of the microbial metabolism leading to a higher CO_2 conversion efficiency. Furthermore, utilization of biomass to produce biofuels and biochemicals seems to be a sustainable approach due to its abundant availability. However, high cost associated with the treatment and processing of the biomass remains a challenge for its upscaling.

Notably harvesting cost of algal or cyanobacterial biomass accounts for more than 30% of total production cost, which needs to be minimized by exploring cost-effective and energy-efficient harvesting methods for profitable and sustainable production of third-generation biofuels or biochemicals via these photosynthetic microbes [93]. Similarly, the recalcitrant nature of the LCB due to the presence of lignin adds to the pretreatment cost that needs to be minimized. Genetic modification or controlling the lignin synthesis pathway of trees or plants can be implemented for reducing the pretreatment cost for removal of lignin from LCB. Furthermore, the high moisture content in the CW biomass adding to the cost of pyrolysis can be overcome by utilizing the heat from local waste for pyrolysis, which can be an economical and sustainable way to reduce cost. In addition, the high capital and operation cost of bioelectrochemical processes, namely MCC and MES, for valorization of CO_2 is a limitation for their field-scale applicability. This can be tackled by introducing low-cost or waste-derived material for electrodes and PEMs, which will curtail the cost to a certain extend. Also, use of renewable energy for fulfilling the energy demand can be a pathbreaker to reduce the cost.

Furthermore, lower power output from CW-MFC can be increased by optimizing the various parameters, such as electrode material, inter-electrode spacing, and culturing efficient macrophyte species, which significantly affects the performance of CW-MFC. Also, to enhance the power output from CW-MFC, the interaction of the electrogens with the anode surface is needed to be understood. Other than this, enzymatic-based methods for the conversion of CO_2 also requires efforts to eliminate the limitation of this method for its field-scale application. For this, improvement in the stability and reusability of the enzymes should be targeted to increase the sustainability of this method. Also, co-factor independent enzymes need to be explored to reduce the cost, so that this method can be implemented at a pragmatic scale [94]. Similarly, the generation of biofuel via gas fermentation is also a sustainable method for the valorization of CO_2 and mitigating GHG. However, lower yield and mass transfer of this process remain major hurdles for the commercialization of this process. These limitations can be eliminated by gene modification of microbe or enhancing the mass transfer by using conductive materials, such as activated carbon and biochar for attaining the higher yield required for its commercial viability [78].

8.5 Conclusion

Carbon dioxide is abundantly available that can be effectively utilized as the raw material for synthesizing biocommodities, such as biofuels and biochemicals. Biological methods for transforming CO_2 into valuables are a sustainable alternative than conventional methods, such as chemical and electrochemical, which demand toxic chemicals and energy-intensive operating conditions. Different biological routes or technology, such as the application of photosynthetic microbes, CW, MES, MCC, gas fermentation, and enzymatic method, possess the potential of biotransformation of CO_2 into different biocommodities, such as bioethanol, biomethane, bioplastic, and biohydrogen. Moreover, CW-MFC is an innovative approach for sequestering CO_2 and simultaneously generating bioelectricity that can fulfill the power requirement of low energy-consuming devices. Although these methods are futuristic solutions for switching to bioeconomy, the lower yield of these technologies or routes is still a major bottleneck, which restricts their commercialization. Hence, further research efforts are needed to overcome the techno-economic barriers of these methods to tap the full potential of these innovative technologies, so that these can be scaled up at the industry level and can compete with the existing methods for the valorization of CO_2.

References

1. Kamkeng ADN, Wang M, Hu J, Du W, Qian F (2021) Transformation technologies for CO_2 utilisation: current status, challenges and future prospects. Chem Eng J 409:128138
2. Okoye-Chine CG et al (2022) Conversion of carbon dioxide into fuels—a review. J CO_2 Util 62:102099
3. Gupta R, Mishra A, Thirupathaiah Y, Chandel AK (2022) Biochemical conversion of CO_2 in fuels and chemicals: status, innovation, and industrial aspects. Biomass Convers Biorefinery
4. Salehizadeh H, Yan N, Farnood R (2020) Recent advances in microbial CO_2 fixation and conversion to value-added products. Chem Eng J 390:124584
5. Bhatia SK, Bhatia RK, Jeon J-M, Kumar G, Yang Y-H (2019) Carbon dioxide capture and bioenergy production using biological system—a review. Renew Sustain Energy Rev 110:143–158
6. Bisbis MB, Gruda N, Blanke M (2018) Potential impacts of climate change on vegetable production and product quality—a review. J Clean Prod 170:1602–1620
7. Hamim (2005) Photosynthesis of C3 and C4 species in response to increased CO_2 concentration and drought stress. HAYATI J Biosci 12:131–131
8. Devi A, Singh A, Bajar S, Pant D, Din ZU (2021) Ethanol from lignocellulosic biomass: an in-depth analysis of pre-treatment methods, fermentation approaches and detoxification processes. J Environ Chem Eng 9:105798
9. Okolie J, Nanda S, Dalai A, Kozinski J (2021) Chemistry and specialty industrial applications of lignocellulosic biomass. Waste Biomass Valorization 12:2145–2169
10. Mariano APB, Unpaprom Y, Ponnusamy VK, Ramaraj R (2021) Bioethanol production from coconut pulp residue using hydrothermal and postalkaline pretreatment. Int J Energy Res 45:8140–8150

11. Pratto B, Chandgude V, de Sousa Junior R, Cruz AJG, Bankar S (2020) Biobutanol production from sugarcane straw: defining optimal biomass loading for improved ABE fermentation. Ind Crops Prod 148:112265
12. Bhatia SK et al (2019) Bioconversion of plant biomass hydrolysate into bioplastic (polyhydroxyalkanoates) using *Ralstonia eutropha* 5119. Bioresour Technol 271:306–315
13. Gonzales RR, Kumar G, Sivagurunathan P, Kim S-H (2017) Enhancement of hydrogen production by optimization of pH adjustment and separation conditions following dilute acid pretreatment of lignocellulosic biomass. Int J Hydrog Energy 42:27502–27511
14. Mehariya S et al (2023) Biopolymer production from biomass produced by Nordic microalgae grown in wastewater. Bioresour Technol 376:128901
15. Umamaheswari J, Kavitha M, Shanthakumar S (2020) Outdoor cultivation of *Chlorella pyrenoidosa* in paddy-soaked wastewater and a feasibility study on biodiesel production from wet algal biomass through in-situ transesterification. Biomass Bioenerg 143:105853
16. Tsolcha ON et al (2021) Utilization of biomass derived from cyanobacteria-based agro-industrial wastewater treatment and raisin residue extract for bioethanol production. Water 13:486
17. Neethu B, Tholia V, Ghangrekar M (2020) Optimizing performance of a microbial carbon-capture cell using Box-Behnken design. Process Biochem 95:99–107
18. Varanasi JL, Prasad S, Singh H, Das D (2020) Improvement of bioelectricity generation and microalgal productivity with concomitant wastewater treatment in flat-plate microbial carbon capture cell. Fuel 263:116696
19. Li S et al (2022) Solid neutral red/Nafion conductive layer on carbon felt electrode enhances acetate production from CO_2 and energy efficiency in microbial electrosynthesis system. Bioresour Technol 363:127983
20. Kiran R, Yadav R, Sathe D, Patil SA (2023) Halophilic CO_2-fixing microbial community as biocatalyst improves the energy efficiency of the microbial electrosynthesis process. Bioresour Technol 371:128637
21. Tahir K et al (2022) Development of a three-dimensional macroporous sponge biocathode coated with carbon nanotube–MXene composite for high-performance microbial electrosynthesis systems. Bioelectrochemistry 146:108140
22. Nevin KP, Woodard TL, Franks AE, Summers ZM, Lovley DR (2010) Microbial electrosynthesis: feeding microbes electricity to convert carbon dioxide and water to multicarbon extracellular organic compounds. MBio 1:e00103-00110
23. Liu D, Zou C, Xu M (2019) Environmental, ecological, and economic benefits of biofuel production using a constructed wetland: a case study in China. Int J Environ Res Public Health 16:827
24. Lin Y et al (2020) The potential of constructed wetland plants for bioethanol production. BioEnergy Res 13:43–49
25. Zhang K et al (2021) Effects of plant location on methane emission, bioelectricity generation, pollutant removal and related biological processes in microbial fuel cell constructed wetland. J Water Process Eng 43:102283
26. Wang Y, Zhang X, Yu G, Yao Y, Lin H (2023) Effects of flow pattern, *Leersia hexandra*, and circuit mode on the Cr (VI) removal capacity, electricity generation performance, and microbial community of constructed wetland-microbial fuel cells. Fuel 338:127326
27. Katakojwala R, Tharak A, Sarkar O, Mohan SV (2022) Design and evaluation of gas fermentation systems for CO_2 reduction to C2 and C4 fatty acids: non-genetic metabolic regulation with pressure, pH and reaction time. Bioresour Technol 351:126937
28. Tharak A, Katakojwala R, Kajla S, Mohan SV (2023) Chemolithoautotrophic reduction of CO_2 to acetic acid in gas and gas-electro fermentation systems: enrichment, microbial dynamics, and sustainability assessment. Chem Eng J 454:140200
29. Zhang Z et al (2022) Novel artificial ionic cofactors for efficient electro-enzymatic conversion of CO_2 to formic acid. J CO_2 Util 60:101978
30. Parmar A, Singh NK, Pandey A, Gnansounou E, Madamwar D (2011) Cyanobacteria and microalgae: a positive prospect for biofuels. Bioresour Technol 102:10163–10172

31. Mathimani T, Pugazhendhi A (2019) Utilization of algae for biofuel, bio-products and bioremediation. Biocatal Agric Biotechnol 17:326–330
32. Nagappan S, Devendran S, Tsai P-C, Dahms H-U, Ponnusamy VK (2019) Potential of two-stage cultivation in microalgae biofuel production. Fuel 252:339–349
33. Shin YS et al (2018) Multilateral approach on enhancing economic viability of lipid production from microalgae: a review. Bioresour Technol 258:335–344
34. Trivedi N, Gupta V, Reddy C, Jha B (2013) Enzymatic hydrolysis and production of bioethanol from common macrophytic green alga *Ulva fasciata* Delile. Bioresour Technol 150:106–112
35. Sadvakasova AK et al (2020) Bioprocesses of hydrogen production by cyanobacteria cells and possible ways to increase their productivity. Renew Sustain Energy Rev 133:110054
36. Kumar M et al (2020) Algae as potential feedstock for the production of biofuels and value-added products: opportunities and challenges. Sci Total Environ 716:137116
37. Mathimani T et al (2019) Review on cultivation and thermochemical conversion of microalgae to fuels and chemicals: process evaluation and knowledge gaps. J Clean Prod 208:1053–1064
38. Devadas VV et al (2021) Algae biopolymer towards sustainable circular economy. Bioresour Technol 325:124702
39. Singh AK, Sharma L, Mallick N, Mala J (2017) Progress and challenges in producing polyhydroxyalkanoate biopolymers from cyanobacteria. J Appl Phycol 29:1213–1232
40. Das S, Das S, Das I, Ghangrekar M (2019) Application of bioelectrochemical systems for carbon dioxide sequestration and concomitant valuable recovery: a review. Mater Sci Energy Technol 2:687–696
41. Zhou M, He H, Jin T, Wang H (2012) Power generation enhancement in novel microbial carbon capture cells with immobilized *Chlorella vulgaris*. J Power Sources 214:216–219
42. Pandit S, Nayak BK, Das D (2012) Microbial carbon capture cell using cyanobacteria for simultaneous power generation, carbon dioxide sequestration and wastewater treatment. Bioresour Technol 107:97–102
43. Saravanan A, Deivayanai VC, Senthil Kumar P, Rangasamy G, Varjani S (2022) CO_2 bio-mitigation using genetically modified algae and biofuel production towards a carbon net-zero society. Bioresour Technol 363:127982
44. Das S, Diels L, Pant D, Patil SA, Ghangrekar MM (2020) Review—microbial electrosynthesis: a way towards the production of electro-commodities through carbon sequestration with microbes as biocatalysts. J Electrochem Soc 167:155510
45. Bian B, Bajracharya S, Xu J, Pant D, Saikaly PE (2020) Microbial electrosynthesis from CO_2: challenges, opportunities and perspectives in the context of circular bioeconomy. Bioresour Technol 302:122863
46. Mohanakrishna G, Vanbroekhoven K, Pant D (2018) Impact of dissolved carbon dioxide concentration on the process parameters during its conversion to acetate through microbial electrosynthesis. React Chem Eng 3:371–378
47. Sharma M et al (2019) Electrode material properties for designing effective microbial electrosynthesis systems. J Mater Chem A 7:24420–24436
48. Zhang X, Li X, Zhao X, Li Y (2019) Factors affecting the efficiency of a bioelectrochemical system: a review. RSC Adv 9:19748–19761
49. Das S, Das I, Ghangrekar M (2020) Role of applied potential on microbial electrosynthesis of organic compounds through carbon dioxide sequestration. J Environ Chem Eng 8:104028
50. Enzmann F, Holtmann D (2019) Rational scale-up of a methane producing bioelectrochemical reactor to 50 L pilot scale. Chem Eng Sci 207:1148–1158
51. Yu L, Yuan Y, Tang J, Zhou S (2017) Thermophilic moorella thermoautotrophica-immobilized cathode enhanced microbial electrosynthesis of acetate and formate from CO_2. Bioelectrochemistry 117:23–28
52. LaBelle EV, May HD (2017) Energy efficiency and productivity enhancement of microbial electrosynthesis of acetate. Front Microbiol 8:756
53. Arends JB, Patil SA, Roume H, Rabaey K (2017) Continuous long-term electricity-driven bioproduction of carboxylates and isopropanol from CO_2 with a mixed microbial community. J CO_2 Util 20:141–149

54. Das S, Chakraborty I, Das S, Ghangrekar MM (2021) Application of novel modular reactor for microbial electrosynthesis employing imposed potential with concomitant separation of acetic acid. Sustain Energy Technol Assess 43:100902
55. Pahunang RR et al (2021) Advances in technological control of greenhouse gas emissions from wastewater in the context of circular economy. Sci Total Environ 792:148479
56. Rhee JS, Iamchaturapatr J (2009) Carbon capture and sequestration by a treatment wetland. Ecol Eng 35:393–401
57. Anna Kajumulo Tibaijuka AD (2008) Constructed wetlands manual 1-88. UN-Habitat, Kathmandu, Nepal
58. Liu D et al (2012) Constructed wetlands as biofuel production systems. Nat Clim Change 2:190–194
59. Li J, Fan J, Zhang J, Hu Z, Liang S (2018) Preparation and evaluation of wetland plant-based biochar for nitrogen removal enhancement in surface flow constructed wetlands. Environ Sci Pollut Res 25:13929–13937
60. Fan X et al (2021) Increasing plant diversity to mitigate net greenhouse effect of wastewater treatment in floating constructed wetlands. J Clean Prod 314:127955
61. Pinho HJO, Mateus DMR (2023) Bioenergy routes for valorizing constructed wetland vegetation: an overview. Ecol Eng 187:106867
62. Gizińska-Górna M et al (2016) The possibility of using plants from hybrid constructed wetland wastewater treatment plant for energy purposes. Ecol Eng 95:534–541
63. Du Y et al (2018) Plant diversity decreases net global warming potential integrating multiple functions in microcosms of constructed wetlands. J Clean Prod 184:718–726
64. Zhang K et al (2021) Effects of the bioelectrochemical technique on methane emission and energy recovery in constructed wetlands (CWs) and related biological mechanisms. Environ Technol 44:540–551
65. Das S, Raj R, Das S, Ghangrekar MM (2023) Evaluating application of photosynthetic microbial fuel cell to exhibit efficient carbon sequestration with concomitant value-added product recovery from wastewater: a review. Environ Sci Pollut Res 30:98995–99012
66. Xu F et al (2018) Electricity production and evolution of microbial community in the constructed wetland-microbial fuel cell. Chem Eng J 339:479–486
67. Kataki S et al (2021) Constructed wetland, an eco-technology for wastewater treatment: a review on various aspects of microbial fuel cell integration, low temperature strategies and life cycle impact of the technology. Renew Sustain Energy Rev 148:111261
68. Gupta S, Srivastava P, Patil SA, Yadav AK (2021) A comprehensive review on emerging constructed wetland coupled microbial fuel cell technology: potential applications and challenges. Bioresour Technol 320:124376
69. Bengelsdorf FR, Durre P (2017) Gas fermentation for commodity chemicals and fuels. Microb Biotechnol 10:1167–1170
70. Liew FE et al (2022) Carbon-negative production of acetone and isopropanol by gas fermentation at industrial pilot scale. Nat Biotechnol 40:335–344
71. Karekar S, Stefanini R, Ahring B (2022) Homo-acetogens: their metabolism and competitive relationship with hydrogenotrophic methanogens. Microorganisms 10:397
72. Bengelsdorf FR, Dürre P (2017) Gas fermentation for commodity chemicals and fuels. Microb Biotechnol 10:1167–1170
73. Marcellin E et al (2016) Low carbon fuels and commodity chemicals from waste gases—systematic approach to understand energy metabolism in a model acetogen. Green Chem 18:3020–3028
74. Mohammadi M, Younesi H, Najafpour G, Mohamed AR (2012) Sustainable ethanol fermentation from synthesis gas by *Clostridium ljungdahlii* in a continuous stirred tank bioreactor. J Chem Technol Biotechnol 87:837–843
75. Diender M, Stams AJ, Sousa DZ (2016) Production of medium-chain fatty acids and higher alcohols by a synthetic co-culture grown on carbon monoxide or syngas. Biotechnol biofuels 9:1–11

76. Tang Y et al (2021) Ethanol production from gas fermentation: Rapid enrichment and domestication of bacterial community with continuous CO/CO_2 gas. Renew Energy 175:337–344
77. Jack J, Lo J, Maness P-C, Ren ZJ (2019) Directing *Clostridium ljungdahlii* fermentation products via hydrogen to carbon monoxide ratio in syngas. Biomass Bioenerg 124:95–101
78. He Y, Kennes C, Lens PN (2022) Enhanced solventogenesis in syngas bioconversion: role of process parameters and thermodynamics. Chemosphere 299:134425
79. Kit McDonnell FB, El-Sharkawy O (2023) ArcelorMittal and LanzaTech announce first ethanol samples from commercial flagship carbon capture and utilisation facility in Ghent, Belgium. https://corporate.arcelormittal.com/media/news-articles/arcelormittal-and-lanzatech-announce-first-ethanol-samples-from-commercial-flagship-carbon-capture-and-utilisation-facility-in-ghent-belgium
80. Shen Y, Brown R, Wen Z (2014) Syngas fermentation of Clostridium carboxidivoran P7 in a hollow fiber membrane biofilm reactor: evaluating the mass transfer coefficient and ethanol production performance. Biochem Eng J 85:21–29
81. Shi J et al (2015) Enzymatic conversion of carbon dioxide. Chem Soc Rev 44:5981–6000
82. Unlu A, Duman-Ozdamar ZE, Caloglu B, Binay B (2021) Enzymes for efficient CO_2 conversion. Protein J 40:489–503
83. Marpani F, Pinelo M, Meyer AS (2017) Enzymatic conversion of CO_2 to CH_3OH via reverse dehydrogenase cascade biocatalysis: quantitative comparison of efficiencies of immobilized enzyme systems. Biochem Eng J 127:217–228
84. Liao Q, Liu W, Meng Z (2022) Strategies for overcoming the limitations of enzymatic carbon dioxide reduction. Biotechnol Adv 60:108024
85. Yu Y et al (2012) Enzymatic conversion of CO_2 to bicarbonate in functionalized mesoporous silica. Microporous Mesoporous Mater 153:166–170
86. Wu W, Lin K-H, Chang J-S (2018) Economic and life-cycle greenhouse gas optimization of microalgae-to-biofuels chains. Bioresour Technol 267:550–559
87. Huang R et al (2022) Comparative life-cycle assessment of microalgal biodiesel production via various emerging wet scenarios: energy conversion characteristics and environmental impacts. Energy Convers Manage 257:115427
88. Ramirez JA, Rainey TJ (2019) Comparative techno-economic analysis of biofuel production through gasification, thermal liquefaction and pyrolysis of sugarcane bagasse. J Clean Prod 229:513–527
89. Okoroafor T, Haile S, Velasquez-Orta S (2021) Life cycle assessment of microbial electrosynthesis for commercial product generation. J Hazard Toxic Radioact Waste 25:04020062
90. Christodoulou X, Okoroafor T, Parry S, Velasquez-Orta SB (2017) The use of carbon dioxide in microbial electrosynthesis: advancements, sustainability and economic feasibility. J CO_2 Util 18:390–399
91. Fang Y-K et al (2023) Life cycle assessment of integrated bioelectrochemical-constructed wetland system: environmental sustainability and economic feasibility evaluation. Resour Conserv Recycl 189:106740
92. CO2ValueEurope (2023) CCU projects. https://database.co2value.eu/
93. Ogbonna CN, Nwoba EG (2021) Bio-based flocculants for sustainable harvesting of microalgae for biofuel production. A review. Renew Sustain Energy Rev 139:110690
94. Long NVD, Lee J, Koo K-K, Luis P, Lee M (2017) Recent progress and novel applications in enzymatic conversion of carbon dioxide. Energies 10:473

Chapter 9
CO_2-Sourced Polymers: Synthesis, Property, Application

Hui Li, Fengyu Zhao, and Haiyang Cheng

Abstract The polymeric materials could be synthesized directly from CO_2 as a monomer, which are named as CO_2-sourced polymers, mainly including polyurethane, polyurea, polycarbonate, and polyester. The CO_2-sourced polymers are of some advantages compared to their conventional counterparts, such as environmental benignity without using toxic reagents, carbon reduction, degradability, and reproducibility. This chapter will review the advancements achieved in the CO_2-sourced polymer synthesized directly and indirectly by using CO_2 as a monomer over the past five years (from 2019 to present). Herein, we pay main attention to the synthetic methodologies, the structure, properties, and possible applications of several kinds of important CO_2-sourced polymers such as polyurethanes, polyureas, polycarbonates, and polyesters.

9.1 Introduction

Polymeric materials such as polyurethanes (PUs), polyureas (PUas), polycarbonates (PCs), and polyesters (PEs) have broad applications in the fields of coatings, biomedicine, membranes, adhesives, packaging materials, and 3-D printing [1–3]. However, the polymer industry is highly reliant on petrochemicals. Besides, the industrial production of PUs and PCs usually involves highly toxic reagents such as

H. Li · F. Zhao (✉) · H. Cheng
State Key Laboratory of Electroanalytical Chemistry, Changchun Institute of Applied Chemistry, Chinese Academy of Sciences, Changchun 130022, China
e-mail: zhaofy@ciac.ac.cn

H. Cheng
e-mail: hycyl@ciac.ac.cn

H. Li · F. Zhao
School of Applied Chemistry and Engineering, University of Science and Technology of China, Hefei 230026, China

H. Li · F. Zhao · H. Cheng
Jilin Province Key Laboratory of Green Chemistry and Process, Changchun Institute of Applied Chemistry, Chinese Academy of Sciences, Changchun 130022, China

G. Zhang et al. (eds.), *Advances in CO_2 Utilization*, Green Chemistry and Sustainable Technology, https://doi.org/10.1007/978-981-99-8822-8_9

isocyanate, phosgene, and bisphenol A (BPA) [4, 5], posing a severe threat to the environment and human health. Consequently, the quest for more sustainable and renewable raw materials in the production of polymers has been drawing increasing attention in recent years. To our relief, a lot of qualified and promising resources that originated either from the nature or from the industry have already been incorporated in the formulation of polymers, including CO_2, vegetable oil, carbohydrate, fatty acid, and crude glycerol, as concluded in the previous reviews [6, 7]. Among all the resources mentioned above, CO_2 stands out as a recyclable, low cost, green, and simple C1 source. The upcycling of CO_2 into polymers can not only reduce greenhouse gas emission but also turn waste into treasure. Thus, using CO_2 to prepare polymeric materials is a prospective and feasible route. The simultaneous utilization of CO_2 with bio-derived raw materials together can even afford 100% renewable polymers which are able to biodegrade or hydrolyze.

The CO_2-derived polymers do not always exhibit comparable properties to their conventional counterparts, which limits their commercialization and application [8]. The polymerization that results in well-defined block polymers is expected to be a powerful toolbox to make up for the lack of performance of CO_2-sourced polymers [9]. Besides, introducing more economic and readily catalyst, broadening the scope of novel monomers, as well as understanding the relationship of structure–performance are also important. As the CO_2-sourced polymers have been comprehensively reviewed in 2019 by Christophe et al. [10], this chapter mainly summarizes the latest important advancements which have been achieved in CO_2-sourced PU, PUa, PC, and PE since 2019. The synthetic method, structural characteristics, properties, and applications of CO_2-sourced PU and PUa are described. Specially, the relationship of structure and property is thoroughly discussed, aiming to provide a deep understanding of the relation between the monomer structure and the mechanical and thermal properties of the final polymers. In the end, the prospectives and challenges are proposed.

9.2 CO_2-Sourced Polyurethanes (PUs)

PU, synthesized conventionally via the step-growth polymerization of diisocyanates with bi- or multi-functional polyols, is one kind of the most important and versatile polymeric materials. Low molecular weight diol or diamine is often used as chain extender in order to produce high molecular weight PU. Because of their profound physicochemical and mechanical properties, flexible and rigid PUs have been widely applied in the fabrication of coatings, adhesives, elastomers, and rigid foams [2]. However, the development of PUs has been hindered by the largely petroleum-derived isocyanate/alcohol monomers and the hazardous isocyanates or phosgene used in the conventional synthetic methods. CO_2 is regarded as a satisfying alternative to fossil fuels, the preparation of PUs using CO_2 directly or indirectly is a green and sustainable method (Scheme 1). The versatility of CO_2-sourced PUs is demonstrated by the prosperous advancements of not only their wide applications

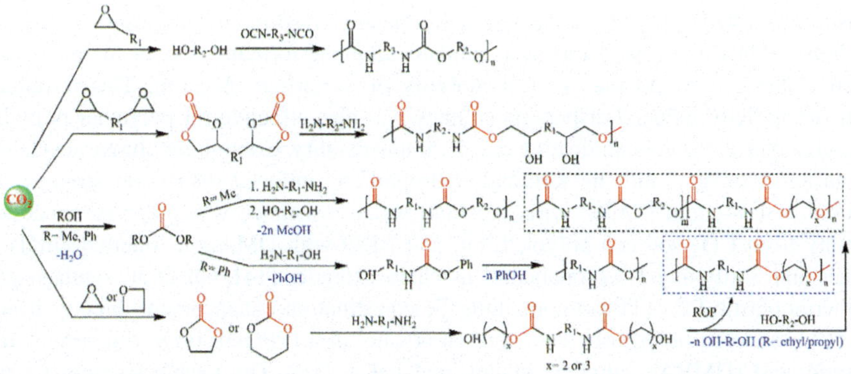

Scheme 1 Synthetic routes for CO_2-sourced PUs

in the coatings, adhesives, hydrogels, foams, and insulation materials, but also their recyclable properties as when there are dynamic moieties in their molecular chain. The development of CO_2-sourced PUs and PU hybrids are summarized according to the synthetic methods.

9.2.1 *CO_2-Sourced Polyol as Building Block*

For the CO_2-sourced PUs derived from CO_2-sourced polyols, the replacement of petroleum resources-based polyols by CO_2-sourced polyols still matters despite the use of isocyanate in the synthesis, owning to the sustainability and greenness of this pathway. Generally, CO_2-sourced polyols are essentially hydroxyl telechelic poly(ether carbonate) polyols with low molecular weights based on epoxide and CO_2. These CO_2-sourced polyols combine the advantages of polycarbonate polyol and polyether polyol, their polyaddition with diisocyanates usually produce PUs with good performances such as high mechanical properties and high hardness due to the coexistence of carbonate and ether units in the main polymeric chain [11, 12]. The properties of PUs can be well regulated by tuning the content of carbonate units in CO_2-sourced polyols.

Wang et al. reported that CO_2-polyols could be utilized in the preparation of waterborne PU coatings with the assistance of an internal emulsifier [11–13]. A good emulsification performance of the internal emulsifier was essential to prepare cationic waterborne polyurethane (CWPU) with improved water resistance and mechanical properties. They found that the emulsifier's incorporation position in the CWPU chain (backbone, side chain, or terminal of the prepolymer chain) had an impact on the emulsification efficiency, which was attributed to the difference in the steric hindrance and mobility of the tertiary *N* group. The PU prepolymer was formed by reacting CO_2-sourced polyol, isophorone diisocyanate (IPDI), and 1,4-butanediol (BDO). Then, the prepolymer reacted with the internal emulsifier and acetic acid to form CWPU [13]. Terminal internal emulsifier 3-(dimethylamino)propane-1,2-diol (t-DMAD) showed the best emulsification performance owing to the least steric hindrance and the greatest mobility of its tertiary *N* atom. This terminal hydrophilicity-induced dispersion strategy was also effective for polyether polyols and polyester polyols, indicating excellent universality. Combining ultraviolet (UV) curing chemistry and the terminal hydrophilicity-induced dispersion strategy, a kind of UV-cured CWPU with excellent water resistance was prepared successfully from CO_2-sourced polyol, IPDI, and BDO with DMAD as internal emulsifier and 3-allyloxy-1,2-propanediol as chain extender [14]. Gu et al. synthesized thermoplastic PUs (TPUs) containing Fe-coordination enhancement directly from a CO_2-sourced poly(propylene ether carbonate) diol (PPCD), IPDI, dimethylolbutyric acid (DMBA), ethylene glycol, and $FeCl_3$ [12]. The tensile strength (σ_m), Young's modulus (E), and the glass transition temperature (T_g) of TPU increased while its thermal stability and elongation at break (ε_b) decreased with the increase of $FeCl_3$ content, which was attributed to the increased density of cross-links. Liu and co-workers conducted the polymerization of a PPCD-derived PU prepolymer and slow-reacting polyaspartic ester (PAE) with 4,4′-dicylocyhexylmethane diisocyanate as chain extender, producing brushable polyurethane-urea (PPCD-PUU) coatings [15]. At an –NCO content of 8% and an –NCO/–NH molar ratio of 1.15, PPCD-PUU exhibited a σ_m of 20.5 MPa, a E of 168.2 MPa, and an ε_b of 352%. Isocyanate-free (poly)urethane polyols with various urethane content and hydroxyl equivalent weights were synthesized by Mannari et al. via cyclic carbonate/amine chemistry [16]. The thermal curing of (poly)urethane polyols by varying amounts of hexa(methoxymethyl) melamine cross-linker (20, 25, 30 wt%) then produced PU coatings. Excellent dry adhesion was observed for all coatings; however, urethane polyols-based coatings lost their adhesion while those polyurethane polyols-based coatings showed better adhesion capacity after being immersed in deionized water for one day. A good balance between cross-linker level and urethane/OH content of the cured films is necessary for obtaining properties which are superior to their commercial polyols-based counterparts (i.e., adhesion, impact resistance, and flexibility).

9.2.2 *CO_2-Sourced (A)cyclic Carbonate as a Building Block*

Dialkyl carbamates generated from dimethyl carbonate (DMC) or diphenyl carbonate (DPC), together with dihydroxyethyl dicarbamates produced from ethylene carbonate (EC), are important intermediates in the synthesis of PU from (a)cyclic carbonates. Generally, in the first step, dicarbamates are synthesized via the reaction of (a)cyclic carbonate with diamine. In the second step, the dicarbamates undergo either self-polycondensation or transurethane polycondensation with diols to furnish the final PUs.

Schimpf et al. [17] synthesized methacrylated glycerol carbonate (GCMA) via the carbonation of glycidyl methacrylate (GMA). The aminolysis of GCMA by diamines afforded a series of liquid hydroxyurethane methacrylate (HUMA) oligomers containing 3.2 ~ 16.9 wt% CO_2. These HUMA oligomers were later incorporated in the formulation of UV-curable acrylic resins, among which the cured acrylic resin from 1,8-diamino-3,6-dioxaoctane had a E of 3600 MPa and a σ_m of 85 MPa, and these performances are much better than that of the commercial Laromer UA 9089. The esterification of HUMA by methacrylic acid anhydride resulted in a significant increment of both the E and T_g of the final cured resin due to an increased acrylate functionality. The HUMA-based acrylic resins were potential 3-D printing materials. Chen et al. prepared a series of 3-D printable PUs from trimethylolpropane allyl ether-cyclic carbonate (TMPMEC), cadaverine, and multi-thiol [18]. TMPMEC was synthesized from trimethylolpropane monoallyl ether and DMC. Firstly, TMPMEC was ring-opened by cadaverine to give the diallyl diurethane prepolymer (DEDU). Then, DEDU cross-linked with multi-thiol to produce the final PUs. Functional PUs including foams, coating, and elastomers could also be prepared from CO_2-sourced (a)cyclic carbonates. PU rigid foams derived from bio-based alcohols, DMC, and hexamethylene diamine (HMDA) were fabricated with the following two steps: first, synthesize the PU resins and then transform the PU resins to PU foams [19, 20]. The glucose-based PU resin was transformed into a self-blowing foam with maleic acid as an initiator and glutaraldehyde as a cross-linker [19]. As for the tannin-based PU resin, a mixture of citric acid and glutaraldehyde was used as blowing agent [20]. The morphology and properties of the prepared foams were affected by the amount of cross-linker and foaming reagents. Nasreddine and co-workers prepared photo-cross-linked PU-acrylate coatings with a prepolymer strategy [21]. NIPU oligomers with acrylate chain ends were synthesized via two steps: the transurethane polycondensation of poly(tetramethylene oxide) (PTMO) and dimethyl dicarbamates originated from hexamethylenediamine, 4,4′-methylenebis (cyclohexylamine), priamine 1074 with or without a diol chain extender, and the subsequent end-functionalization of NIPU oligomers by acryloyl chloride. The acrylate chain ends of NIPU oligomers were then used to prepare NIPUA coatings. These coatings showed a $T_{5\%} > 155$ °C, a σ_m up to 11.8 MPa, an E up to 25.5 MPa, and an ε_b in the range of 20% to 520%. In addition, poly(carbonate-urethane)s (NIPCUs) were prepared from the melt polycondensation of bis(hydroxyalkyl carbamate) (BHEC) with oligo(decamethylene carbonate) diol [22], in which BHEC was obtained via the ring opening of EC by Priamine 1075. NIPCUs showed exceptional mechanical properties where the sample with a molecular weight (M_n) of 43,100 g/mol showed a σ_m of 20 MPa and an ε_b of 1800%, and such exceptional mechanical properties were ascribed to the strong H-bonding interactions among urethane and carbonate moieties. Later, NIPCU elastomers with a M_n ranging from 10 to 34 kg/mol were produced by the polycondensation of bis(hydroxyalkyl carbamate)s (BHACs) with oligocarbonate diols [23]. BHACs synthesized from EC and α,ω-diamine served as the hard segment while oligocarbonate diols served as the soft segment. NIPCUs displayed comparable mechanical properties to conventional isocyanate-based PUs: a σ_m of 32 MPa and an ε_b of 800%. The PU elastomer (TPCUE) with a $M_n > 30$ kg/mol reported by Yang et al. was synthesized from the melt polycondensation of 1,6-bis(hydroxyethyloxycarbonylamino)hexane (BHCH) and polycarbonate diols

(PCDLs) [24]. BHCH was obtained from the reaction of EC and HMDA while PCDLs were synthesized from DMC and 1,6-hexanediol via a two-step method. TPCUEs exhibited excellent thermal and mechanical properties with a $T_{5\%} > 240$ °C, a T_g ranging from − 18 to − 1 °C, a melting point (T_m) ranging from 93 to 122 °C, a σ_m ranging between 28.1 and 47.3 MPa, a E ranging between 13.8 and 32.7 MPa, and an $\varepsilon_b > 1000\%$.

9.2.3 CO_2-Sourced Bis-, Tri-, Multi-Functional Cyclic Carbonates as Building Blocks

CO_2-sourced multi-functional cyclic carbonates (CCs), which are easily accessible from the carbonation reaction of epoxy compounds, are one kind of the most important intermediates in the preparation of CO_2-sourced PU materials. The powerful 100% atom economic ring-opening reaction between polyfunctional CCs and amines produces isocyanate-free poly(hydroxyurethane)s (PHUs) with –OH groups in their backbone. Effort in recent years has been put into the development of two sorts of CO_2-sourced PUs: (1) functional CO_2-sourced PUs in the form of coatings, adhesives, hydrogels, and so on, (2) recyclable PU systems with dynamic bonds. Green biomass-derived PUs from bio-derived epoxy/multi-functional CC monomers are also paid more attention. The CO_2-sourced PUs and their hybrids synthesized from CO_2-sourced bis-, tri-, multi-functional CCs will be described according to their functions.

9.2.3.1 Coatings

Mannari et al. prepared non-isocyanate urethane acrylate (NIPU-AC) oligomers from polyurethane polyamines (PUPAs) which were synthesized by the polymerization between bis- or tri-cyclic carbonate and diamine. Four routes have been reported: (1) First, non-isocyanate polyurethane polyols (PUPOs) were synthesized by either the ring-opening reaction of bis-CCs with monoethanolamine or the reaction of PUPA with ethylene carbonate (EC). Then, PUPOs were functionalized by methacrylic anhydride (MAAH). (2) NIPU-ACs were synthesized from the reaction of PUPAs and glycidyl methacrylate. (3) NIPU-ACs were synthesized directly from PUPAs and MAAH. (4) NIPU-ACs were synthesized from the Michael addition reaction between PUPAs and hexanediol diacrylate [25]. NIPU-AC oligomers from the direct reaction between PUPA and MAAH were further formulated for UV-curable NIPU aerospace coatings preparation. These coatings contained a lower level of volatiles by comparison to the traditional aerospace coatings and exhibited good comprehensive properties of fine water resistance, good low temperature flexibility, and acceptable chemical resistance.

9.2.3.2 Adhesives

Hybridization methodology can overcome several limitations of the cyclic carbonate/amine chemistry used in the synthesis of PHUs, that is, the low reactivity of CC toward aminolysis, poor solubility, or high prices of more reactive monomers. PHU-siloxane hybrid adhesives were prepared from the polyaddition reaction among trimethylolpropane tris-carbonate (TMPTC), poly(propylene glycol) dicyclic carbonate (PPGdiCC), and m-xylylene diamine (MXDA) with dopamine and amino-propyl trimethoxysilane (APTMS) serving as adhesion and fast curing promoter, respectively [26]. Dopamine and APTMS formed a synergetic effect on the rheological and adhesive properties of PHUs: Dopamine increased the adhesion strength of PHUs by increasing the lap shear adhesion value while APTMS endowed the adhesives with fast curing properties, better cohesive strength, and thermal resistance. Afterward, PPGdiCC was incorporated as a soft segment in the formulation of adhesives in order to obtain PHU with optimal properties. As the TMPTC content increased from 30 to 70 mol%, the lap shear strength of PHU increased from 3.1 to 19.2 MPa. PHUs with a TMPTC content > 50 mol% could preserve their adhesive properties even at high temperature ($T > 200$ °C). Zhang et al. took the advantage of hyper-branched, rigid, and bio-based CC (Ec-MTDA) with four CC groups, which was synthesized from 1,8-menthane diamine, to develop high-performance NIPUs (ECMP) with Priamine 1074 as curing agent [3]. Upon decreasing the molar ratio of amino/CC from 1.4 to 1.0, the σ_m, E, and T_g of ECMP increased from 3.77 to 34.9 MPa, 6.06 to 462.29 MPa, and 52.71 to 81.31 °C, respectively, as a result of increased hyper-branched and rigid six-membered ring structures. The hyper-branched structure and rigidity of Ec-MTDA could limit the movement of polymeric chains and enhance the thermal stability. As a result, ECMP-1.0 exhibited the poorest thermal stability because it owned the lowest proportion of Ec-MTDA. Besides, ECMP-1.4 showed the highest catalyst-free relaxation rate (5.16 min at 120 °C) in the stress relaxation test because of the decrease of the cross-linking density and the increase of the flexibility since Priamine 1074 accelerated the dynamic transesterification and relaxation of polymer chains. In addition, ECMP-1.0 displayed satisfying shape memory, self-healing, and chemical reprocessing properties due to the transesterification reaction between the carbamate groups and the –OH groups. The healing efficiency was 91% at 160 °C for 8 h, 66% of the σ_m was recovered, and the mechanical properties were only slightly decreased after three chemical cycles. Moreover, the abundant polar groups endowed ECMP-1.0 with excellent adhesion properties (an initial shear strength of 7.09 MPa).

9.2.3.3 Hydrogels

Christophe et al. [27] reported a one-step synthetic method of isocyanate-free PU hydrogels in one pot with poly(ethylene glycol) dicyclic carbonate (PEG-di5CC) and polyethylene imine (PEI) as raw materials. The pH of the media was adjusted within a range of 10.5 ~ 11.5 to limit the CC hydrolysis where the best mechanical performances (E = ~ 233 kPa, stress at break = ~ 222 kPa) of PU hydrogels were achieved at a pH of 10.5. When the gel content was identical (85%), the equilibrium water absorption of the prepared PU hydrogels was six times larger than that of the previously reported PHU hydrogels prepared in the bulk. The robustness of this process was exemplified by the manufacturing of PHU/clay (Cloisite Na or Laponite S182) nanocomposites and PHU/gelatin double network hydrogels with improved and modulative mechanical properties.

9.2.3.4 Foams

Self-blowing non-isocyanate PU foams containing both urethane and thioether linkages were prepared by the DBU-catalyzed selective addition of a diamine (1,2-bis(2-aminoethoxy)ethane, **A1**) and a dithiol (2,2′-(ethylenedioxy)diethanethiol, **T1**) to trimethylolpropane tris-carbonate (**C2**) [28]. The aminolysis of **C2** by **A1** produced PHU matrix while the decarboxylation of **C2** with **T1** generated the blowing agent (CO_2) and thioether linkages. This new synthetic process for isocyanate-free PU foams synthesis demonstrated its robustness through its tolerance to additives (i.e., clay, PDMS) as well as the diversity and accessibility of its monomers. The prepared foams exhibited a density as low as 0.166 g/cm^3 and a $T_{5\%}$ in the range of 220 and 248 °C. The thioether linkages in foams contributed to the polymer matrix construction and improved their thermal properties. Later, the foam formulation was further adjusted to TMPTC and an aliphatic or aromatic diamine [29]. The in situ release of CO_2 was realized through the DBU-catalyzed partial hydrolysis of TMPTC. The nature (rigid or flexible), the morphology, and the properties of the prepared foams could be tuned via changing the diamine or water content and adding clay. Importantly, the fabrication of foams in 30 min was realized after applying this water-induced foaming technology to the reaction injection molding process under the help of some foaming acceleration procedures. This is a significant improvement in foaming technologies in which the foaming usually takes several hours or even days.

9.2.3.5 Insulation Materials

PHU-mica powder composites based on inorganic silylated mica, bis-cyclocarbonate of bisphenol A (CCGEBA), and isophorone diamine (IPDA) were fabricated by Ubirajara et al. [30]. An excess of CCGEBA over the carbonate/amine stoichiometric ratio lowered the thermal stability of the composite. The mechanical, thermal, and electrical properties of the PHUs were improved by the addition of silylated mica, which was attributed to the higher cross-linking density in the organosilicate network. The optimized PHU-mica powder composite showed a $T_{10\%}$ of 309 °C, a hardness of 0.41 GPa, an elastic modulus of 5.04 GPa, a dielectric strength of 20.2 kV/mm, a dielectric constant of 10.4 at 1 kHz, and a dissipation factor of 0.12. As a result, these PHU-mica powder composites could replace the traditional epoxy-mica paper as insulating materials.

9.2.3.6 CO_2-Sourced Recyclable PUs

Covalent adaptable networks (CANs) refer to a type of thermoset-resembling polymeric materials with dynamic bonds such as boronic ester, imine bond, disulfide bond, urethane bond, silyl ether bond, and so on [31]. CANs can go through dynamic and reversible bond exchange under external stimuli of heat, light, or pH value. Generally, CANs exhibit similar high mechanical properties and stability like classic thermosets while could be softened, recycled, repaired, and reshaped like classic thermoplastics.

Xu et al. synthesized a kind of recyclable PHUs through the polyaddition of cyclic-carbonated soybean oil (CSBO) with various diamines [31, 32]. CSBO was produced with two steps: Firstly, epoxidation of soybean oil was carried out to form epoxidized soybean oil (ESBO), and then, ESBO reacted with CO_2 to produce CSBO. CSBO was cured by 1,8-menthane diamine (MDA) to produce fully bio-based PHUs under solvent-free and catalyst-free conditions [32]. The highly active –NH_2 groups of rigid MDA could auto-catalyze the reaction system, facilitating the curing of CSBO and the bond exchange in PHUs. Three kinds of PHUs with different amino/CC molar ratios (CSBO-MDA$_x$, x is 1.0, 1.2, 1.8) were synthesized, and among them, CSBO-MDA$_{1.2}$ exhibited the best mechanical and thermal properties (σ_m, 2.2 $\pm$ 0.12 MPa; $T_{10\%}$, 307 °C; T_g, 30 °C), which was attributed to its best stable cross-linking network and the highest cross-linking density (607.77 mol/m^3). In addition, CSBO-MDA$_{1.0}$ exhibited the highest relaxation rate because the introduction of an appropriate amount of MDA could increase the number of carbamate bonds and accelerate the transesterification reaction via auto-catalysis without hindering the mobility of polymer chains. Afterward, PHUs (CSBO-DDS) with dual dynamic covalent bonds were developed from CSBO and 4,4′-diaminodiphenyldisulfide (DDS) with dynamic disulfide bonds (S–S) [31]. Two control samples that contained either single S–S dynamic system (ESO-DDS) or single carbamate groups dynamic system (CSBO-DDM) were also prepared. CSBO-DDS with dual dynamic systems exhibited faster relaxation rate than the single dynamic systems, which was attributed to the enhanced bond exchange reaction by the S–S/carbamate dual dynamic networks. CANs of PHU could also be fabricated using bis(6-membered cyclic carbonate) of 4,4′-biphenol (BCC-BP) which originated from bi(1,3-diol) and CO_2 [33]. Either 2,2′-(ethylenedioxy)bis(ethylamine) (EGDA) or 4,7,10-trioxa-1,13-tridecane diamine (DEGDA) was used with tris(2-aminoethyl)-amine (TAEA) to cure BCC-BP to produce cross-linked PHUs (CLPHUs). CLPHUs exhibited the T_g in a range of 60 ~ 90 °C, $T_{5\%}$ > 320 °C, σ_m in the range of 41 ~ 65 MPa, E in the range of 719 ~ 1233 MPa, and ε_b in the range of 42 ~ 452%. EGDA-cured CLPHUs displayed faster stress relaxation compared to DEGDA-cured CLPHUs, which was attributed to the higher density of exchangeable bonds in the latter resulting from the shorter molecular chain of EGDA. CLPHU with a lower cross-link density exhibited a higher recovery efficiency, but its recovery efficiency decreased with increasing of the remolding cycles due to the reduced –OH density. CLPHUs showed excellent shape memory properties with the shape fixity and shape retention reaching 99% and 93%, respectively. Besides, the chemical recycling of CLPHUs was feasible (purity is 99.62%, recycling rate is 87%).

9.2.4 Other Methodologies

Several particular methodologies have been developed for the synthesis of CO_2-sourced PUs in the past few years. In 2022, Maria's group synthesized poly(acyl-urethane) (PUAcyl) with a M_n of 13.5 kg/mol from urea and castor oil in the presence of a Lewis acid promoter ($BF_3{\cdot}OEt_2$) [34]. This method can be applied also to other oils containing ester and –OH groups. Meanwhile, Zhang et al. prepared PU vitrimers successfully from alkylene bisurea (1,10-decylenebisurea (DBU), 1,6-hexylene bisurea (HBU)) and epoxide (1,4-butanediol diglycidyl ether, BDE) [35]. Bisureas (BUs) were synthesized from urea and diamines (1,10-diaminodecane (DA10), HMDA). A eutectic system was formed in the presence of $ZnCl_2$ which promoted the melting of BUs, in which PUs were produced at a relatively lower temperature of 130 °C. PU produced from DBU gave a σ_m of ~ 17 MPa, a E of ~ 90 MPa, and an ε_b of ~ 200%, which could be retained after reprocessing due to the existence of the dynamic exchange reactions of urethane/urea/carbonate cross-linkers.

9.3 CO_2-Sourced Polyureas (PUas)

Polyureas (PUas) are a category of PUs carrying urea bond (–HN–C(=O)–NH–) in their backbones which possess some advantages, for example, fast setting nature, excellent thermal and mechanical properties, good anti-abrasion properties [36]. Traditionally, PUas are prepared via the catalyst-free, rapid polyaddition between isocyanate and amines, contrary to which, the use of CO_2 to replace isocyanate in PUa preparation is a green process. CO_2 can be incorporated in PUas either directly as a monomer or indirectly in the form of CO_2-sourced building blocks including methanol, (cyclic) organic carbonate, and urea (Scheme 2).

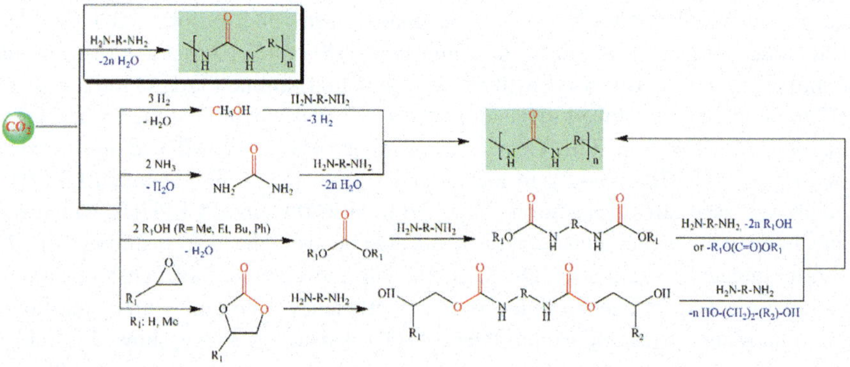

Scheme 2 Synthetic routes for CO_2-sourced PUas

9.3.1 PUas Synthesized from CO_2 Directly

Our group has made some contributions to the development of CO_2-sourced PUas in the past few years [37–41]. We developed a two-step method to get rid of water in the synthesis of PUas [42], and the first series of PUas ($DAD_{100-x}TTD_x$) (x represents the weight ratio of diamines) with measurable mechanical properties were synthesized from CO_2 with 4,7,10-trioxa-1,13-tridecanediamine (TTD) and 1,12-diaminododecane (DAD) [37]. $DAD_{100-x}TTD_x$ ($0 \leq x \leq 75$) exhibited a $T_{5\%} \geq$ 295 °C and a T_m at a range of 90 ~ 207 °C. The thermal and mechanical properties of $DAD_{100-x}TTD_x$ depended on its H-bond, crystallinity, and chain flexibility. Upon varying the feed ratio of two diamines and the chain length or molecular weight of diamine, the H-bond and the flexibility of PUas could be modulated, leading to PUas with adjustable properties. For example, as the TTD content increased from 0 to 100 wt%, the T_m, crystallizing temperature (T_c) and T_g of $DAD_{100-x}TTD_x$ decreased gradually, which was because that the network structure of homo-PUa (DAD_{100}) was gradually destroyed as the TTD content increased, causing a reduction in the strength and/or the density of H-bonds as well as the crystallinity due to the increase of head-to-head (–(C=O)– and –(C=O)–) and tail-to-tail (–NH and –NH) structures. Besides, $DAD_{90}EDR\text{-}104_{10}$ (EDR-104: 1,5-diamino-3-oxapentane) exhibited a E of 1190 MPa, a σ_m of 57.4 MPa, and an ε_b of 71%, which are superior to those of non-isocyanate PU and polyamide 6. Other than linear CO_2-sourced PUas, we have also synthesized macrocyclic oligourea (OUa) unexpectedly from the post-polymerization of TTD-derived linear OUa in N-methyl pyrrolidone (NMP) [43]. Efficient separation and purification methods were developed for isolation of macrocyclic OUa out from its linear one [40].

CO_2-sourced PUas and PUa hybrids imparted with self-healing properties were also produced [38, 39, 41]. The combination of TTD and IPDA in different proportions resulted in the first self-healing PUas directly from CO_2 ($TTD_x\text{-}IPDA_{100-x}$) [38]. TTD-based PUa moieties owned robust and ordered H-bonds while IPDA-based PUa moieties had irregular arranged molecular chains caused by the steric hindrance of the bulky fatty ring of IPDA. Both the E and the σ_m of $TTD_x\text{-}IPDA_{100-x}$ decreased while the ε_b and toughness increased with increasing the IPDA content from 0 to 43 wt%. Meanwhile, the T_m decreased and then disappeared, suggesting a transformation from semi-crystalline to amorphous. Among the prepared PUas, $TTD_{57}\text{-}IPDA_{43}$ (M_n = 38 kDa, polydispersity ($Đ$) = 1.58) exhibited competitive mechanical properties and self-healing properties at room temperature. Afterward, a self-healing polyurethane-urea (PUUa) based on TTD-derived OUa and a furfuryl amine-derived bis(cyclic carbonate) (BCCFA) and its Diels–Alder adduct (PUUa-DA) were synthesized [39]. The highlight of this strategy is that both OUa and BCCFA are CO_2-sourced, and the furan ring in BCCFA imparted PUUa with the ability to undergo Diels–Alder reaction. PUUa behaved like a tough thermoplastic (σ_m, 15.9 MPa; ε_b, 46%; E, 206 MPa) compared to the fragile PUa polycondensed from OUa. As for PUUa-DA, the E decreased to 151 MPa, but the σ_m increased to 18.5 MPa and the ε_b increased to 136% due to the formation of a cross-linking structure. Thermosetting PUUa-DA could not only be self-healed with a healing efficiency of 94%, but also be recycled via hot-pressing. 1,3-bis(3-aminopropyl)-tetramethyldisiloxane-derived OUa was polymerized with terephthalaldehyde (TA) with tris(2-aminoethyl)amine (TAEA) as cross-linker, yielding a type of self-healing poly(urea-imine) thermoset (PUIs) [41]. Similarly, the properties of PUIs could be tuned upon varying the OUa/TAEA molar ratio because OUa affected the H-bonds and TAEA controlled the cross-linking density. Benefiting from the H-bonds among urea groups and the dynamic imine bonds, the PUI from a OUa/TAEA molar ratio of 2:3 not only exhibited good self-healing property with a healing efficiency of 92%, but also could be recycled by physical or chemical processes although it was a thermosetting material.

Xiao et al. prepared polyurea-multi-block-poly(propylene carbonate) (PUa-mb-PPCs) via the polyaddition of CO_2-derived PUa oligomer (PUa) with CO_2-derived poly(propylene carbonate) diols carrying terminal –NCO groups (PPC-MDI) [44]. PUa synthesized from the polycondensation of CO_2, TTD, and HMDA acted as the hard segment, and PPC-MDI synthesized from propylene epoxide, CO_2, and methylene diisocyanate was the soft segment. PUa-mb-PPCs based on a TTD:HDA ratio of 1:1 exhibited a σ_m up to 30 MPa, an ε_b of ~ 700%, and a E′ of 142.1 MPa at ambient temperature, which are comparable to the conventional thermoplastic PU elastomers.

Altering the structure of diamines gives access to PUas with a diversity of structure and property. Starting from diamine with POSS cages (3,13-dianilino DDSQ) and amino-terminated polyether (Jeffamine D-2000), Zheng et al. synthesized the first group of novel organic–inorganic siloxane-PUas [45]. The use of NaH as water scavenger ensured the high M_ns of the siloxane-PUas (11,200 ~ 21,200 Da). With the introduction of POSS, the product transformed from viscous liquid to elastic solids due to the formation of physically cross-linked networks. Recently, Yan et al. synthesized a CO_2-sourced ionic polyurea (CIPUa) via the polycondensation of CO_2 and an amino-functionalized ionic liquid (BPY-NH_2) [36]. The developed CIPUa exhibited super strong and reversible adhesion toward a variety of substrates (e.g., rubber, metal) under various environments. The synergistic effect of the rich urea groups and the ionic species (bipyridyl cations and PF_6 anions) in CIPUa where they interacted with the substrate, respectively, via H-bonds and electrostatic interactions provided the reinforced interfacial adhesion of CIPUa.

9.3.2 PUas Synthesized Indirectly from CO_2-Derived Monomers

CO_2-sourced PUas can also be fabricated indirectly from CO_2-sourced building blocks such as (poly)carbonate [46], urea [47, 48], and methanol [49]. The aminolysis of CO_2-derived (a)cyclic carbonates (i.e., DMC, DPC, EC) by amines produces dialkyl carbamates which are used directly in the fabrication of PUas via transurethane polycondensation, affording PUas with high molecular weight under mild conditions by removing the alcohol by-products (i.e., methanol, phenol, catechol) and organic carbonates.

9.3.2.1 (Poly)carbonate

Pan et al. reported a kind of biomass-based PUas with a M_n ranging at 2800 ~ 16,700 Da from the catalytic polycondensation of novel bio-based dicarbamates (DIFFA-DC, isomannide) and diamines [46]. The two dicarbamates were synthesized by reacting difurfurylamine (DIFFA) or diaminoisoidide (DAIID) with DMC. Under catalyst-free and solvent-free conditions, PUas (M_n = 1030 ~ 3200 Da) with a $T_{5\%}$ ranging from 283 to 305 °C and a high T_m ranging from 278 to 375 °C were synthesized by the polymerization of catechol carbonate with diamines [50]. Eugenio et al. reported the first synthesis of PUas with a yield about 78 ~ 99% from the chemical recycling poly-(bisphenol a carbonate) (PBAC) with diamines [51]. PBAC could be produced by the transesterification and polycondensation of BPA and DPC. The formed PUas exhibited a $T_{5\%}$ of 212 ~ 270 °C and a $T_{d,max}$ at 323 ~ 366 °C.

9.3.2.2 Urea

PUas elastomers were synthesized from the melt polycondensation of poly(tetramethylene oxide) (PTMO, Jeffamine® THF-170), 2,2-(ethylenedioxy)bis(ethylamine) (EBA), and urea [47]. Urea polymerized with PTMO to form the soft segment and polymerized with EBA to form the hard segment. To ensure the elastomeric properties of PUa, the hard segment content should be no more than 30 wt%. PUas with 5 ~ 30 wt% hard segment exhibited the $T_{5\%}$ ranging from 311 to 384 °C, the ε_b in the range of 340% ~ 770%, the σ_m around 1.1 ~ 14 MPa, and an E in a range of 0.76 ~ 29.5 MPa. Specially, the PUa containing 20 wt% hard segments possessed the highest dielectric permittivity at 10.6 (1 kHz, 300 K) and exhibited a relatively high dielectric loss of 3.2, making it a suitable candidate for dielectric elastomer applications. Thereafter, Tang et al. prepared thermoplastic PUas via the copolymerization of DA10 or/and HMDA with urea. The mechanical properties of the PUas could be adjusted by changing the diamine feed ratio [48]. The polycondensation of urea with diamines (i.e., triethylenetetramine) or (branched) polyamines with multiple amine groups (i.e., TAEA) at 115 ~ 140 °C under solvent and catalyst-free conditions could provide formaldehyde-free PUa wood adhesives with superior lap shear strengths up to 2.64 MPa [52, 53]. PUa adhesives with higher cross-linking degree, which were prepared from polyamines with more terminal amine groups, exhibited better bonding properties [53].

9.3.2.3 Methanol

Amit et al. synthesized PUas (M_n, 2214 ~ 5500 Da; $Đ$, 1.03 ~ 1.32) from the dehydrogenative coupling of methanol and diamines in the presence of a Ru pincer catalyst [49]. The incorporation of 2 mol% KO^{t}Bu could facilitate the dehydrogenation of MeOH by promoting the C-H cleavage or by scavenging impurities. The synthesized PUas exhibited a $T_{5\%}$ at 230 ~ 299 °C. Moreover, 100% renewable PUas, a chiral PUa, and the first ^{13}C-labeled PUa were synthesized by incorporating bio-sourced diamines, a chiral diamine, and $^{13}CH_3OH$. Mn complex/KO^{t}Bu catalytic system was also found active for the polymerization of diamine and methanol [54]. This catalytic method was suitable to a variety of diamines, giving PUas in good yields (68 ~ 88%). According to the mechanism, the Mn-catalyzed polymerization proceeded via an in situ formed isocyanate intermediate [54, 55]. First, methanol dehydrogenated to formaldehyde which reacted with amine to form formamide. Then, the formamide dehydrogenated to form the isocyanate. Mn catalyst is of more advantageous compared to Ru catalyst as Mn is an earth-abundant metal.

9.4 CO_2-Sourced Polycarbonates (PCs)

PCs are a class of polymer materials with carbonate linkages (–OC(O)O–) in their main chain, having a wide range of applications as foams, adhesives, coatings, and packaging materials [1]. The traditional production of (aromatic) PCs relies on toxic phosgene and BPA which is a suspect endocrine disruptor, a neurotoxic and carcinogenic agent [56]. Thus, the quest for greener alternatives to BPA and safer synthetic process has pushed the development of CO_2-sourced PCs. As shown in Scheme 3, there are mainly three ways to synthesize PCs from CO_2 based on the polymerization mechanism: (1) the direct ring-opening copolymerization (ROCOP) of CO_2 with cyclic ethers (epoxides), (2) the step-growth polymerization (SGP) of diol with acyclic carbonate (or dialkyl carbonates)/bis(α-alkylidene cyclic carbonates) /CO_2 (and dihalide), (3) the ring-opening polymerization (ROP) of CO_2-sourced five-membered or six-membered CCs. The CO_2/diols/dihalide polycondensation and the direct polymerization of diol and CO_2 (equilibrium yields < 1%) [57] suffer from uncontrollable polymerization and harsh reaction conditions [58], as well as the use of hazardous halides. ROP of CCs for the fabrication of PCs is effective and gives PCs with good reproducibility, high quality (high M_n and low $Đ$), and a greater functional diversity [59]. However, the ROCOP of epoxide and CO_2 is the most well-studied synthetic method of CO_2-sourced PCs owning to the high activity and accessibility of epoxides [58].

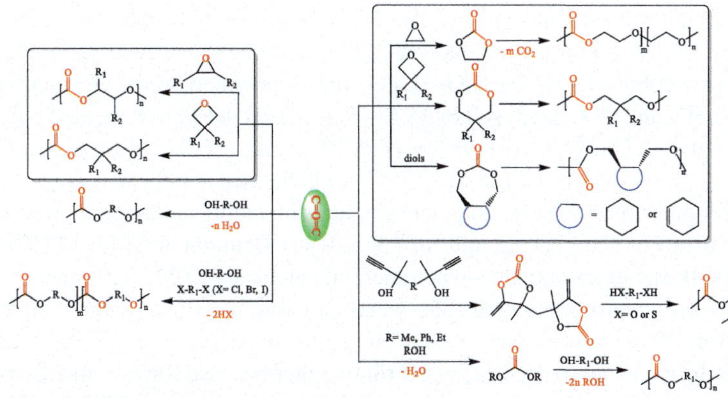

Scheme 3 Synthetic routes for CO_2-sourced PCs

9.4.1 The ROCOP of Epoxide with CO_2

The first example of ROCOP of epoxide/CO_2 was demonstrated by Inoue and co-workers in 1969 where CO_2 was polymerized with propylene oxide (PO) to give poly(propylene carbonate) (PPC), catalyzed by ($ZnEt_2$)/H_2O [60]. Lewis acid sites (a metal center, i.e., Zn(II), Co(III), Al(III)) and Lewis base sites are always used together in the most of the catalytic systems, where the Lewis acid sites activate epoxides and the Lewis base sites act as nucleophiles to open the epoxide ring [61]. According to the mechanism, the epoxide ring is activated by the Lewis acid metal center and then attacked by the Lewis base sites, leading to the formation of a metal alkoxide intermediate. Next, the metal alkoxide intermediate can either attack CO_2 to form a carbonate intermediate or continue to insert epoxide repeatedly, resulting in the formation of (poly)ether units. If CO_2 attack occurs, the carbonate intermediate can either form undesired CCs by backbiting reaction or continue to undergo repeated alternating epoxide/CO_2 insertion, leading to PCs with perfectly alternating nature [62]. The dissociation of the carbonate with epoxide binding and ring opening by the growing anionic polymer chain is generally accepted as the rate-determining step instead of the following CO_2 insertion step [63]. The selectivity is affected by a variety of parameters, including the nature of the substrate, the reaction conditions, and the catalyst [7].

Degradable PC analogs of commercial poly(alkyl acrylate)s were synthesized from the copolymerization of CO_2 and either alkyl glycidate or alkyl glycidyl epoxides, catalyzed by rac-[SalcyCo^{III}DNP]DNP [64]. Lower TOF and M_n were observed for the glycidate/CO_2 system compared to the glycidyl/CO_2 system, which resulted from the higher steric hindrance during the polymerization in the glycidate/CO_2 system and the more reactive methine carbon of the glycidate epoxide. Besides, poly(M acrylate carbonate)s exhibited higher T_g (M = ethyl) and $T_{d,50\%}$ (M = ethyl, butyl) compared to their poly(glycidyl M ester carbonate) counterparts, which was attributed to a higher backbone rigidity. The PC analogs not only inherited adhesive properties from the non-degradable poly(alkyl acrylate)s, but also were able to degrade into alcohol, acids, and CO_2. Later, the terpolymerization of glycidyl butyrate (GB), PO, and CO_2 catalyzed by [*S,S*]-[SalcyCo$_{III}$DNP]/DNP yielded poly (propylene-co-glycidyl butyrate carbonate)s (PPGBC)s ($M_n = 6.0 \sim 13.6$ kg/mol, $Đ = 1.08 \sim 1.12$, $T_g = -30 \sim 0$ °C) [65]. PPGBCs could be used as adhesive, showing varied peel strengths toward a wide range of substrates including metal, glass, wood, and Teflon® surfaces. The peel strength of 4.9 ± 0.41 N/cm obtained for PPGBC with 56 mol% GB was superior to that of commercial Ducttape® (3 M 2929). In order to prepare PCs with controllable molecular weight and unimodal molecular weight distribution from PO or cyclohexene oxide (CHO), triisobutylaluminum (TiBA) was employed as a drying agent to dry CO_2 to get rid of water because water could act as a chain transfer agent [66]. As a result, poly(cyclohexene carbonate)s (PCHCs) ($M_n = 68 \sim 450$ kg/mol, $Đ = 1.09 \sim 1.37$) and poly(propylene carbonate)s (PPCs) ($M_n = 58 \sim 203$ kg/mol, $Đ = 1.14 \sim 1.29$) with monomodal GPC traces were obtained. A degree of polymerization as high as 16,000 was achieved in both cases, and the absolute molar mass of PCHC even reached 1030 kg/mol. Di-block copolymers poly(propylene carbonate-b-cyclohexene carbonate)s (PPC-b-PCHCs) with a M_n up to 310 kg/mol were also fabricated.

Biomass-derived epoxide monomers have also been exploited for the synthesis of PCs. Poly-(limonene carbonate)-block-poly(cyclohexene carbonate) di-block copolymers (PLimC-b-PCHCs) with a M_n up to 12.1 kg/mol ($Đ$, $1.08 \sim 1.18$) were synthesized from limonene-derived trans-limonene oxide, CHO, and CO_2 by sequential epoxide addition [67]. The similarity of PCHC and PLimC homopolymers in the structure and T_gs resulted in a single T_g of PLimC-b-PCHCs. Upon varying the composition and M_n of PLimC-b-PCHCs, varied morphologies including cylindrical, lamellar, and hexagonally perforated lamellar were observed. Glycerol can not only be utilized to synthesize CC for PU synthesis, but also be transformed into potential epoxide monomers for PC production. Cui et al. synthesized an epoxide from fatty acids residues waste (CG-FAW) that derived from crude glycerol refining process. Epichlorohydrin, which could also be derived from crude glycerol, was employed to functionalize CG-FAW with an oxirane group, producing the desired epoxide. The copolymerization of this epoxide and CO_2 afforded bio-based PC with 98.1% carbonate content [68]. Karen et al. synthesized a group of bio-based PCs from a D-xylose-derived oxetane and CO_2 through a dual strategy [69]. The first pathway was the direct ROCOP of oxetane with CO_2 catalyzed by (salen)CrCl/PPNCl. The second one consisted of two steps where oxetane was first transformed into a six-membered CC using ZnI_2/n-Bu_4NI, followed by the ROP of six-membered CC catalyzed by TBD. PCs from the two pathways exhibited similar thermal properties.

Additionally, CO_2-sourced PCs with novel architectures were synthesized. Benefiting from the Diels–Alder reaction of maleic anhydride (MA) and furfuryl glycidyl ether (FGE), PPCs containing cross-linked networks (PPC-MFs) (M_n, 84,525 ~ 99,367 g/mol; $Đ$, 3.7 ~ 4.5) were developed by Feng et al. via a one-pot polymerization of PO, CO_2, MA, and FGE [70]. PPC-MFs exhibited $T_{5\%}$ at a range of 249.7 ~ 271.3 °C which was improved by comparison to that (217.1 °C) of PPC. Both the $T_{5\%}$

and gel content of PPC-MFs increased as the MA/FGE loading increased. Besides, PPC-MFs displayed good elasticity and dimensional stability with the minimum permanent deformation of 6.5%, which is much lower than that of PPC (157.2%). However, the mechanical properties of PPC-MFs (σ_m, 6.6 ~ 8.3 MPa, ε_b, 273 ~ 513%) were weakened as compared to PPC (σ_m, 11 MPa, ε_b, 559%) due to increased flexibility. Recently, the copolymerization of *meso*-CHO and CO_2 afforded unprecedentedly recyclable cyclic PCHCs containing a tricarbonate linkage (M_w, 27.9 ~ 33.5 KDa, $Đ$, 1.02 ~ 1.17) [71]. All the PCHCs exhibited similar M_ns (~30 KDa) regardless of the monomer feed ratio, which implied that chain transfer reaction had occurred during the polymerization. However, there is no chain transfer reagent in the polymerization system. The most PCHCs contain an end group with a M_n of 88 which is a result of the continuous insertion of two CO_2 molecules, suggesting the formation of PCHCs with a cyclic structure (tricarbonate linkage). This explained why the most PCHCs with various monomer ratios showed similar M_ns since the ring closure is a kind of chain transfer reactions. By mechanism, one CO_2 would be trapped between two Zn centers before the initiation of copolymerization, and it would later bridge the nucleophilic attack of CO_2 on the chain end by the last insertion of CO_2 onto the propagation site, which resulted in the closure of the polymer chain to form a cyclic polymer.

9.4.2 The SGP of Diol with Carbonates/CO_2

The polycondensation between diol and CO_2 has been a promising way for the synthesis of CO_2-sourced PCs not only because water is the only by-product, but also because diols are easily available [72, 73]. Moreover, it is possible to acquire 100% renewable resource-based PCs using biomass-derived diols, which will contribute to carbon neutrality [57]. The SGP between diol and acyclic carbonate or bis(α-alkylidene cyclic carbonates) (bis-αCCs) for PC synthesis is also promising because it provides the access to a broad range of both aliphatic and aromatic PCs with tunable properties from a diversity of CO_2-sourced monomers. Besides, the higher reactivity of bis-αCCs allows to synthesize CO_2-sourced PCs efficiently.

Keiichi et al. synthesized PCs with a M_n up to 5 kg/mol ($Đ$, 1.54) in good selectivity (> 99%) from the direct polymerization of 1,6-hexanediol and CO_2 under a CeO_2 and 2-furonitrile catalysis system [72]. The M_n of PCs could be controlled by changing the amount of 2-furonitrile and diols. Compared to CeO_2 and 2-cyanopyridine catalysis system that produced PCs with low M_ns (900 ~ 1700 g/mol), the CeO_2 and 2-furonitrile catalysis system was more superior due to (1) lower reactivity of 2-furonitrile-derived 2-furamide toward the produced PC diols, which led to lower selectivity toward the ester-capped PCs, (2) weak adsorption of 2-furonitrile on CeO_2 which led the steric hindrance to decrease at the active sites of CeO_2. Under the catalysis of organic bases such as tetramethylguanidine (TMG), DBU, or DBN, diol reacted with CO_2 and allyl bromide to form diverse α,ω-diene functionalized carbonate monomers which were subsequently polymerized for the synthesis of poly(thioether carbonates) (M_n, 13.8 ~ 45.5 kg/mol, Đ, 1.9 ~ 2.5) and unsaturated PCs (M_n, 2.5 ~ 37.6 kg/mol, Đ, 1.2 ~ 2.1) [73]. The post-polymerization modification of the obtained poly(thioether carbonates) and PCs via oxidization and hydrogenation reactions was also conducted successfully.

Mono- and bis-αCCs derived from CO_2 and propargylic alcohols have appeared as a class of important and powerful synthons for the synthesis of PCs. These bis-αCCs exhibited higher reactivity than traditional five-membered CCs owing to the keto-enol equilibrium and the presence of an exocyclic olefinic group [74, 75]. Regioregular poly(monothiocarbonates) and poly(thioether-co-cyclic carbonate)s (M_n, 9200 ~ 54,000 g/mol, Đ, 1.37 ~ 2.58, T_g, 9 ~ 126 °C) were prepared from the ROP of bis-αCCs and dithiols catalyzed by DBU with or without fluorinated alcohol [75]. An on/off switchable domino process discovered in the synthesis of linear thiocarbonates or elusive tetrasubstituted ethylene carbonates from mono-αCC and thiol also occurred in the polymerization of bis-αCCs and dithiols. The transformation of the thiocarbonate linkages in poly(monothiocarbonates) to the thioether-co-cyclic carbonate linkages in poly(thioether-co-cyclic carbonate)s via rearrangement was observed. As the thioether-co-cyclic carbonate linkages content increased, the T_g of the obtained polymer increased. Despite the exciting synthetic possibilities endowed by this domino process in organic chemistry and polymer design, the underlying mechanism has not been explained yet. Christophe et al. reported another domino approach for the one-pot cascade synthesis of poly(β-oxo-carbonate)s from CO_2, propargylic alcohols, and diols [76]. A highly selective catalytic system [TBA][OPh]/AgI was identified in the synthesis of oxo-alkyl carbonate via two cascade reactions: the cycloaddition of CO_2 with propargylic alcohol to selectively yield *in* situ αCC and the alcoholysis of αCC to furnish oxo-alkyl carbonate. Upon employing [TBA][Oph]/AgI as catalyst, the terpolymerization of 4,4′-(ethane-1,2-diyl)bis(4-methyl-5-methylene-1,3-dioxolan-2-one), diol, and CO_2 under 1 ~ 15 bar CO_2 at 60 ~ 80 °C led to the formation of the target poly(β-oxo-carbonate)s with M_w up to 4300 g/mol. Thomas et al. synthesized PCs (M_n, 1905 ~ 2113 g/mol, Đ, 1.01 ~ 1.03) from CO_2-based bis-αCCs at a conversion of 99% [74]. Initially, readily available bis-epoxides occurred a ring-opening reaction with 1,4-butynediol to produce primary propargylic alcohols. Then, the propargylic alcohols were carbonated with CO_2 to produce bis-αCCs. Finally, poly(β-oxo-carbonates)s were afforded from the copolymerization of bis-αCCs and BDO.

Biorenewable glucose-derived isosorbide (ISB), a diol containing an endo-OH and an exo-OH, is the ideal alternative to BPA. Poly(isosorbide carbonates) (PICs) could be synthesized from ISB and CO_2-derived dialkyl carbonate (DMC, DPC). However, the low reactivity of the endo-OH in ISB and the possible alkylation side reaction (e.g., methylation by DMC) have hindered the synthesis of high molecular weight PIC [77]. Catalyst that can efficiently activate the endo-OH of ISB is the key to achieve high molecular weight PICs. Zhang et al. have developed various catalysts for the synthesis of ISB-derived PICs [77–80]. A series of tetrabutylphosphonium-based ionic liquid (IL) catalysts were synthesized and presented high effective for the synthesis of PICs from ISB and DPC via two successive stages (transesterification and polycondensation). Among these IL catalysts, $[P_{4444}]^+[CH_3COO]^-$ was identified as the most efficient IL catalyst for PIC synthesis due to two well-balanced activation effects: the activation of C=O in DPC by $[P_{4444}]^+$ and the activation of endo-OH in ISB by $[CH_3COO]^-$ through disrupting the intramolecular H-bonding between endo-OH and the furan ring [78]. Notably, the ratio of endo-OH/exo-OH of the obtained PIC (M_w up to 66.9 kg/mol, T_g, 169 °C) was as low as 0.25. Furthermore, isosorbide-based copolycarbonates (CPICs) with adjustable flexibility (M_w, 14.7 ~ 110 kg/mol, $Đ$, 1.62 ~ 1.88) were also prepared from the terpolymerization of DPC, ISB, and diols. All CPICs and PIC started to degrade from 270 °C to 320 °C, and their dynamic mechanical properties varied with their structures. Li and co-workers applied another IL catalyst, 1-ethyl-3-methylimidazole lysine ([Emim][Lys]) in the synthesis of PIC from ISB and DPC [81]. Impressively, PIC with a M_w up to 150 kg/mol and a T_g up to 174 °C was obtained. [Emim][Lys] stood out among all the amino acid ionic liquids (AAILs) screened due to its suitable electronegativity, and the suitable basicity, steric hindrance, and H-bond donating ability of $[Lys]^-$. The tetrapolymerization of ISB, DPC, BDO, and 1,4-cyclohexanedimethanol was conducted in order to improve the mechanical properties of PICs. As a result, copolymers (PIBCCs) with the E up to 979 MPa, the σ_m up to 57 MPa, and the ε_b up to 145% were obtained, which were comparable to commercial petroleum-based PCs. Then, the phenolic IL catalysts were employed to promote the reaction between ISB and CO_2-sourced DMC to furnish PICs with tunable molecular weights [80].

9.4.3 The ROP of CO_2-Sourced CCs

There are fewer reports on the synthesis of PCs from the ROP of CCs compared to other methodologies discussed above due to the limited scope of polymerizable CC monomers. Compared with the diol-related polycondensation method for PC synthesis, the ROP of CCs offers the possibility of controlling the M_n and distribution, the microstructure of the polymer, and the nature of the end groups [82]. Eco-friendly trimethyl glycine (TMG) was proved to be an efficient organocatalyst for the controllable synthesis of poly(trimethylene carbonate)s (PTMCs) (M_n, 1200 ~ 4000 g/mol, $Đ$, 1.12 ~ 1.22) via the ROP of TMC [83]. The intermolecular transesterification side reaction that led to cyclic oligomers became obvious when the conversion of TMC was over 80%. Interestingly, TMG analogs (i.e., *N*,*N*-dimethylglycine) also showed catalytic activity toward the ROP of TMC. In the results reported by Zhu et al., two novel seven-membered CCs with trans-cyclohexyl-fused rings, trans-octahydrobenzo[e][1,3]dioxepin-3-one (**M1**), and trans-1,5,5a,6,9,9a-hexahydrobenzo[e][1,3]dioxepin-3-one) (**M2**) were synthesized through cyclization of diols and CO_2 in good yields [82]. The enantiopure monomers, (R,R)-**M1** and (S,S)-**M1,** were also prepared. Under the catalysis of a yttrium alkyl complex, the room temperature ROP of **M1**, **M2**, (R,R)-**M1,** or (S,S)-**M1** yielded PCs with a M_n up to 673 kg/mol and a TOF up to 6×10^5 h^{-1}. The PCs based on **M1** or **M2** exhibited a $T_{5\%}$ in the range of 307 ~ 325 °C and a single T_g in the range of 32 ~ 39 °C while the PCs based on (R,R)-**M1** or (S,S)-**M1** exhibited a T_g in the range of 34 ~ 39 °C. Excitingly, the PCs are able to depolymerize to their corresponding cyclic dimers that could re-polymerize to synthesize the initial PCs in moderate yields over the catalyst of MeOK.

Another novel type of six-membered bicyclic CCs was fabricated in a yield up to 96% from the coupling of CO_2 and a variety of anti-configured β-epoxy alcohols in the presence of a binary Al^{III} aminotriphenolate complex/DIPEA catalyst [84]. Two of these six-membered CCs were subjected to the ROP to produce PCs, further broadening the scope of polymerizable CC substrates. Recently, in order to modify the properties of PCHC, the terpolymerization of CHO, CO_2, and six-membered CC was conducted in the presence of a dissymmetric Cr complex, affording terpolycarbonates with various sequence distributions (random, gradient, and block) [85]. The reaction conditions and the reactivity of the six-membered CC affected the sequence distribution of the obtained PC. Taking the CHO/CO_2/1,3-dioxan-2-one (TMC) system (PCs: $M_{n,}$18.4 ~ 31.6 kg/mol, $Đ$, 1.13 ~ 1.38) as an example, the poly(trimethylene carbonate) (PTMC) segment from the ROP of TMC increased when the cocatalyst amount decreased. Besides, the increase in temperature (60 ~ 100 °C) also led to an increase in the PTMC content (15.9 ~ 25.3%). What's more, increasing the CO_2 pressure (< 2 MPa) caused the formation of random PC while 4 MPa CO_2 pressure hindered the ROP of TMC. The substitutes of six-membered CC influenced its reactivity directly, and its reactivity significantly influenced the sequence distributions of the synthesized terpolymer. Upon decreasing the reactivity of the six-membered CC, the ROP of the six-membered CC was gradually restrained in the initial stage of terpolymerization, leading to the translation of sequence distribution from random to gradient and block-like. For terpolymers based on CO_2, CHO, and TMC, its T_g (69.5 ~ 115.5 °C) decreased with the increase of PTMC segment (1.2 ~ 38.6%).

9.5 CO_2-Sourced Polyesters (PEs)

PEs are a family of polymeric materials with –C(=O)–O– linkages in the backbone. The representative synthetic methods of PEs include the ROP of lactones/cyclic esters [86], the polycondensation of dicarboxylic acids with diols [87], and the self-polycondensation of renewable resources-derived α-hydroxyacids [88]. The ROP synthetic method of PEs is limited by possible side reactions such as transesterification as well as the poor functional diversity of available lactones while the polycondensation method requires high temperature to remove the by-product water [89], and long reaction time to obtain high molecular weight PEs. The synthesis of CO_2-sourced PEs has attracted increasing interest because of its sustainability and versatility. Generally, CO_2 can be incorporated into PEs directly or indirectly with the help of olefins or alkyne. The recent development on CO_2-sourced PEs is centered on a disubstituted δ-valerolactone, ethylidene-6-vinyltetrahydro-2H-pyran-2-one (EVL) synthesized from CO_2, and 1,3-butadiene.

Ni et al. copolymerized EVL with β-butyrolactone (BBL) in the presence of [Sc(OTf)$_3$] [90]. Notably, this is the first successful ROP of EVL and produced functional poly(EVL-*r*-BBL)s (M_w, 1.5 ~ 4.1 kg/mol, *Đ*, 1.86 ~ 3.66) containing 16 ~ 51 mol% EVL. The difficulty in the ROP of EVL is: (1) the α, β-conjugated C=C on the six-membered ring is a typical Michael acceptor that may react with potential chain-end alkoxides to terminate the chain-propagating processes in anionic ROPs, (2) the limited M_w and broad dispersity were attributed to the chain transfer reactions of the BBL-terminated polymer chains resulting from the steric hindrance of the allyl group [91]. The copolymerization of EVL and BBL followed an alternating tendency because the homopropagation of BBL or EVL was not favored. Additionally, unsaturated groups in poly(EVL-*r*-BBL)s allowed the post-polymerization functionalization by mercapto-ended PSar via thiol-ene click reaction, leading to the formation of an amphiphilic grafting copolymer poly(EVL-*r*-BBL)-g-PSar. A novel strategy named "scrambling polymerizations" was designed for the synthesis of PEs with a M_n up to 29 kg/mol and a EVL content up to 50 mol% from EVL and ε-caprolactone (CL) in the presence of sodium 2,6-ditert-butyl-4-methylphenolate (NaOAr) [92]. The "scrambling" refers to the intermolecular transesterification in the anionic ROP of CL. CL was much more active than tri-EVL or oEVL, and the latter was the oligomer of EVL with three kinds of lactone rings. During the copolymerization of CL and EVL, the anionic ROP of CL and the C=C addition oligomerization of EVL occurred first, leading to the formation of PCL and oEVL, respectively. Then, oEVL propagated onto the anionic PCL active chain via the ring opening of ring A after CL was consumed. Afterward, the slow ROP of ring B of oEVL occurred and generated branching of P(CL-co-EVL). The active chain end of either oEVL or CL always prefers transferring to the other chain end, leading to random P(CL-co-EVL). Specially, the topology of P(CL-co-EVL) is tunable to cyclic, branched, and cross-linked by extending the polymerization time. Cross-linked P(CL-co-EVL) obtained with a [EVL]/[CL] feed ratio of 1.5/1 and a reaction time of 36 h exhibited good elastic behavior: an ε_b of 608 ± 36%, a σ_m of 1.46 ± 0.71 MPa, an elastic recovery > 97% at 200% elongation.

In the work reported by Ian et al., EVL (denoted as EVP here) was either semi-hydrogenated or exhaustively hydrogenated, respectively, in the synthesis of EtVP (86% yield) or DEP (77% yield) [93]. Then, the organocatalyzed ROP of EtVP (DEP) by TBD (NaOMe/CO(NHCy)(NHPh)) afforded poly(EtVP) (M_n,13.6 kg/mol, *Đ*, 1.32) with 28.5 wt% CO_2 or poly(DEP) (M_n,9.7 kg/mol, *Đ*, 1.27) with 28.2 wt% CO_2. The T_g (− 42 °C) of poly(DEP) was lower than that (− 38.8 °C) of poly(EtVP), which may be resulted from the further saturation in DEP. Furthermore, poly(EtVP) could be chemically recycled. In addition, the post-polymerization modification of poly(EtVP) could be conducted via performing thiol-ene reaction, quaternization, and cross-linking on its alkene side chain.

9.6 Summary

The recycling of CO_2 into polymers including PUs, PUas, PCs, and PEs is of great importance and owns a methodological superiority since it achieves the recycle of CO_2 and the transformation of CO_2 into value-added chemicals simultaneously. Recently, the CO_2-sourced polymers have attracted much attention, and some achievements have already been obtained in the synthetic methods, the structure and properties of functional materials, and their potential applications. Both the direct utilization of CO_2 as a monomer and the indirect utilization of CO_2 with CO_2-derived compounds (e.g., CC, lactone) as monomers are depicted. This chapter focus is put on the structure–property relationships of CO_2-sourced polymers including PUs, PUas, PCs, and PEs, "how the thermodynamic properties alter with the unit structures" is mainly described and explained.

There still remain uncountable challenges to achieve large-scale production of CO_2-sourced polymers despite some achievements have been made on the properties, the versatility of monomers, and the catalysts. First of all, the dilemma of the lack of performances for most of the CO_2-sourced polymers compared to their commercial counterparts has to be overcome. The synthesis of a copolymer with di-block or even tri-block by incorporating a third monomer (or more) might be an efficient strategy because it integrates the advantages of multiple blocks. The most typical example is poly(carbonate-co-ester) synthesized from the terpolymerization of epoxide, anhydride, and CO_2, as reviewed lately in 2021 [94]. However, the use of metal catalysts, the low M_n of copolymers produced, the poor compatibility between blocks, and the insufficient understanding in the structure–performance relationships have limited the development of CO_2-sourced polymers [95]. Another strategy to obtain CO_2-sourced polymers with superior performance is the post-polymerization via versatile synthetic chemistry such as thiol-ene click chemistry and Diels–Alder reaction. In these strategies, various functional groups can be introduced into the polymers to provide a wealth of properties for the CO_2-sourced polymers (i.e., hydrophilicity, recyclability). The second challenge lies in the catalysts. Catalysts are important in the synthesis of both CO_2-sourced synthons (e.g., CCs, carbamates) and well-defined CO_2-sourced polymers without undesired by-products (i.e., CC in PU synthesis, ether segments in PC synthesis). Nevertheless, high catalyst loading, metal residue in the final products, and the requirement of toxic cocatalysts could cause troubles. As a result, organic catalysts have emerged to become a research hot spot in the past few years. The third challenge is the limited scope of monomers where the most monomers used in the synthesis of CO_2-sourced polymers are derived from fossil feedstock. Consequently, the utilization of biomass-resourced monomers (e.g., glycerol) instead of fossil fuels-derived ones is under development. All in all, the utilization of CO_2 in the synthesis of polymers is of great significance, and we cannot deny that it is highly promising in future with the developments in both the catalyst and monomers.

Acknowledgements This work was partially supported by the National Key Research and Development Program of China (2022YFA1504901)

References

1. Mandal M (2020) Group 4 complexes as catalysts for the transformation of CO_2 into polycarbonates and cyclic carbonates. J Organomet Chem 907:121067
2. Cornille A, Auvergne R, Figovsky O, Boutevin B, Caillol S (2017) A perspective approach to sustainable routes for non-isocyanate polyurethanes. Eur Polym J 87:535–552
3. Zhang B et al (2023) High-strength, self-healing, recyclable, and catalyst-free bio-based non-isocyanate polyurethane. ACS Sustain Chem Eng 11:6100–6113
4. Niesiobędzka J, Datta J (2023) Challenges and recent advances in bio-based isocyanate production. Green Chem 25:2482–2504
5. Garrison MD et al (2021) BPA-free high-performance sustainable polycarbonates derived from non-estrogenic bio-based phenols. Green Chem 23:8016–8029
6. Hai TAP et al (2021) Renewable polyurethanes from sustainable biological precursors. Biomacromol 22:1770–1794
7. Kamphuis AJ, Picchioni F, Pescarmona PP (2019) CO_2-fixation into cyclic and polymeric carbonates: principles and applications. Green Chem 21:406–448
8. Taherimehr M, Pescarmona PP (2014) Green polycarbonates prepared by the copolymerization of CO_2 with epoxides. J Appl Polym Sci 131:41141
9. Zhang Y-Y, Wu G-P, Darensbourg DJ (2020) CO_2-based block copolymers: present and future designs. Trends Chem 2:750–763
10. Grignard B, Gennen S, Jerome C, Kleij AW, Detrembleur C (2019) Advances in the use of CO_2 as a renewable feedstock for the synthesis of polymers. Chem Soc Rev 48:4466–4514
11. Zou CY, Zhang HM, Qiao LJ, Wang XH, Wang FS (2020) Near neutral waterborne cationic polyurethane from CO_2-polyol, a compatible binder to aqueous conducting polyaniline for eco-friendly anti-corrosion purposes. Green Chem 22:7823–7831
12. Gu G, Dong J, Duan Z, Liu B (2021) Construction of mechanically reinforced thermoplastic polyurethane from carbon dioxide-based poly (ether carbonate) polyols via coordination cross-linking. Polymers 13:2765
13. Gong R et al (2020) Terminal hydrophilicity-induced dispersion of cationic waterborne polyurethane from CO_2-based polyol. Macromolecules 53:6322–6330
14. Gong RN, Cao H, Zhang HM, Qiao LJ, Wang XH (2021) UV-curable cationic waterborne polyurethane from CO_2-polyol with excellent water resistance. Polymer 218:123536
15. Huang Z et al (2023) Manual applied polyurethane-urea: high performance coating based on CO_2-based polyol and polyaspartic ester. Prog Org Coat 181:107580
16. Asemani HR, Mannari V (2019) Synthesis and evaluation of non-isocyanate polyurethane polyols for heat-cured thermoset coatings. Prog Org Coat 131:247–258
17. Schimpf V, Asmacher A, Fuchs A, Bruchmann B, Muelhaupt R (2019) Polyfunctional acrylic non-isocyanate hydroxyurethanes as photocurable thermosets for 3D printing. Macromolecules 52:3288–3297
18. Warner JJ et al (2019) 3D printable non-isocyanate polyurethanes with tunable material properties. Polym Chem 10:4665–4674
19. Chen X et al (2020) Preparation and characterization of condensed tannin non-isocyanate polyurethane (NIPU) rigid foams by ambient temperature blowing. Polymers 12:750
20. Xi X et al (2019) Preparation and evaluation of glucose based non-isocyanate polyurethane self-blowing rigid foams. Polymers 11:1802
21. Boisaubert P, Kébir N, Schuller A-S, Burel F (2020) Photo-crosslinked non-isocyanate polyurethane acrylate (NIPUA) coatings through a transurethane polycondensation approach. Polymer 206:122855
22. Wołosz D, Parzuchowski PG (2022) Biobased non-isocyanate poly (carbonate-urethane) s of exceptional strength and flexibility. Polymer 254:125026
23. Wołosz D, Parzuchowski PG, Rolińska K (2022) Environmentally friendly synthesis of urea-free poly (carbonate-urethane) elastomers. Macromolecules 55:4995–5008
24. Liu T et al (2022) Synthesis and properties of high performance thermoplastic polycarbonate polyurethane elastomers through a non-isocyanate route. RSC Adv 12:30167–30173

25. Zareanshahraki F, Asemani HR, Skuza J, Mannari V (2020) Synthesis of non-isocyanate polyurethanes and their application in radiation-curable aerospace coatings. Prog Org Coat 138:105394
26. Gomez-Lopez A, Grignard B, Calvo I, Detrembleur C, Sardon H (2021) Synergetic effect of dopamine and alkoxysilanes in sustainable non-isocyanate polyurethane adhesives. Macromol Rapid Commun 42:2000538
27. Bourguignon M, Thomassin J-M, Grignard B, Jerome C, Detrembleur C (2019) Fast and facile one-pot one-step preparation of nonisocyanate polyurethane hydrogels in water at room temperature. ACS Sustain Chem Eng 7:12601–12610
28. Monie F et al (2020) Chemo-and regioselective additions of nucleophiles to cyclic carbonates for the preparation of self-blowing non-isocyanate polyurethane foams. Angew Chem Int Ed 59:17033–17041
29. Bourguignon M, Grignard B, Detrembleur C (2022) Water-induced self-blown non-isocyanate polyurethane foams. Angew Chem Int Ed 61:e202213422
30. Santos JJ et al (2023) Hybrid bisphenol a non-isocyanate polyurethane composite with mica powder: a new insulating material. J CO_2 Util 67:102303
31. Yang XX et al (2021) Preparation of non-isocyanate polyurethanes from epoxy soybean oil: dual dynamic networks to realize self-healing and reprocessing under mild conditions. Green Chem 23:6349–6355
32. Liu X et al (2021) Fully bio-based polyhydroxyurethanes with a dynamic network from a terpene derivative and cyclic carbonate functional soybean oil. ACS Sustain Chem Eng 9:4175–4184
33. Miao P et al (2022) Mechanically robust and chemically recyclable polyhydroxyurethanes from CO_2-derived six-membered cyclic carbonates. ACS Appl Mater Interfaces 15:2246–2255
34. Rodrigues JDO, Andrade CKZ, Quirino RL, Sales MJA (2022) Non-isocyanate poly (acyl-urethane) obtained from urea and castor (Ricinus communis L.) oil. Prog Org Coat 162:106557
35. Xie K, Tang D, Zhang G (2022) A nonisocyanate strategy toward polyurethane vitrimers from alkylene bisurea and epoxide through eutectic-assisted melting. Macromol Chem Phys 223:2100452
36. Ou X et al (2023) Recyclable, fire-resistant, superstrong, and reversible ionic polyurea-based adhesives. Chem Mat 35:1218–1228
37. Jiang S et al (2019) Direct synthesis of polyurea thermoplastics from CO_2 and diamines. ACS Appl Mater Interfaces 11:47413–47421
38. Wu PX et al (2020) Basic approaches to the design of intrinsic self-healing polymers for triboelectric nanogenerators. ACS Sustain Chem Eng 8:12677–12685
39. Wu PX et al (2021) A self-healing and recyclable polyurethane-urea Diels–Alder adduct synthesized from carbon dioxide and furfuryl amine. Green Chem 23:552–560
40. Shi RH et al (2022) Cyclic oligourea synthesized from CO_2: purification, characterization and properties. Green. Energy Environ 7:477–484
41. Wu PX, Wang XC, Shi RH, Cheng HY, Zhao FY (2022) A self-healing and recyclable poly (urea-imine) thermoset synthesized from CO_2. Green Chem 24:1561–1569
42. Jiang S, Shi R, Cheng H, Zhang C, Zhao F (2017) Synthesis of polyurea from 1, 6-hexanediamine with CO_2 through a two-step polymerization. Green Energy Environ 2:370–376
43. Ying Z et al (2016) Carbon dioxide as a sustainable resource for macrocyclic oligourea. Green Chem 18:2528–2533
44. Qin J et al (2019) A novel thermoplastic elastomer from double CO_2-route oligomers. Adv Ind Eng Polym Res 2:186–195
45. Zhao B, Mei H, Wang H, Li L, Zheng S (2021) Organic–inorganic polyureas with POSS cages in the main chains via polycondensation of diamines with carbon dioxide. ACS Appl Polym Mater 4:509–520
46. Jiang Y, Yang S, Pan X (2022) Biomass-based polyureas derived from rigid furfurylamine and isomannide. ACS Appl polym mater 4:2197–2204
47. White BT et al (2020) Isocyanate-and solvent-free synthesis of melt processible polyurea elastomers derived from urea as a monomer. RSC Adv 10:18760–18768

48. Lin CX, Xie KZ, Tang DL (2022) High-performance thermoplastic polyureas via a non-isocyanate route from urea and aliphatic diamines. J Appl Polym Sci 139:e52513
49. Kumar A, Armstrong D, Peters G, Nagala M, Shirran S (2021) Direct synthesis of polyureas from the dehydrogenative coupling of diamines and methanol. Chem Commun 57:6153–6156
50. Soccio M et al (2020) Urea and polyurea production: An innovative solvent-and catalyst-free approach through catechol carbonate. ACS Sustain Chem Eng 8:15640–15650
51. Quaranta E, Dibenedetto A, Nocito F, Fini P (2021) Chemical recycling of poly-(bisphenol A carbonate) by diaminolysis: a new carbon-saving synthetic entry into non-isocyanate polyureas (NIPUreas). J Hazard Mater 403:123957
52. Li T et al (2020) Novel highly branched polymer wood adhesive resin. ACS Sustain Chem Eng 8:5209–5216
53. Yang H et al (2021) Superstrong adhesive of isocyanate-free polyurea with a branched structure. ACS Appl Polym Mater 3:1638–1651
54. Owen AE et al (2022) Manganese-catalyzed dehydrogenative synthesis of urea derivatives and polyureas. ACS Catal 12:6923–6933
55. Guo J, Tang J, Xi H, Zhao S-Y, Liu W (2023) Manganese catalyzed urea and polyurea synthesis using methanol as C1 source. Chinese Chem Lett 34:107731
56. Parrino F et al (2048) Polymers of limonene oxide and carbon dioxide: polycarbonates of the solar economy. ACS Omega 3:4884–4890
57. Tamura M, Nakagawa Y, Tomishige K (2022) Direct CO_2 transformation to aliphatic polycarbonates. Asian J Org Chem 11:e202200445
58. Liang JX, Ye SX, Wang SJ, Xiao M, Meng YZ (2021) Design and structure of catalysts: syntheses of carbon dioxide-based copolymers with cyclic anhydrides and/or cyclic esters. Polym J 53:3–27
59. Xu J, Feng E, Song J (2014) Renaissance of aliphatic polycarbonates: new techniques and biomedical applications. J Appl Polym Sci 131:39822
60. Inoue S, Koinuma H, Tsuruta TJ (1969) Copolymerization of carbon dioxide and epoxide. Polym Sci Part B Polym Lett 7:287–292
61. Kathalewar MS, Joshi PB, Sabnis AS, Malshe VC (2013) Non-isocyanate polyurethanes: from chemistry to applications. RSC Adv 3:4110–4129
62. Andrea KA, Kerton FM (2021) Iron-catalyzed reactions of CO_2 and epoxides to yield cyclic and polycarbonates. Polym J 53:29–46
63. Bhat GA, Darensbourg DJ (2022) Progress in the catalytic reactions of CO_2 and epoxides to selectively provide cyclic or polymeric carbonates. Green Chem 24:5007–5034
64. Beharaj A, Ekladious I, Grinstaff MW (2019) Poly (alkyl glycidate carbonate) s as degradable pressure-sensitive adhesives. Angew Chem Int Ed 58:1407–1411
65. Beharaj A, McCaslin EZ, Blessing WA, Grinstaff MW (2019) Sustainable polycarbonate adhesives for dry and aqueous conditions with thermoresponsive properties. Nat Commun 10:5478
66. Jia MC, Hadjichristidis N, Gnanou Y, Feng XS (2019) Monomodal ultrahigh-molar-mass polycarbonate homopolymers and diblock copolymers by anionic copolymerization of epoxides with CO_2. ACS Macro Lett 8:1594–1598
67. Bailer J et al (2019) Synthesis and self-assembly of biobased poly(limonene carbonate)-block-poly(cyclohexene carbonate) diblock copolymers prepared by sequential ring-opening copolymerization. Green Chem 21:2266–2272
68. Cui SQ, Borgemenke J, Liu Z, Keener HM, Li YB (2019) Innovative sustainable conversion from CO_2 and biodiesel-based crude glycerol waste to bio-based polycarbonates. J CO_2 Util 34:198–206
69. Tran DK, Rashad AZ, Darensbourg DJ, Wooley KL (2021) Sustainable synthesis of CO_2-derived polycarbonates from d-xylose. Polym Chem 12:5271–5278
70. Gao LJ et al (2019) A novel one-pot synthesis of poly(propylene carbonate) containing cross-linked networks by copolymerization of carbon dioxide, propylene oxide, maleic anhydride, and furfuryl glycidyl ether. Polymers 11:881

71. Liao X et al (2022) A sustainable approach for the synthesis of recyclable cyclic CO_2-based polycarbonates. Chem Sci 13:6283–6290
72. Gu Y et al (2019) Highly efficient synthesis of alkyl *N*-arylcarbamates from CO_2, anilines, and branched alcohols with a catalyst system of CeO_2 and 2-cyanopyridine. ACS Sustain Chem Eng 7:6304–6315
73. Gong ZJ, Li YR, Wu HL, Lin SD, Yu WY (2020) Direct copolymerization of carbon dioxide and 1, 4-butanediol enhanced by ceria nanorod catalyst. Appl Catal B Environ 265:118524
74. Dabral S et al (2020) Synthesis and polymerisation of α-alkylidene cyclic carbonates from carbon dioxide, epoxides and the primary propargylic alcohol 1, 4-butynediol. Green Chem 22:1553–1558
75. Ouhib F et al (2019) A switchable domino process for the construction of novel CO_2-sourced sulfur-containing building blocks and polymers. Angew Chem Int Ed 58:11768–11773
76. Tounzoua CN et al (2020) A catalytic domino approach toward oxo-alkyl carbonates and polycarbonates from CO_2, propargylic alcohols, and (mono- and di-)alcohols. ACS Sustain Chem Eng 8:9698–9710
77. Yang ZF et al (2020) Cost-effective synthesis of high molecular weight biobased polycarbonate via melt polymerization of isosorbide and dimethyl carbonate. ACS Sustain Chem Eng 8:9968–9979
78. Zhang ZC et al (2019) Synthesis of high-molecular weight isosorbide-based polycarbonates through efficient activation of endo-hydroxyl groups by an ionic liquid. Green Chem 21:3891–3901
79. Qian W et al (2020) Efficient synthesis of bio-derived polycarbonates from dimethyl carbonate and isosorbide: regulating exo-OH and endo-OH reactivity by ionic liquids. Green Chem 22:5357–5368
80. Qian W et al (2020) Synthesis of bioderived polycarbonates with adjustable molecular weights catalyzed by phenolic-derived ionic liquids. Green Chem 22:2488–2497
81. Zhang ZC et al (2020) A non-phosgene process for bioderived polycarbonate with high molecular weight and advanced property profile synthesized using amino acid ionic liquids as catalysts. Green Chem 22:2534–2542
82. Zhang W et al (2022) Highly reactive cyclic carbonates with a fused ring toward functionalizable and recyclable polycarbonates. ACS Macro Lett 11:173–178
83. Saito T et al (2019) Trimethyl glycine as an environmentally benign and biocompatible organocatalyst for ring-opening polymerization of cyclic carbonate. ACS Sustain Chem Eng 7:8868–8875
84. Qiao C et al (2022) A novel catalytic route to polymerizable bicyclic cyclic carbonate monomers from carbon dioxide. Angew Chem Int Ed 61:e202205053
85. Gao Z et al (2023) Synthesis of random, gradient, and block-like terpolycarbonates via one-pot terpolymerization of epoxide, CO_2, and six-membered cyclic carbonates. Macromolecules 56:2062–2069
86. Zhu JB, Chen EY (2023) Living coordination polymerization of a six-five bicyclic lactone to produce completely recyclable polyester. Angew Chem Int Ed 57:12558–12562
87. Diaz A, Katsarava R, Puiggali J (2014) Synthesis, properties and applications of biodegradable polymers derived from diols and dicarboxylic acids: from polyesters to poly (ester amide)s. Int J Mol Sci 15:7064–7123
88. Shuai XT, Jedlinski Z, Kowalczuk M, Rydz J, Tan HM (1999) Enzymatic synthesis of polyesters from hydroxyl acids. Eur Polym J 35:721–725
89. Longo JM, Sanford MJ, Coates GW (2016) Ring-opening copolymerization of epoxides and cyclic anhydrides with discrete metal complexes: structure–property relationships. Chem Rev 116:15167–15197
90. Yue S et al (2021) Ring-opening polymerization of CO_2-based disubstituted δ-valerolactone toward sustainable functional polyesters. ACS Macro Lett 10:1055–1060
91. Lou YJ, Xu LY, Gan NL, Sun YY, Lin BL (2022) Chemically recyclable polyesters from CO_2, H_2, and 1, 3-butadiene. Innovation 3:100216

92. Chen KH et al (2022) A topology-defined polyester elastomer from CO_2 and 1,3-butadiene: a one-pot-one-step "scramblingpolymerizations" strategy. Angew Chem Int Ed 61:e202213028
93. Rapagnani RM, Dunscomb RJ, Fresh AA, Tonks IA (2022) Tunable and recyclable polyesters from CO_2 and butadiene. Nat Chem 14:877–883
94. Liang X, Tan F, Zhu YQ (2021) Recent developments in ring-opening copolymerization of epoxides with CO_2 and cyclic anhydrides for biomedical applications. Front Chem 9(20):647245
95. Li Y et al (2018) Carbon dioxide-based copolymers with various architectures. Prog Polym Sci 82:120–157

Chapter 10
Plasma-Based CO_2 Conversion

Annemie Bogaerts, Xin Tu, and Tomohiro Nozaki

Abstract This chapter discusses the applications of plasma and plasma catalysis for CO_2 conversion. After a general introduction on plasma technology and why it is interesting for CO_2 conversion, we briefly explain the most common plasma reactor types used for this application. Subsequently, we present the state-of-the-art on plasma-based CO_2 conversion including dry reforming of methane (CH_4) (DRM) in these different types of plasma reactors. This will be followed by the state-of-the-art on plasma catalysis for CO_2 conversion, including CO_2 splitting, CO_2 hydrogenation, DRM, and CO_2 reduction with water.

10.1 What Is Plasma, and Why Is It Interesting for CO_2 Conversion?

Plasma is a (partially) ionized gas, also called the "fourth state of matter", next to solid, liquid, and gas. It consists of molecules, but also electrons, various types of ions, radicals, and excited species. This "cocktail of species" makes plasma of interest for many applications [1], like in the semiconductor industry and materials technology (for thin film deposition, surface modification, etching,...), for medical applications, as light sources, chemical reactors, and for environmental and energy

A. Bogaerts (✉)
Research Group PLASMANT, Department of Chemistry, University of Antwerp, Universiteitsplein 1, 2610 Antwerp, Belgium
e-mail: annemie.bogaerts@uantwerpen.be

X. Tu
Department of Electrical Engineering and Electronics, University of Liverpool, Liverpool L69 3GJ, UK
e-mail: xin.tu@liverpool.ac.uk

T. Nozaki
Department of Mechanical Engineering, Tokyo Institute of Technology, O-Okayama, Tokyo 152-8550, Japan
e-mail: nozaki.t.ab@m.titech.ac.jp

G. Zhang et al. (eds.), *Advances in CO_2 Utilization*, Green Chemistry and Sustainable Technology, https://doi.org/10.1007/978-981-99-8822-8_10

applications, e.g., air pollution control, N_2 fixation, as well as CO_2 conversion into value-added chemicals and fuels.

Plasma occurs in nature, and above 99% of the visible matter in the universe is in plasma state, e.g., our Sun. Other examples closer to Earth are lightning and the Auroras. Plasma can also be created, by heating a gas, or by inserting electrical energy. High-temperature plasmas (with temperature of 10^6 K and more) are fully ionized and used for fusion research, while low-temperature plasmas (or gas discharges) are partially ionized, and used for the many applications listed above, including CO_2 conversion. Hence, they will be the subject of this chapter.

As mentioned above, plasma contains many different species. If they all have the same temperature (typically 10,000 K or more), the plasma is in "thermal equilibrium" ("thermal plasma"). However, often the temperature of the (light) electrons is much higher than for the other (heavy) species (i.e., several 10,000 K (or a few eV) vs. room temperature up to 1000 K), and then the term "non-thermal plasma" is used. If the gas temperature is still quite high (several 1000 K), the plasma is called "warm plasma".

In non-thermal or warm plasmas, the electrical energy supplied to the gas will selectively heat the (light) electrons, after which they will activate the gas molecules (e.g., CO_2), forming excited species, ions and/or radicals, which will quickly react into new molecules. Hence, the gas must not be heated as a whole for the chemical conversions to take place, and thermodynamically unfavorable chemical reactions, such as CO_2 splitting, or dry reforming of methane (DRM) can be performed at mild operating conditions (typically atmospheric pressure and near room temperature).

Depending on the reduced electric field in the plasma (i.e., ratio of electric field over gas number density: *E/N*) and the gas temperature, the vibrationally excited molecules can also have a higher temperature than the other plasma species, i.e., neutral gas molecules and ions. This means that the vibrational levels are overpopulated, compared to a Boltzmann distribution (representative for that gas temperature). This causes the most energy-efficient CO_2 dissociation, by so-called vibrational ladder climbing [2]. This strong vibrational excitation typically occurs when the *E/N* is around 50 Td (Townsend; $1\ \text{Td} = 10^{-21}\ \text{V m}^2$). At higher *E/N* values, the electron energy is too high for efficient vibrational excitation, and it goes into electronic excitation and ionization [2]. However, even when vibrational excitation is dominant, if the gas temperature is too high (several 1000 K), collisions with the gas molecules cause depopulation of the vibrational levels, so-called vibrational–translational (VT) relaxation.

The major reason why plasma is so appealing for CO_2 conversion (and gas conversion more in general) is because it is almost instantaneously switched on/off, making it suitable for combination with (fluctuating) renewable electricity (e.g., wind or solar energy), for energy storage in a chemical form.

Other advantages of plasma technology are its process versatility, because different types of reactions can be carried out, such as pure CO_2 splitting, but also the combined CO_2 conversion with a H-source, like CH_4, H_2 or H_2O. Furthermore, it does not make use of rare earth metals, it has a low investment cost, i.e., no expensive equipment is needed, and there is no economy of scale, which makes it useful for local, small-scale on-demand production of fuels from renewable energy. Finally, various types of renewable electricity can be applied, which increases its application versatility, as it can be operated independent of the availability of solar radiation.

In the following sections, we will first present a short description of the most common plasma reactors used for CO_2 conversion (Sect. 10.2). In Sects. 10.3 and 10.4, we will provide the state-of-the-art on plasma-based CO_2 conversion and dry reforming of methane, respectively, in these different types of plasma reactors, mostly without catalysts. Subsequently, in Sect. 10.5, plasma catalysis will be introduced, followed with the state-of-the-art on plasma-catalytic CO_2 conversion, including CO_2 splitting, CO_2 hydrogenation, DRM, and CO_2 reduction with water.

10.2 Most Common Plasma Reactor Types for CO_2 Conversion

The plasma reactor types most often studied for CO_2 conversion are a dielectric barrier discharge (DBD), microwave (MW) plasma and gliding arc (GA) discharge (see Fig. 10.1), although other plasma types are also gaining interest for CO_2 conversion, such as atmospheric pressure glow discharge (APGD) and nanosecond (ns) pulsed plasmas. Here, we briefly present their main features. The reduced electric field (*E/N*) is a very important characteristic in the plasma, as it defines the importance of vibrational excitation. The different plasma types are characterized by different *E/N* values. A DBD is a typical non-thermal plasma, with *E/N* above 100–200 Td, while MW, GA, ns-pulsed plasmas and APGDs typically operate around 50 Td and are defined as warm plasmas.

10.2.1 Dielectric Barrier Discharge (DBD)

In a DBD, an (alternating current) potential difference is applied between two electrodes, of which at least one is covered by a dielectric barrier. Most commonly, two concentric cylindrical electrodes (cf. Fig. 10.1a) are used, in which the inner electrode is surrounded by a dielectric tube, around which a mesh or foil electrode is wrapped. One of the electrodes is powered by high voltage, while the other is grounded. The gap between inner electrode and dielectric tube is typically a few millimeters and allows the gas to flow through.

DBDs can be easily scaled up, by placing a large number of reactors in parallel, as demonstrated already many years ago for ozone synthesis [3]. However, they have a quite low energy efficiency for CO_2 conversion (typically around 10% [2]). The reason is their high *E/N*, making the electrons too energetic for efficient vibrational excitation of CO_2. The energy efficiency (or performance in general) can be improved by inserting a dielectric packing in the DBD reactor. Indeed, the applied electric

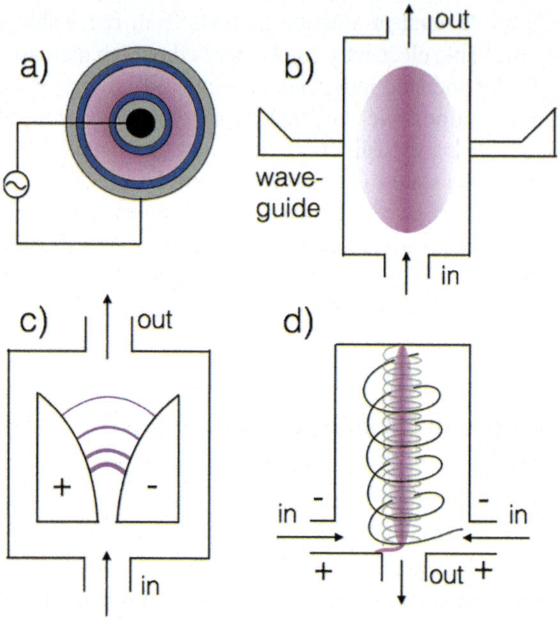

Fig. 10.1 Schematic illustration of the three plasma reactors most often used for CO_2 conversion, i.e., **a** dielectric barrier discharge (DBD), **b** microwave (MW) plasma, and gliding arc (GA) discharge, in **c** classical, flat 2D configuration and **d** cylindrical geometry, like a gliding arc plasmatron (GAP). In **a** the inner and outer electrode are indicated in gray, while the dielectric barriers are in blue. In this example, there are two dielectric barriers, but often there is only one, i.e., the outer electrode (metallic mesh or foil) is wrapped around a dielectric tube. The plasma is indicated in each panel in purple. Adopted from [6] with permission

field causes polarization of the dielectric packing beads, resulting in electric field enhancement, and thus also higher electron energy, near the contact points of the beads, for the same applied power [4], yielding more CO_2 conversion. Furthermore, such a packed-bed DBD is the most logical setup for plasma catalysis (see Sect. 10.5). On the other hand, the packing causes loss of reactive plasma species at its surface, and a shorter residence time at the same flow rate (due to the smaller discharge volume). Therefore, the CO_2 conversion and energy efficiency are not always enhanced in a packed-bed DBD [5].

10.2.2 Microwave (MW) Plasma

In a microwave (MW) plasma, electromagnetic radiation with frequency between 0.3 and 10 GHz is applied to a gas, without using electrodes. For CO_2 conversion, the most common configuration is the surface wave discharge; see Fig. 10.1b. Here, a rectangular waveguide intersects with a quartz tube transparent to MW radiation, through which the gas flows, and this causes plasma ignition. This configuration has the advantage of being electrode-free, thus avoiding electrode degradation over time.

MW plasmas can operate in a wide pressure range. At low pressure, the most energy-efficient CO_2 conversion has been reported, i.e., up to 90% for supersonic gas flow and pressures around 100–200 Torr [7], which was attributed to efficient vibration-induced CO_2 dissociation, as explained above, while at atmospheric pressure, which is more suitable for industrial implementation, the vibrational levels are depopulated by VT relaxation [8] due to the high gas temperature, and the dissociation occurs by thermal reactions.

A characteristic feature of MW plasmas for CO_2 conversion is that the plasma is radially expanded and filling up most of the quartz tube at low pressure (below 100 mbar), i.e., the so-called diffuse mode, while at higher pressure, the plasma is confined in a narrow filament in the reactor center, i.e., the so-called contracted mode [9]. Experiments revealed that the optimal conditions for CO_2 conversion coincide with the transition from diffuse to contracted mode [10, 11]. This was attributed to optimized transport of O atoms from the hot plasma to the cold edges, where they can react with unconverted CO_2, forming more CO, instead of recombining with CO in the hot core, forming again CO_2 [12].

Furthermore, recent insights indicate that at sub-atmospheric pressure, in both modes, but especially in the contracted mode, the CO_2 conversion proceeds by thermal dissociation [13]. Indeed, gas temperatures of 3000–5000 K were measured in the hot core of MW plasmas at sub-atmospheric pressure, with a sudden rise to above 6000 K when approaching atmospheric pressure [9, 11, 14]. The abrupt increase in temperature is accompanied by plasma contraction [10, 11, 15], also called power concentration, to indicate that it is more than just a shrinkage of the plasma parameters [14].

10.2.3 Gliding Arc (GA) Plasma

A classical (2D) gliding arc (GA) plasma (Fig. 10.1c) is created by applying a potential difference between two flat diverging electrodes. The arc is formed at the shortest interelectrode distance, but due to the gas flow, it "glides" along the electrodes toward larger interelectrode distance, until it extinguishes, and a new arc is formed at the shortest interelectrode distance. This 2D design, however, has some drawbacks, because a large fraction of the gas does not pass through the arc, and the gas flow must be high to have smooth arc gliding. Both effects limit the conversion.

To overcome these limitations, other (3D) GA designs were developed, which are based on cylindrical electrodes. Examples are the so-called gliding arc plasmatron (GAP; see Fig. 10.1d) [16], but also various other types of (rotating or magnetically stabilized) GA reactors [17]. In such 3D designs, the gas typically enters tangentially, and an arc is created around the reactor axis, in the longitudinal direction. In the ideal case, all the gas passes through the arc for maximum conversion.

GA plasmas typically operate at atmospheric pressure, and exhibit fairly good energy efficiency, i.e., around 20–40% for CO_2 splitting [16, 18] and 60% for DRM [19] (see more details in Sect. 10.4). Due to the high gas temperature (around 3000 K [20]), the CO_2 conversion again largely proceeds thermally. The conversion and energy efficiency, both in a GA and MW plasma, can be further enhanced by fast quenching (gas cooling) immediately after the plasma, as shown in Sect. 10.3 below.

10.2.4 Other Plasma Reactors Used for CO_2 Conversion

Other plasma reactor types are also gaining increasing interest for CO_2 conversion or DRM. Examples are (i) ns-pulsed plasmas [21], in which a strong thermal non-equilibrium can be created, due to the high power applied in short pulses, and the fact that the gas can drop again in between pulses, (ii) spark discharges [22], which are based on streamers between two electrodes, developing into highly energetic spark channels that extinguish and reignite periodically, as well as (iii) corona discharges [23], which are created near sharp electrode edges, but just like DBDs, they are typically not very efficient for CO_2 conversion due to too high *E/N* values.

Finally, atmospheric pressure glow discharges (APGDs) are also particularly interesting, for both CO_2 conversion [24] and DRM [25], due to their simple design and homogeneous glow, so that most of the gas passes through the plasma. They exhibit a gas temperature of around 2500 K [24], so that the CO_2 conversion is also dominated by thermal dissociation.

10.3 State-of-the-Art on Plasma-Based CO_2 Conversion

Figure 10.2 gives an overview of the energy efficiency as a function of CO_2 conversion, grouped per plasma type, for the three main plasma reactors used in literature, i.e., DBD (note: "other" mainly stands for corona discharges), MW plasmas (including low-pressure RF plasmas), and GA plasmas. This figure was first constructed in [2], based on a critical analysis of all data reported in literature up to 2017, and was recently updated with more recent publications [26]. The figure also indicates the efficiency target of 60%, that was defined in [2] to be competitive with classical and other emerging technologies.

DBD plasmas yield a reasonable conversion (up to 40%), but the energy efficiency is low, typically around 5–10%, with some exceptions up to 15–20%, i.e., by applying the power in so-called burst mode [27]. This is, however, still a factor 3–4 too low to reach the 60% efficiency target (see Fig. 10.2). As mentioned in Sect. 10.2.1 above, packed-bed DBD reactors can sometimes improve the performance, as demonstrated, e.g., in [28, 29], where a simultaneous rise by a factor 2 in both conversion and energy efficiency was obtained, attributed to the electric field enhancement due to polarization of the packing beads. Furthermore, TiO_2 and $BaTiO_3$ packings can also yield

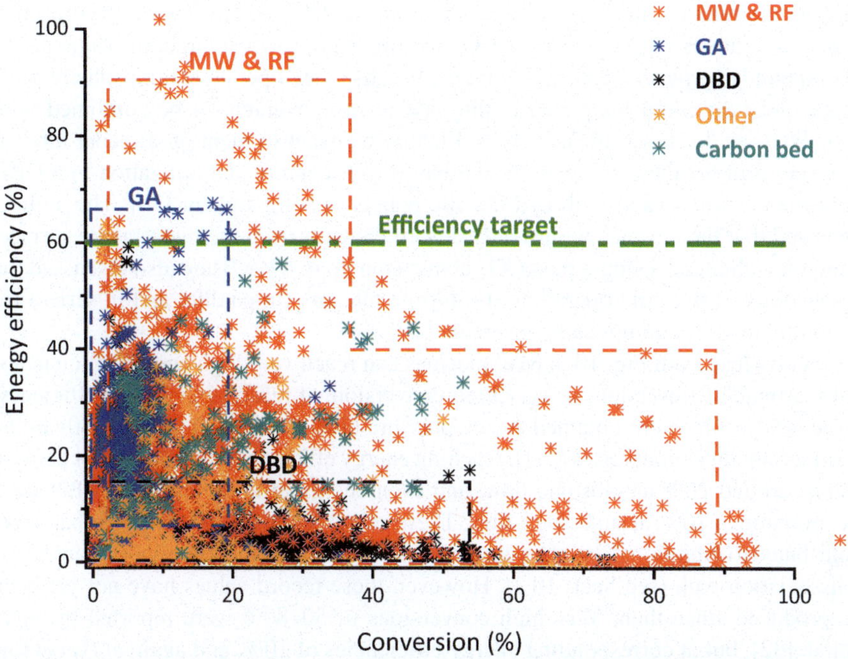

Fig. 10.2 Comparison of all data collected from literature for CO_2 splitting in the most common plasma types, showing the energy efficiency as a function of the CO_2 conversion. The 60% efficiency target, defined in [2], is also indicated. The figure was originally made in [2], and later updated in [26] with more recent data. Adopted from [26] with permission

a (photo)catalytic effect, by the formation of electron–hole pairs due to the plasma electrons, causing oxygen vacancies, thus acting as active sites for adsorption and activation of the reactants, as discussed in [29]. Nevertheless, as also mentioned in Sect. 10.2.1 above, packed-bed DBD reactors do not always give rise to improvement (e.g., [5, 28]), because the packing also causes a smaller plasma volume, and thus it reduces the residence time at constant flow rate, which negatively affects the conversion. Many factors, such as discharge gap, packing material, packing (bead) size, as well as applied power and flow rate, determine whether the positive effect of electric field enhancement or the negative effect of smaller plasma volume will be dominant, resulting in either better or worse performance compared to an empty reactor. Furthermore, the influence of the packing material is very complex, and not fully understood. The dielectric constant of the packing is thought to play a major role, as it affects the polarization, and thus the electric field enhancement, but it is clearly not the only parameter affecting the performance, because the trends in CO_2 conversion and energy efficiency do not entirely follow the trends in dielectric constant [5]. In general, it is clear from Fig. 10.2 that DBDs, even with a packing, exhibit too low energy efficiency to be competitive with other emerging technologies, or even with other plasma reactor types, at least for pure CO_2 splitting, despite their inherent strengths, like ease of operation and scalability. However, DBD reactors are more interesting for DRM or CO_2 hydrogenation, when targeting value-added compounds by plasma catalysis. Indeed, because they operate at atmospheric pressure and near room temperature, they are most convenient to be combined with catalysts, as discussed in Sect. 10.5. Finally, we want to mention a recent innovative concept for the combined plasma-based CO_2 capture and utilization in a DBD plasma reactor packed with hydrotalcite pellets as solid sorbent [30]. The authors showed that the plasma can instantly desorb the CO_2 from the hydrotalcite surface, and they obtained a single pass CO_2 conversion up to 67%. They also discussed the possibility of periodic operation of multiple reactors (in parallel and in series) for continuous CO_2 capture and conversion [30].

As is clear from Fig. 10.2, MW plasmas can reach very high energy efficiencies, at reasonable conversion, or very high conversion, at reasonable energy efficiency. The best results were obtained at low pressure (100–200 mbar) in the 1980s in the former Soviet Union (e.g., [7, 31], reaching energy efficiencies up to 80%, for conversions around 20% at subsonic flow, and even energy efficiencies up to 90%, for a conversion of 10%, at supersonic flow. These values, obtained at 100–200 mbar, were attributed to vibrational-induced dissociation, which is the most energy-efficient CO_2 dissociation path (see Sect. 10.2). However, these record values have not yet been reproduced since then. Very high conversions of 50–80% were reported by Silva et al. [32], but at corresponding energy efficiencies of 10%, and again at (very) low pressure (1–10 mbar). Research at DIFFER (the Netherlands) for MW plasmas at (sub)atmospheric pressure achieved maximum energy efficiencies around 40–50%, for a conversion around 10–20%, at 100–200 mbar [10], and up to 600 mbar [33], or maximum conversions up to 80% for energy efficiencies of 20%, in quenching experiments at 200 mbar [33]. At (or near) atmospheric pressure, however, more typical values for conversion around 10% at energy efficiencies around 30% were reported, both at DIFFER and IPP (Germany) [11, 34]. However, recent work at both groups showed that a quenching nozzle can improve the conversion. A so-called converging–diverging nozzle yielded 21% relative rise in both energy efficiency and CO_2 conversion at 700 mbar, to values of 31% energy efficiency and 13% conversion, while at 300 mbar, 71% relative increase was achieved, yielding an energy efficiency of 23% and a conversion of 28% [35]. Likewise, Hecimovic et al. [36] reported significant improvements in conversion by a factor 7 upon adding a quenching nozzle, at 5 L/min, 1500 W and 900 mbar, i.e., from 5% without nozzle, up to 35% for a 2.5 mm nozzle (at an energy efficiency of about 20%). In all these cases, as the gas temperature is typically 3000–5000 K (and even up to 6000 K), there is no overpopulation of the vibrational levels, and the conversion is dominated by thermal dissociation, as also confirmed by computer modeling, where at atmospheric pressure and corresponding high temperature, the vibrational distribution function (VDF) is nearly thermal [8]. Indeed, a strong non-equilibrium VDF with pronounced overpopulation of the higher vibrational levels, can only be obtained at low pressure and low gas temperature [8].

Finally, GA plasmas exhibit energy efficiencies for CO_2 splitting of 20–40% (e.g., [16, 17, 37–39]) with some exceptions above the efficiency target of 60%. Just like in MW plasmas, vibrational excitation is important, but due to the high temperature (calculated to be around 3000 K [20]), the vibrational levels are not overpopulated, and thermal chemistry again dominates. The conversion is typically limited to about 10%, with some exceptions up to 20% (see Fig. 10.2). This might be attributed to the fact that only a limited fraction of gas passes through the active arc plasma [20], but also by recombination of the products after the plasma (see below). Smart reactor design, to enhance the processed gas fraction, might lead to improved performance. This was demonstrated for a so-called dual vortex plasmatron [40], enhancing the energy efficiency, but the conversion was still limited to about 10%, although this was attributed to limitations of the power supply. A successful example of improved conversion, due to smart reactor design, was however illustrated for an APGD reactor [24]. Indeed, a so-called confined APGD gave three times higher conversion than a basic APGD design, because all the gas was allowed to pass through the plasma.

On the other hand, the reactor design of some cylindrical GA plasma reactors seems already quite optimized, like for the GAP [16], because design modifications only gave rise to minor improvements [26]. Therefore, efforts are being made already for GA upscaling, currently handling CO_2 feed flow rates of 100 L/min (i.e., 100 ton/year), as a first step toward industrial exploitation.

The fact that the CO_2 conversion of both atmospheric pressure MW plasmas and GA plasmas (as well as other atmospheric pressure plasmas, such as APGDs) seems mostly limited to the order of 10%, with energy efficiency around 30%, for various reactor configurations, suggests that the conversion inside the reactor might be much higher (even up to 100%, at the high temperatures occurring in these reactors), but that limitations are caused by recombination of CO back into CO_2, once the gas slowly cools down after leaving the plasma:

$$CO + O + M \rightleftarrows CO_2 + M$$

$$CO + O_2 \rightleftarrows CO_2 + O$$

Hence, further improvements are possible by limiting these recombination reactions after the plasma. Fast cooling of the gas after the plasma can quench the recombination reactions, thereby “freezing” the conversion [34]. Modeling revealed that this can enhance the conversion and energy efficiency by up to a factor 3 [41]. In addition, next to this so-called “ideal quenching”, it is also possible to even further dissociate CO_2 upon quenching, in so-called “super-ideal quenching”, when vibrational non-equilibrium upon a fast drop in gas temperature promotes further CO_2 dissociation [41]:

$$CO_2(v) + O \rightleftarrows CO + O_2,$$

in which $CO_2(v)$ represents a vibrationally excited CO_2 molecule.

As mentioned above, the beneficial effect of quenching was recently demonstrated for MW plasmas [35, 36], and the significant improvements of the latter experiments (i.e., sevenfold increase in conversion) were also explained by modeling [42].

In addition, we believe that further improvements in plasma-based CO_2 conversion can also be made by focusing on the gas flow dynamics inside the reactor (e.g., by smart reactor design), to have most of the gas passing through the active plasma, and also to improve mixing between hot and cold gas, for bringing energetic O atoms in contact with unconverted CO_2, producing more CO, as mentioned in Sect. 10.2.2 above. van de Steeg et al. [12] indeed suggested that the theoretical efficiency limit of thermal CO_2 dissociation (~52%) may be increased to ca. 70%, by controlling the transport of O atoms from the hot plasma core toward the colder edges, reacting with unconverted CO_2, instead of recombining with CO into CO_2. The same authors also reported that the enhanced performance in their MW plasma reactor is a consequence of thermal conversion in the plasma core, and vibrational non-equilibrium in the immediate surroundings, allowing this super-ideal quenching reaction (see above) [12]. The plasma core is stated to be in chemical non-equilibrium, due to transport as a result of the steep gradients between the hot plasma core and the cooler periphery [12]. Fluid dynamics simulations would be very useful to get more insights into this transport. For a classical GA, model calculations demonstrated that an improvement in treated gas fractions could be realized by a higher relative velocity between arc and gas flow [37, 38].

Other solutions to limit the recombination of O/O_2 with CO after the plasma, besides fast quenching, could be to remove the O/O_2 from the gas mixture, e.g., by scavenging materials or chemicals [43], by membranes [44], in a hybrid reactor composed of a DBD with solid oxide electrolysis cell [45], or by means of a carbon bed. The latter has indeed been successfully applied for various plasma reactors [46–48]. On one hand, the O atoms and O_2 molecules react with the C atoms, to form extra CO, and they are consumed, hence not available anymore for recombination reactions with CO. On the other hand, if the temperature in the carbon bed is high enough, the reverse Boudouard reaction can take place, in which unconverted CO_2 reacts with C, forming two CO molecules. A thermal arc plasma reactor with post-plasma carbon bed yielded record values for the CO_2 conversion of 67–94% with corresponding energy efficiency of about 70%, hence much higher than ever reported for other (non-thermal or warm) plasmas [46]. Likewise, a post-plasma carbon bed after a GAP reactor could enhance both the overall CO_2 conversion and energy efficiency by a factor 2, while the CO concentration was increased by even a factor 3, and nearly all O_2 was removed from the product mix, thus greatly reducing separation costs [48]. However, more research is needed to maintain this improved performance over a long time-scale. Note that the O_2 removal also has great potential in combination with gas recirculation, or placing plasma reactors in series. It was demonstrated in [49] that a recycle flow can drastically increase the CO_2 conversion (e.g., by a factor 3 compared to a single-pass plasma treatment), but O_2 removal in combination with recycling further improves the performance, because it eliminates recombination of O/O_2 with the formed CO in the next pass, and shifts the chemical balance toward the products [49].

A similar effect as in a post-plasma carbon bed was recently demonstrated by plasma-assisted chemical looping [50]. A nanostructured CeO_2/Fe_2O_3 oxygen scavenger, pre-reduced by H_2 plasma, was placed post-plasma, to capture the produced oxygen species and suppress CO + O recombination, reaching an overall CO_2 conversion at the reactor outlet of ca. 29%. Such conversion can only be achieved at temperatures above 2775 K according to chemical equilibrium calculations, hence indicating that post-plasma scavenging by chemical looping materials can drastically overcome chemical equilibrium limits of CO_2 conversion.

10.4 State-of-the-Art on Plasma-Based Dry Reforming of Methane (DRM)

Plasma-based DRM is gaining more and more interest, as it can valorize biogas for the production of syngas (CO/H_2 mixture), used for the further synthesis of value-added chemicals and fuels, through the Fischer–Tropsch process or for methanol synthesis (and subsequently methanol or ethanol to olefin synthesis). Figure 10.3 summarizes the energy cost versus total conversion, for all types of (major) plasma reactors, collected from all results published in literature. Note that the total conversion is calculated as the sum of the effective conversions of both CO_2 and CH_4 (hence accounting for the fraction of these components in the gas mix), to have only one parameter and allow comparison for different CO_2/CH_4 ratios in one figure. The original figure was made in [2], but was updated in 2021 with more recent data [25]. Note that we plot the energy cost now, instead of the energy efficiency like in Fig. 10.2 for pure CO_2 splitting, because the energy efficiency can only be determined if all possible reaction products are taken into account. However, most papers in literature only report the syngas components (CO and H_2) as products, and sometimes the light hydrocarbons. This makes it not possible to calculate the true energy efficiency, while the energy cost can be calculated without knowledge of all reaction products. To allow easy comparison with Fig. 10.2 for pure CO_2 splitting, we have reversed the y-axis in Fig. 10.3.

The green dash-dotted line in Fig. 10.3 again indicates the efficiency target. We assume the same 60% efficiency target, which corresponds to an energy cost of 4.27 eV/molecule when syngas is the main product of DRM [2]. Note however, that the conversion of syngas into value-added chemicals and fuels through the Fischer–Tropsch process or for methanol synthesis also requires a lot of energy. Thus, if plasma (catalysis) can directly produce these value-added compounds in one step, with sufficient yields, the efficiency target can be drastically reduced, because the energy-intensive step of processing syngas into the desired products can be avoided. However, major research efforts are needed to better understand how these desired compounds can be selectively produced, especially in terms of catalyst choice and design. This will be discussed in more detail in Sect. 10.5.

Just like for CO_2 splitting, most research on DRM was carried out with DBDs, especially in combination with (catalytic) packing materials. More details on the performance of plasma-catalytic DRM will be given in Sect. 10.5, but the data points in Fig. 10.3 comprise both the results without and with catalysts. As is clear from

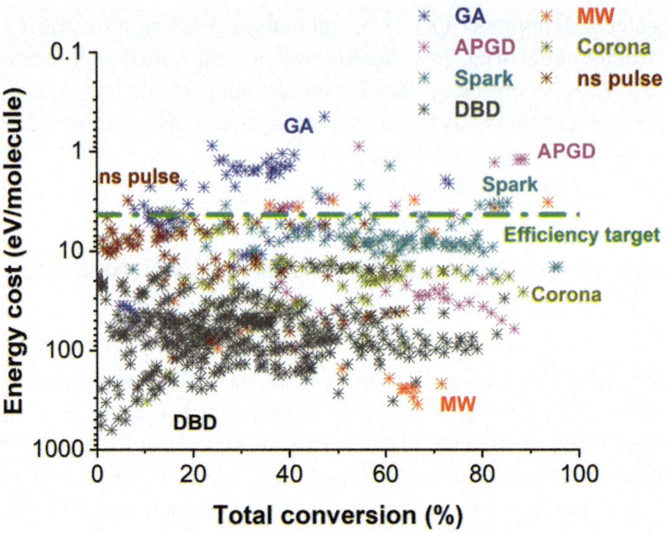

Fig. 10.3 Comparison of all data collected from literature for DRM in the most common plasma types, plotting the energy cost vs total conversion. The target energy cost of 4.27 eV/molecule is also indicated. It corresponds to the 60% efficiency target, in case of syngas production. Note that the y-axis is reversed to be more comparable with Fig. 10.2. The figure was originally made in [2], and later updated in [25] with more recent data. Adopted from [25] with permission

Fig. 10.3, a DBD provides reasonable conversion (up to 60% and up to 80% for non-packed and packed-bed DBD reactors, respectively), but the energy cost is around 20–100 eV/molecule (and even above), corresponding to an energy efficiency of 12.8–2.6%, when only syngas is produced. This energy cost is at least a factor 5 too high (or the efficiency a factor 5 too low) compared to the target mentioned above. Modeling for a very wide range of conditions predicted that by careful selection of the operating conditions, 16.9 eV/molecule would be the lowest energy cost that can be reached [51]. This is still clearly too high for being competitive with other (emerging) technologies, unless of course suitable catalysts can be found for the direct production of value-added compounds with high enough yields, which would drastically lower the efficiency target, as explained above.

For MW plasmas, only a few experiments on DRM have been reported in literature up to now; they typically present a high conversion, but also a high energy cost, with only a few exceptions (see Fig. 10.3). This is in stark contrast to the results for pure CO_2 splitting in MW plasmas, where very high energy efficiency is typically reported (see previous section). Some recent papers, however, show that MW plasmas are also very promising for DRM. For instance, in [52], energy costs around 3.6 eV/molecule, with CO_2 and CH_4 conversions of 68% and 97% were achieved. In [53], energy costs of 3.2 eV/molecule were reported, with CO_2 and CH_4 conversions both above 90%, and in [54], energy costs of 2.8–3.0 eV/molecule were obtained, with CO_2 and CH_4 conversions up to 50 and 67%, respectively.

For GA plasmas, very promising results have been reported for DRM, with total conversions up to 30–50% and energy costs down to 1–2 eV/molecule (see Fig. 10.3), hence clearly below the energy cost target of 4.27 eV/molecule (or energy efficiency target of 60%) for syngas production. Some examples can be found in [19, 55–61]. In pure CH_4/CO_2 mixtures, it is often difficult to reach stable plasma conditions at high CH_4 fractions, due to coking [19]. Adding N_2 [58] and O_2 [59] helps to get higher CH_4/CO_2 ratios, and also improves the performance, as they assist in the conversion process. In [58], CO_2 and CH_4 conversions of 29% and 36% were obtained, with an energy cost of 2.2 eV/molecule, while in [59], CO_2 and CH_4 conversion above 50% and up to 99% were reported, with total energy cost of 3.4 eV/molecule. Note that this number is higher than the record values of 1–2 eV/ molecule, because part of the applied power is also used for N_2 (and O_2) activation. In [60] adding O_2, as well as a Ni-based catalyst after the GA, in a heat-insulated reactor, yielded a spectacular enhancement in energy efficiency (up to 86%), with absolute CH_4 and CO_2 conversions of 92% and 23%, respectively. A similar setup provided an energy efficiency of 79% at an absolute CH_4 and CO_2 conversion of 99% and 79%, respectively [61]. This study was based on an interesting concept on the combination of GA, for plasma-catalytic reforming, with water electrolysis, to produce high-quality syngas from CH_4, CO_2, and H_2O.

Also other plasma types have been investigated for DRM, such as ns-pulsed plasmas [21], spark discharges [22] and APGDs [25], yielding very promising results. For instance, in [25] CO_2 and CH_4 conversions of 64 and 94% were obtained, at an energy cost of 3.5–4 eV/molecule, hence better than the efficiency target. However, more research is needed in these plasma types to further explore their potential.

In general, plasma-based DRM seems very promising for syngas production, especially in warm plasmas, reaching good conversion and low energy costs, clearly better than the target defined in [2]. The CH_4/CO_2 ratio in the feed gas has a determining role in both conversion and energy cost, as well as product distribution and syngas ratio. While at low CH_4/CO_2 ratios, part of the H atoms is lost into H_2O formation, at high CH_4/CO_2 ratios, carbon deposition can become a major problem. DBD plasmas also exhibit reasonable conversion, but an energy cost that is about five times too high. However, a DBD is very suitable for integrating catalysts in the reactor, and if catalysts can be found that can selectively produce value-added compounds, the efficiency target could be lowered (or in other words, a higher energy cost would still be competitive), as explained above. This will be further elaborated in Sect. 10.5.

10.5 Plasma-Catalytic CO_2 Conversion

10.5.1 Why Plasma Catalysis is Promising for CO_2 Conversion

Plasma-based technology has shown its promise for CO_2 conversion, as demonstrated in previous sections. However, a wide range of products can be produced, especially in plasma-based DRM and CO_2 hydrogenation, depending on the operating conditions, types of plasma and reactor configurations. Such indiscriminate plasma chemistry does not favor the selective synthesis of fuels and chemicals from CO_2. The combination of plasma and heterogeneous catalysis, known as plasma catalysis, is an emerging solution for the conversion of CO_2 into desirable products. This promising approach combines the highly energetic nature of the plasma, generating energetic electrons and highly reactive species that enables the activation of CO_2 at room temperature and atmospheric pressure, with the selective nature of catalysis to accelerate the reaction rates and improve selectivity toward target fuels and chemicals. Catalysts play a critical role in plasma catalysis and provide an active surface for the adsorption and activation of CO_2 and aid in the subsequent reactions.

Two reactor configurations are typically used in plasma catalytic gas conversion: single-stage plasma catalysis and two-stage plasma catalysis. The single-stage configuration places the catalysts within the plasma discharge region, allowing direct interaction between the reactive species and catalyst surfaces, therefore also known as in plasma catalysis (IPP). The single-stage process has the potential to enhance reaction performance and is widely used for plasma-catalytic CO_2 conversion. In the two-stage plasma catalysis system, catalysts can be placed either before or after the plasma region. Post-plasma catalysis (PPC), where catalysts are typically positioned after the plasma region, is more commonly employed in a two-stage plasma catalysis system. In PPC, reactants undergo plasma gas-phase reactions, followed by catalytic surface reactions. The catalyst bed might be heated to facilitate thermal catalytic reactions. In PPC, only long-lived species and end-products from the plasma interact with the catalyst surface, while short-lived vibrationally excited species or radicals do not significantly contribute, particularly in CO_2 conversion processes. PPC has received less attention in plasma catalytic CO_2 conversion. Therefore, single-stage plasma catalysis or IPP will be the main focus in this section.

The interaction between plasma and the catalyst surface in IPP is a complex process that occurs through multiple pathways (Fig. 10.4) [62–64]. Plasma-generated species, including excited species and radicals, can directly interact with the catalyst surface. These species may undergo adsorption onto reactive sites, such as metal sites and oxygen vacancies, leading to the formation of intermediates. These intermediates play a crucial role in subsequent chemical reactions, following the Langmuir–Hinshelwood (L–H) mechanism. Additionally, plasma-induced gas species, can react with adsorbed species on the catalyst surfaces via the Eley–Rideal (E–R) mechanism. This dynamic equilibrium between plasma-assisted adsorption and desorption

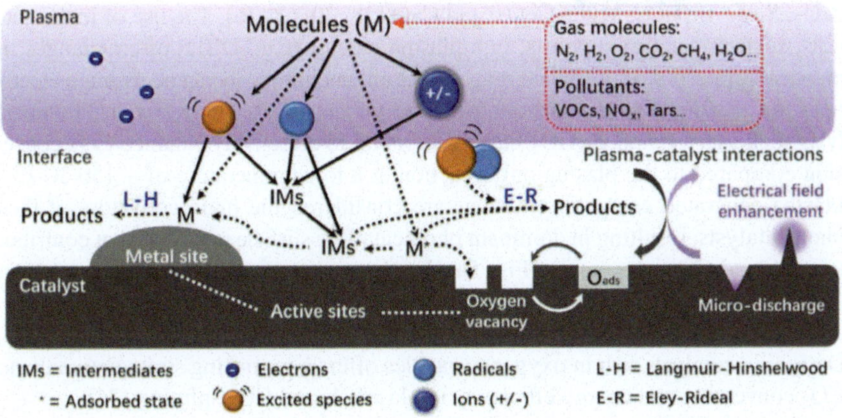

Fig. 10.4 Plasma–catalyst interface and plasma–catalyst interactions. Reproduced from [63] with permission

processes contributes to the overall catalytic activity. The plasma–catalyst interactions have the potential to create an enhanced electrical field, micro-discharges, and hot spots, which further facilitate the desired reaction pathways in both the plasma gas phase and plasma–catalyst interface. The physical and chemical interactions between plasma and catalysts have the potential to generate a synergistic effect, wherein the reaction performance of plasma catalysis can exceed the combined effectiveness of the catalytic process and plasma process alone under same reaction conditions. The synergy of plasma catalysis for CO_2 conversion including CO_2 splitting, CO_2 hydrogenation and DRM has been reported several times [29, 65–67], although there are also examples where plasma catalysis exhibits no synergy. In the following subsections, different pathways for plasma-catalytic CO_2 conversion will be discussed, including CO_2 splitting, CO_2 hydrogenation, DRM, and CO_2 reduction with water, followed by a brief section on limitations of plasma catalysis.

10.5.2 Plasma-Catalytic CO_2 Splitting

Plasma-based CO_2 splitting without using a catalyst typically produces CO and O_2 with a high selectivity of CO (95–100%). Therefore, the use of catalysts in plasma-based CO_2 splitting has limited potential to further improve the selectivity of CO. However, combining plasma with appropriate catalysts has the potential to enhance the CO_2 conversion, CO yield, and energy efficiency. Additionally, plasma catalysis has great potential to overcome the trade-off between CO_2 conversion and energy efficiency observed in plasma-based CO_2 splitting without a catalyst.

Many inert materials and catalysts have been investigated for plasma-based CO_2 splitting, including glass wool, glass beads, quartz wool, Al_2O_3, Ni/Al_2O_3, $CaTiO_3$,

ZrO_2, SiO_2, $BaTiO_3$, MgO, CeO_2, and CaO [28, 29, 68–76]. The use of inert materials, such as catalyst supports, in a plasma reactor (e.g., DBD) often changes the physical properties of the discharge, such as enhancing the local and average electric field [4, 28, 76], which could contribute to enhanced CO_2 conversion. Photocatalysts such as TiO_2 and $BaTiO_3$ have been found to effectively enhance CO_2 conversion compared to the plasma-only reaction at a low temperature of ~ 150 °C [29]. Plasma-generated energetic electrons are considered the main activators of these photocatalysts, resulting in dominant photocatalytic surface reactions that contribute to enhanced CO_2 conversion (Fig. 5a). Furthermore, oxygen vacancies on catalyst surfaces, such as CeO_2-based catalysts (FeO_x/CeO_2 and Mo-doped CeO_2), have been demonstrated to serve as active sites for enhancing CO_2 conversion (Fig. 5b) [74, 75]. Designing catalysts rich in oxygen vacancies offers a promising strategy to enhance CO_2 conversion and energy efficiency in plasma-based CO_2 splitting [74].

Although DBD has been extensively used for plasma-catalytic CO_2 splitting, the combination of GA and MW plasmas with catalysts has also been explored [77, 78]. For example, NiO/TiO_2 was placed in the post-discharge area of an MW plasma reactor, and this post-plasma catalysis system doubled the CO_2 conversion compared to the MW plasma-only system [71]. Zhang et al. found that placing TiO_2 in the discharge region of a GA reactor boosted CO_2 conversion and energy efficiency by 135% compared to the reaction in the absence of TiO_2 or placing TiO_2 downstream of the GA [77].

10.5.3 Plasma-Catalytic DRM

Although plasma-catalytic DRM has been carried out using different plasma systems, including DBD, plasma jet and GA [55, 79, 80], DBD is the most widely used plasma reactor for plasma-catalytic DRM.

Supported metal catalysts have been extensively investigated in plasma-catalytic DRM [81]. Specifically, the related research focuses on three main aspects: active metals, promoters, and supports. Both transition metals (e.g., Ni, Cu, Co, Fe, Mn) and noble metals (e.g., Pt, Pd, Ag) have been studied for the plasma-catalytic DRM process over the past few decades [66, 81–85]. Among them, Ni/Al_2O_3 has emerged as the most commonly used catalyst for the plasma-catalyzed DRM process [81, 84, 86–92]. Different Ni-based bimetallic catalysts have also been designed for plasma-catalytic DRM, including NiCu, NiCo, NiMn, and NiFe [93–95]. The formation of metal alloys and enhanced reducibility of bimetallic catalysts could contribute to the enhanced reaction performance [95]. Additionally, the metal loadings and metal loading ratio are critical to determine the performance of this reaction [94]. Ni-based bimetallic catalysts (NiCu, NiCo, NiMn) have also been shown to tune the selectivity of gas products, favoring the generation of C_3–C_4 hydrocarbons and syngas while reducing the production of C_2 hydrocarbons [95].

Adding promoters to metal-based catalysts has been demonstrated as an effective method for enhancing plasma-catalytic processes by modifying the chemical and

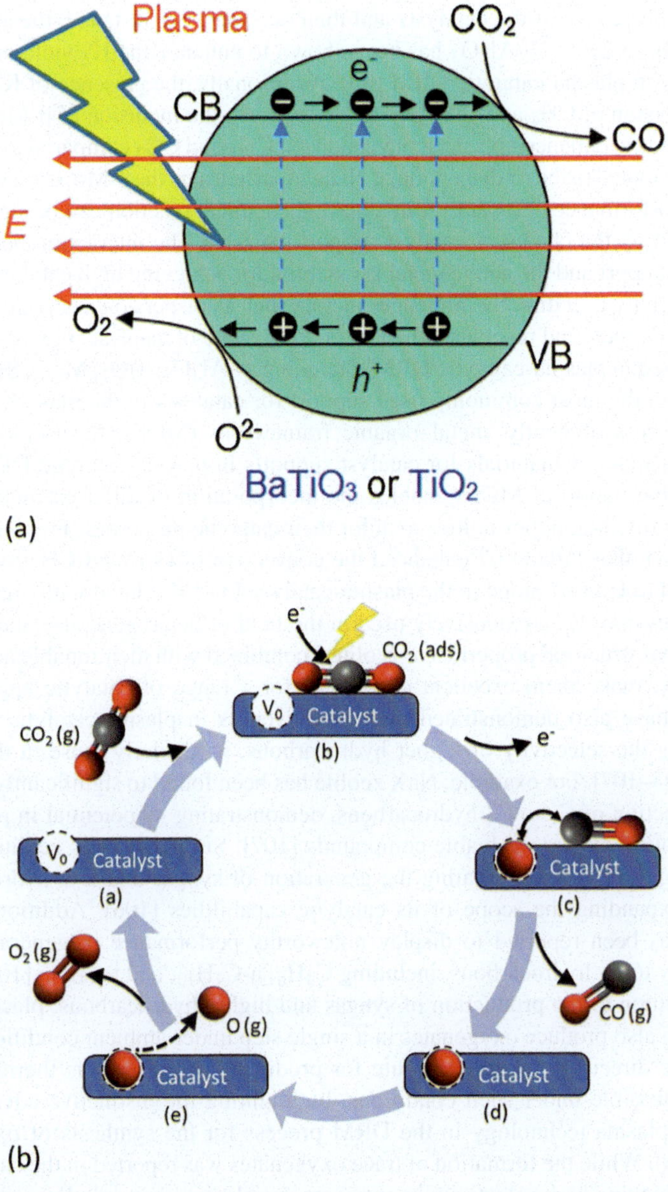

Fig. 10.5 **a** Reaction mechanisms of plasma-catalytic CO_2 splitting on the surface of a photocatalyst. Reproduced from [29] with permission; **b** CO_2 dissociation mechanism via dissociative electron attachment at the catalyst surface. Reproduced from [75] with permission

physical properties of the catalysts and their surfaces. For instance, the addition of K, Ce, Mg or La to Ni/Al_2O_3 has been shown to enhance the H_2 yield and energy efficiency in plasma-catalytic DRM [96]. Additionally, the presence of K promotes the conversion of CH_4 and CO_2 [96–99]. Moreover, the inclusion of β-Mo_2C in Ni/Al_2O_3 leads to enhanced plasma conversions of CH_4 and CO_2 compared to Ni/Al_2O_3 alone, attributed to the re-dispersion of Ni nanoparticles on the β-Mo_2C surface [100].

The performance of a catalyst in a plasma chemical reaction can be significantly influenced by the choice of catalyst support material. The interaction between the catalyst support and the active metal is essential for achieving high catalytic activity and selectivity as it directly influences factors such as dispersion, accessibility, electronic properties, and reactant diffusion. A wide range of catalyst supports has been investigated in plasma-catalytic DRM, including γ- Al_2O_3, TiO_2, MgO, SiO_2 [101]. γ-Al_2O_3 is the most commonly used support for catalysts in the plasma-catalyzed DRM process. Recently, metal–organic frameworks (MOFs) have emerged as a promising class of materials for catalyst supports in plasma-catalytic DRM [102]. The tunable nature of MOFs enables the incorporation of different metal ions or clusters, providing opportunities to tailor their catalytic properties. Previous studies have shown that Pt/UiO-67 enhanced the conversion of CO_2 and CH_4 by 10–20% compared to UiO-67 alone in the plasma-catalyzed DRM diluted with argon [102].

Zeolites have been extensively used in the field of heterogeneous catalysis. The exceptional structural properties of zeolites, combined with their tunable acidity and selectivity, make them excellent candidates for a range of catalytic applications. Zeolites have also demonstrated their effectiveness in plasma-catalytic DRM by enhancing the selectivity of higher hydrocarbons, particularly those in the C_2–C_4 range [103–107]. For example, NaX zeolite has been found to significantly enhance the production of C_2 to C_4 hydrocarbons, demonstrating its potential in promoting the formation of these valuable compounds [107]. Similarly, NaY zeolite exhibits promising activity in facilitating the generation of syngas and C_{5+} hydrocarbons, further expanding the scope of its catalytic capabilities [106]. Additionally, HY zeolite has been reported to display noteworthy performance in generating high selectivity to C_4 hydrocarbons, including C_4H_8, n-C_4H_{10}, and i-C_4H_{10} [105].

In addition to the production of syngas and higher hydrocarbons, plasma-based DRM can also produce oxygenates in a single step under ambient conditions [108–117]. The direct dry reforming route for producing oxygenates is thermodynamically unfeasible under such conditions, highlighting the distinctive advantage of utilizing plasma technology in the DRM process for the synthesis of oxygenates (Fig. 10.6). While the formation of trace oxygenates was reported in the early 2000s [117], achieving high selectivity in oxygenate production has only been reported in recent years [116]. In 2017, Wang et al. reported a direct, one-step plasma process for converting CO_2 and CH_4 into high selectivity oxygenates (e.g., acetic acid, methanol, ethanol, and acetone) at ambient pressure (1 bar) and room temperature (30 °C) using a water-cooled DBD reactor. The total selectivity to oxygenates, which were condensed and measured using off-line gas chromatography (GC), was approximately 50–60% without a catalyst, with acetic acid being the major liquid product [116]. However, the use of a Cu/Al_2O_3 catalyst has been shown to further enhance

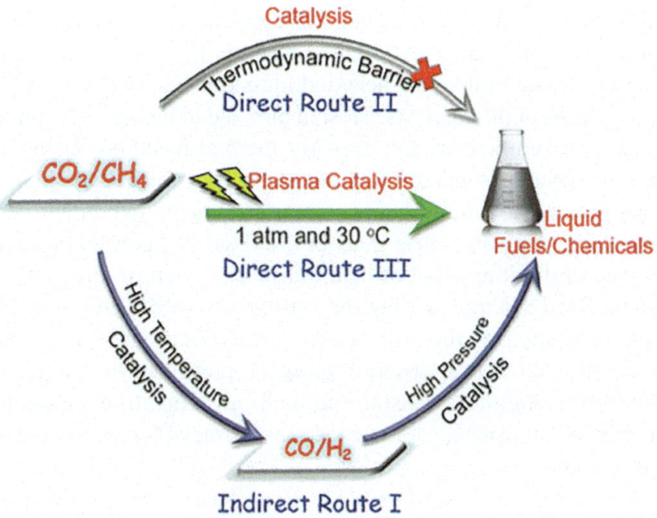

Fig. 10.6 Different routes for the conversion of CO_2 and CH_4 to oxygenate. Reproduced from [116] with permission

the selectivity of acetic acid to 40.2%, while formaldehyde was only formed when using noble metal-based catalysts (Pt/Al_2O_3 and Au/Al_2O_3) [116]. The combination of plasma with catalysts has proven to be effective in tuning the selectivity of oxygenates.

There has been growing interest in this reaction in recent years, and more supported metal catalysts have also been investigated, including $CoFe/SiO_2$, Pt/ZSM-5, Ni/Al_2O_3, Ag/Al_2O_3, Pt/Al_2O_3, NiGa/Nickle foam, Cu/TiO_2, Cu/CeO_2, Cu/HZSM-5, and Cu/SiO_2 [110–116]. Among the active metals, Cu has emerged as the most promising for this reaction. The redox capacity and valence state of Cu species (Cu^{2+} and Cu^{+}) in Cu-based catalysts have been found to manipulate the production of alcohols and acids [115]. For instance, Cu^{2+} species favor the formation of alcohols, while Cu^{+} species contribute to enhanced acetic acid production. Furthermore, the presence of Brønsted acid sites in Cu/HZSM-5 catalysts has been reported to enhance the selectivity of acetic acid. By varying the Cu loading from 1 wt% to 20 wt% in $Cu/Al(OH)_3$ catalysts, the predominant alcohol produced can shift from methanol to ethanol. The optimal Cu loading for methanol and ethanol production falls within the range of 5–15 wt% [115]. Shao and co-workers designed oxygen vacancy-rich Co-based catalysts and found that metallic Co plays a crucial role in enhancing acetic acid generation, while oxygen vacancies (O_v) contribute to the formation of CH_3OH [113].

In addition to considerable efforts on testing different catalysts, innovative reactor design has also been explored to enhance the reaction performance of plasma DRM. For instance, Martin-del-Campo et al. developed a rotating gliding arc-spouted bed hybrid system for plasma-catalytic DRM [118]. Nozaki and co-workers developed a fluidized-bed DBD reactor to enhance the interaction between the plasma and the catalyst surface for plasma-catalytic DRM reaction [97, 119]. They found that the fluidized-bed facilitated improved heat and mass transfer, resulting in an increased specific surface area of the catalysts. This, in turn, led to higher CH_4 conversion rates compared to a packed-bed reactor, even for thermal reactions. Using a powdered catalyst in a fixed bed presents certain limitations. Due to the scaling law (Paschen's law), it is not possible to uniformly generate streamers throughout the dense powder layer. Furthermore, the rate of heat transfer in a fixed-bed powder bed is low, failing to address the rate-limiting issue of heat transfer. In contrast, the fluidized bed has proved to be effective in transporting the plasma-generated active species due to its high transport capacity. As the plasma-driven reaction progressed, the temperature decreased rapidly due to heat absorption, which prevented the reforming reaction from further advancing (heat transfer rate limiting). Therefore, enhancing the heat transport capability of plasma-catalytic reactors is crucial for harnessing the inherent reactivity of plasmas.

10.5.4 Plasma-Catalytic CO_2 Hydrogenation

CO_2 hydrogenation to chemicals and fuels, especially CO, CH_4, and oxygenates, represents a promising approach for CO_2 conversion and utilization, which has been investigated in the field of conventional heterogeneous catalysis for decades. Among the different routes for CO_2 conversion via thermal catalysis, CO and CH_4 can be synthesized from CO_2 thermo-catalytic hydrogenation at atmospheric pressure, via the reverse water–gas shift (RWGS) reaction and methanation (also known as the Sabatier reaction), respectively. Other targeted products, such as C_{2+} hydrocarbons, methanol, and other oxygenates, can only be obtained at high pressures. Meanwhile, the reaction performance of thermal catalytic CO_2 hydrogenation, including activity and selectivity, has been recognized to be strongly related to the catalyst design.

Compared to thermal catalysis, plasma-catalytic CO_2 hydrogenation can be achieved at ambient pressure and low temperatures, even at room temperature [65, 120]. In the absence of a catalyst, plasma-based CO_2 hydrogenation typically generates CO with limited CH_4 [67]. The combination of plasma with a catalyst opens the door to tune the selectivity of target products. CO and CH_4 are the most common gas products in plasma-catalytic CO_2 hydrogenation using a conventional DBD reactor, while the production of methanol, ethanol, and acetic acid was also reported using a water-cooled DBD reactor [65, 120]. A DBD has been widely used for plasma-catalytic CO_2 hydrogenation due to its flexibility to be combined with catalysts for tuning the product selectivity, while some other plasma types, such as GA and MW plasmas, can easily decompose the formed products, such as methane and oxygenates, and thus are not suitable for CO_2 hydrogenation, except for CO production via the RWGS reaction.

10.5.4.1 Reverse Water–Gas Shift Reaction

The RWGS reaction is a chemical process in which CO_2 and H_2 react to produce CO and water. It is an endothermic reaction that typically requires high temperatures ranging from 300 to 800 °C to proceed effectively. One promising application of the RWGS reaction is its combination with Fischer–Tropsch (FT) synthesis, to create a modified FT process using a mixture of CO_2, CO, and H_2 for the production of a wide range of hydrocarbon products including long-chain hydrocarbons such as waxes and liquid fuels [121].

Compared to the extensive investigation of the thermal catalytic RWGS reaction, the use of plasma catalysis for RWGS has received less attention. In the absence of a catalyst, the plasma RWGS reaction typically produces a high selectivity of CO (> 90%) [67, 122, 123]. Therefore, using catalysts has limited potential to further improve the selectivity of CO via the RWGS reaction. However, combining plasma with appropriate catalysts has the potential to enhance the CO_2 conversion and CO yield [67]. For example, Sun et al. reported that the use of a Pd/ZnO catalyst doubled the conversion of CO_2 (36.7%) and the CO yield (35.5%) in the plasma-catalytic RWGS reaction compared to the plasma-only system (21.3% for CO_2 conversion and 19.9% for CO yield). On the other hand, using plasma + ZnO alone showed similar CO_2 conversion and CO yield as those obtained using plasma only [67]. In plasma-catalytic CO_2 hydrogenation over Pd/ZnO, the dominant reaction route for enhanced CO_2 conversion is the hydrogenation of adsorbed surface CO_2 on Pd/ZnO. This can be attributed to the generation of a ZnO_x overlay, resulting from strong metal-support interactions (SMSI) at the Pd–ZnO interface, and the presence of abundant H species on the surface of Pd/ZnO. However, in the plasma + ZnO system, this crucial surface reaction may be limited due to the lack of active H species on the ZnO surface and the absence of SMSI. Instead, CO_2 splitting to CO, both in the plasma gas phase and on the surface of ZnO, is believed to play an important role in the conversion of CO_2 in the plasma + ZnO system [67]. Perovskite catalysts have also been tested for the plasma-catalyzed RWGS reaction. Ni-based perovskite catalysts demonstrated low CO production due to the formation of more methane (6–18%) as Ni-based catalysts favor the methanation pathway over the CO pathway, particularly at higher discharge power levels [124–126]. A few studies reported that alloy-based catalysts are effective for plasma-catalyzed RWGS [123, 124, 127]. Nozaki and his colleagues have developed a fluidized-bed DBD reactor for plasma catalytic RWGS using a Pd_2Ga/SiO_2 alloy catalyst [128]. Under both thermal and non-thermal plasma (NTP) conditions, they observed a remarkable CO selectivity close to 100%. Interestingly, the activation of CO_2 by plasma not only led to a more than twofold increase in CO_2 conversion compared to thermal conditions but also breaks the thermodynamic equilibrium limit (Fig. 10.7). The in situ Fourier transform infrared spectroscopy (FTIR) characterizations of the catalysts during CO_2 hydrogenation revealed that vibrationally excited CO_2 played a significant role in enhancing the catalytic reaction. The DFT calculations identified a key reaction

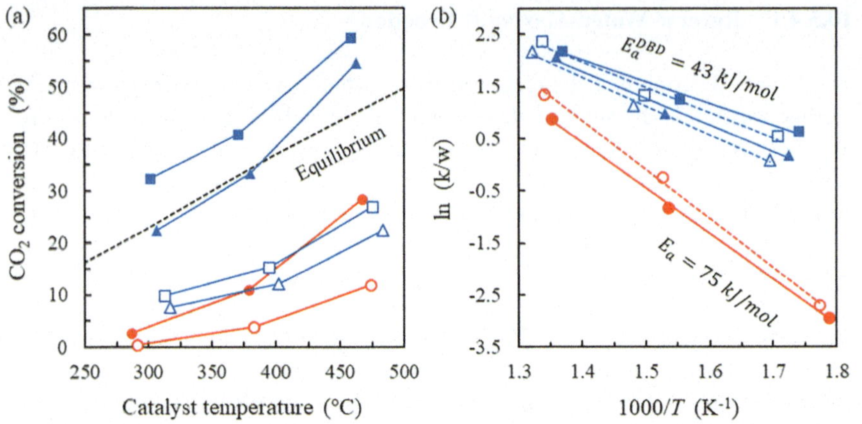

Fig. 10.7 Kinetic studies of reverse water–gas shift (RWGS) reaction in the fluidized-bed DBD reactor: **a** CO_2 conversion, **b** Arrhenius plot: natural logarithm of reaction rate constant per catalyst weight versus the inverse of catalyst temperature [filled square—10 wt% (100 kHz), filled triangle—10 wt% (12 kHz), open square—3 wt% (100 kHz), open triangle—3 wt% (12 kHz), filled circle—10 wt% (Thermal), open circle—3 wt% (Thermal)]. Reproduced from [128] with permission

pathway where the decomposition of monodentate formate was promoted by plasma-activated hydrogen species [128]. Therefore, the choice of the active metal is one of the important factors to consider when designing catalysts for this reaction.

10.5.4.2 CO_2 Methanation

CO_2 methanation, also known as the Sabatier reaction, is a promising process for converting CO_2 into CH_4. Methane is an efficient energy carrier and can serve as a clean energy source that can be easily stored. CO_2 methanation is a highly exothermic reaction and holds potential as a power-to-gas process. In thermal catalytic methanation, nickel-based catalysts are widely recognized as the preferred choice for the active base metal due to the low cost and abundant availability of nickel, particularly at temperatures ranging from 300 to 400 °C.

In recent years, there has been increasing attention on plasma-catalytic CO_2 methanation. Catalysts play a critical role in achieving a high selectivity of CH_4. In the absence of a catalyst, plasma-based CO_2 hydrogenation is more favorable for CO production via the RWGS route rather than methane production via the methanation route [67, 122, 123, 129]. Among the catalysts used for plasma-based CO_2 methanation, Ni-based catalysts are the most widely used, followed by Ru- and Co-based catalysts [130–134]. Commonly used catalyst supports include Al_2O_3 and $CeZrO_2$ [132], which are also commonly used in thermal catalytic CO_2 methanation. Additionally, SiO_2, zeolites (such as ZSM-5 and USY), and MOFs (e.g., UiO-66) have been investigated for plasma-catalytic CO_2 methanation [130, 135–137]. Various promoters, including CeO_2, Mn, Cu, La, Co, have also been explored to enhance

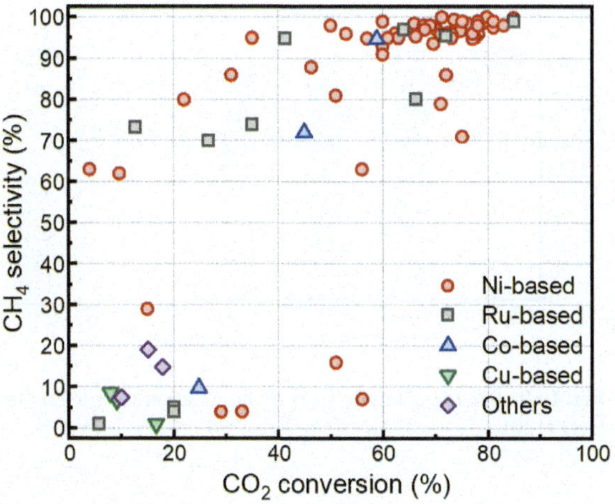

Fig. 10.8 CO_2 conversion versus CH_4 selectivity using different catalysts in plasma-catalytic CO_2 methanation. Data were collected from [140–150]

the activities of methanation catalysts [138, 139]. Among these, CeO_2 is the most commonly considered promoter, especially for Ni-based catalysts.

Figure 10.8 illustrates the relationship between CO_2 conversion and CH_4 selectivity for different metal-based catalysts reported in plasma-catalytic CO_2 methanation. Both Ni and Ru exhibit higher CH_4 selectivity compared to other metals. Nickel is the most active metal for plasma-catalytic CO_2 methanation, achieving CO_2 conversions of 60–90% and CH_4 selectivity of up to 100%. Notably, Ni and Ru-based catalysts can achieve nearly 100% CH_4 selectivity. On the other hand, metals like Cu, Fe, and Mo exhibit much lower CO_2 conversions and low CH_4 selectivity, indicating that they are not effective metals for methanation under plasma conditions.

10.5.4.3 CO_2 Hydrogenation to Methanol

CO_2 hydrogenation to methanol using thermal catalysis typically requires high temperatures (200–400 °C) and high pressures (30–300 bar). $Cu/ZnO/Al_2O_3$ is a well-studied catalyst system widely used for thermal catalytic CO_2 hydrogenation to methanol. However, advancements in plasma-based gas conversion have shown the potential to enable this reaction under milder conditions, such as room temperature and ambient pressure, while achieving higher CO_2 conversion compared to traditional thermal catalysis (which typically achieves around 10–15% single-pass conversion) [65, 120, 151].

Although significant progress has been made in the field of thermal catalytic CO_2 hydrogenation to methanol [152], driven by rational catalyst design, optimization of reaction conditions, and a deeper understanding of reaction mechanisms, the focus on plasma-catalytic CO_2 hydrogenation is still relatively limited. In the late 1990s, Eliasson et al. conducted investigations on plasma-catalytic CO_2 hydrogenation to methanol using a commercial $Cu/ZnO/Al_2O_3$ catalyst at various pressures (1–8 bar)

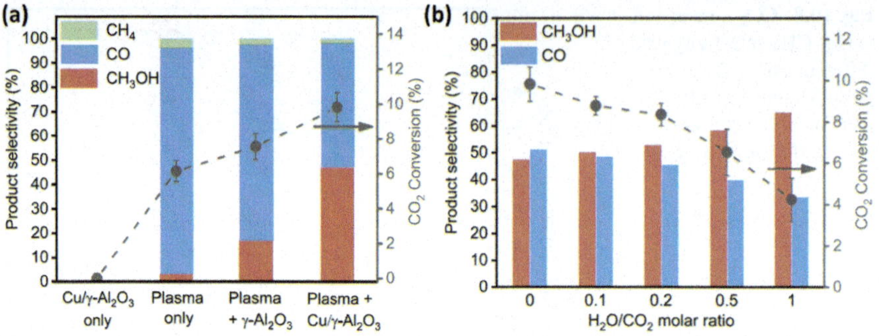

Fig. 10.9 Plasma-catalytic CO_2 hydrogenation over a Cu cluster catalyst **a** without H_2O; **b** with H_2O [120]

[153]. However, the reported methanol yield was very low, reaching up to only 1%, even at higher pressures [153]. In 2018, Wang et al. reported a water-cooled DBD reactor that can enable plasma-catalytic CO_2 hydrogenation to oxygenates at room temperature and atmospheric pressure. A high methanol selectivity of 53.7% and a methanol yield of 11.3% were obtained using a Cu/Al_2O_3 catalyst [65]. Ethanol was also produced alongside methanol, although its selectivity was much lower compared to methanol. Ronda-Lloret et al. investigated the influence of different Co_xO_y loadings of Co_xO_y/MgO catalysts on plasma-catalytic CO_2 hydrogenation [151]. They found that a 15 wt% Co_xO_y/MgO catalyst provided the highest CO_2 conversion (35%) and methanol selectivity (31%). Interestingly, this process also yielded acetic acid [151]. Recently, Cui et al. explored the effect of Cu cluster catalysts on the selectivity of methanol in plasma-catalyzed CO_2 hydrogenation [120]. They achieved a CO_2 conversion of 10% and a methanol selectivity of around 50% using 4 wt% Cu/Al_2O_3 catalysts (Fig. 10.9). Interestingly, the addition of water in a 1:1 ratio with CO_2 further increased the methanol selectivity to 65% [120]. These studies highlight the potential of plasma-catalytic CO_2 hydrogenation to methanol and demonstrate various catalyst systems and conditions that can influence the reaction performance. Note that liquid products were typically condensed and measured using off-line GC in these studies. Other catalysts including MnO_x/ZrO_2, Cu- and Ni-based catalysts have also been tested for plasma-catalyzed CO_2 hydrogenation to methanol [154–156]. Further research and optimization of plasma-catalytic processes hold promise for enhancing the selectivity and efficiency of this conversion, contributing to the development of sustainable and energy-efficient CO_2 utilization technologies.

10.5.5 Plasma-Catalytic Conversion of CO_2 with Water

Plasma-catalytic CO_2 reduction with water for the synthesis of fuels and chemicals is a promising approach for CO_2 conversion. Unlike traditional methods (e.g., CO_2

hydrogenation) that rely on hydrogen as a feedstock, this process utilizes water, which is cost-effective and abundant. While extensive research has been conducted on CO_2 reduction with water using electrocatalysis and photocatalysis [157, 158], limited efforts have been made to explore plasma-based CO_2 reduction with water [159, 160], particularly using plasma catalysis. Currently, the main gas products of plasma-based CO_2 reduction with water are syngas and methane [161–163], although trace amounts of liquid products (e.g., dimethyl ether, methanol, ethanol, formic acid, and oxalic acid) have also been reported [164].

The combination of plasma with a catalyst has been found to enhance the CO_2 conversion and increase the production of syngas and methane. For example, Ma et al. reported that TiO_2 can enhance the CO_2 conversion by up to 15% compared to the plasma-only reaction, while its effect on the water conversion is limited [162]. They also found that 10% Ni/Al_2O_3 can significantly improve the CO_2 conversion, albeit with less impact on water conversion [162]. In another study, $Ni/\gamma\text{–}Al_2O_3$ catalysts, both in unreduced and partially reduced forms, have been reported to enhance plasma-catalytic CO_2 conversion and increase the syngas ratio [163]. Interestingly, the use of reduced $Ni/\gamma\text{–}Al_2O_3$ catalyst leads to the production of carbon nanofibers [163].

Furthermore, studies have shown the production of methane at ppm levels using pulsed corona discharge over a water film [161]. The choice of electrode material, such as NiCr alloy high voltage electrode, has been found to influence methane production, suggesting a potential catalytic effect of the electrode in plasma-based CO_2 reduction with water [161]. Recently, Yang et al. demonstrated the formation of liquid products, including acetic acid, formic acid, and oxalate, through nanosecond discharge in CO_2-saturated water, although the quantification of these products was not reported [164].

Despite the promise of plasma-catalytic reduction of CO_2 with water, there is still limited research focused on this reaction. Only a few catalysts have been tested, and their specific effects on syngas and methane production remain unclear. Nonetheless, these studies emphasize the potential of plasma-catalytic systems for CO_2 conversion and highlight the importance of catalyst design in optimizing process performance. To produce more valuable liquid fuels and chemicals, designing bi-functional catalysts or employing tandem catalysis that enable the simultaneous activation of CO_2 and H_2O, as well as subsequent reactions such as syngas conversion, could be a promising strategy. Further research and investigation is required to fully understand and optimize this innovative approach for sustainable carbon utilization.

10.5.6 Limitations of Plasma Catalysis

Plasma catalysis has exhibited significant potential for activating CO_2 into various fuels and chemicals under mild conditions. In addition, the combination of plasma with catalysts has demonstrated synergistic effects for CO_2 conversion. However, it is important to note that the combination of plasma with a catalyst does not always result in a synergy or benefit the reactions. For instance, Tu et al. observed a decrease in both CO_2 and CH_4 conversion in plasma-catalytic DRM when Ni/γ–Al_2O_3 was packed into the entire discharge area of a DBD reactor, as compared to using plasma alone. They proposed that the plasma-catalytic synergy for DRM depends on the balance between the catalyst-induced change in discharge behavior and the plasma-assisted activation of the catalyst at low temperatures [83]. Andersen et al. investigated the influence of different γ-Al_2O_3-supported metal catalysts (Cu, Ag, and Pt) on plasma-catalytic DRM. They found that the presence of supported Cu and Ag catalysts reduced the conversion of both gases compared to the plasma-only system. On the other hand, the use of Pt/γ–Al_2O_3 did not improve the gas conversion [85]. Similarly, Wang et al. reported that the utilization of ZSM-supported metal catalysts led to a decrease in CH_4 conversion compared to the plasma-only system. Moreover, there was no improvement in CO_2 conversion when considering the measurement errors [165]. In addition to DRM, the negative impact of catalysts on other plasma-based CO_2 conversion reactions has also been reported. Wang et al. also found the use of Cu/Al_2O_3 and Pt/Al_2O_3 catalysts in plasma CO_2 hydrogenation resulted in a slight reduction in methanol selectivity compared to the plasma-only system [65]. Liu and co-workers evaluated the influence of different perovskite catalysts on the plasma-catalytic RWGS reaction and discovered that not all perovskite catalysts can enhance the CO_2 conversion. For instance, the presence of $La_{0.9}Sr_{0.1}FeO_{3-\delta}$ in the plasma-catalytic RWGS reaction showed lower CO_2 conversion compared to the plasma-only system [124]. The underlying mechanism behind the negative effect of catalysts in plasma-catalytic CO_2 conversion is still not fully understood, particularly plasma-assisted surface reactions, due to complicated plasma-catalyst interactions. Very recently, Loenders and co-workers provided in-depth discussion on the various physical and chemical effects that contribute to the inherent complexity of plasma catalysis for gas conversion. They specifically discussed the negative impact of catalysts on plasma-catalytic DRM reactions. Factors such as the properties of catalysts and supports (e.g., size, conductivity, porosity, dielectric constant), the arrangement of catalysts/supports within the plasma reactor, and the configuration of the reactor itself all influence the discharge properties. These properties, in turn, have a direct influence on the physical and chemical interactions between the plasma and catalyst, ultimately affecting the performance of plasma-catalytic reactions. They highlighted that a comprehensive understanding of the physical and chemical interactions occurring within plasma-catalytic systems is crucial for avoiding conditions that could lead to detrimental effects on the reaction performance [64]. In addition, rational catalyst design specifically for plasma catalysis or plasma-catalytic reactors, such as considering plasma–catalyst interactions, is critical to enhance the selectivity of targeted products from plasma-catalytic CO_2 conversion. This could be supported through the development of more advanced in situ catalyst characterization under plasma conditions and plasma catalysis modeling [67, 120, 128, 131], both of which would provide valuable insights into the plasma-assisted surface reactions, including the adsorption and desorption of species on the catalyst surface. These insights, in turn, would provide useful guidance for catalyst design and development dedicated for plasma catalysis.

10.6 Conclusion and Future Research Directions

This chapter presents the possibilities and limitations of plasma-based CO_2 conversion, in various types of plasma reactors. At an industrial level, non-thermal plasma is an excellent means for exploiting non-equilibrium processes that generate reactive species under practical reaction conditions. For CO_2 splitting, DBD plasmas exhibit too low energy efficiency for industrial implementation. The energy efficiency is typically only 5–10%, hence far below the target of 60%, which was defined to be competitive with classical and other emerging technologies [2] (see Fig. 10.2). GA and especially MW plasmas have much higher energy efficiencies for CO_2 splitting. For low-pressure MW plasmas, this can be attributed to the important role of the vibrational kinetics for energy-efficient CO_2 dissociation, but at atmospheric pressure, which is more convenient for industrial use, the conversion proceeds mainly by thermal dissociation in both GA and MW plasmas.

Also for DRM, DBD plasma reactors provide too low energy efficiency, clearly below the target of 60%. For MW plasmas, only limited results have been reported in literature, although recent data show its potential. GA plasmas seem very promising, with energy efficiencies close to or above the efficiency target. The same is true for ns-pulsed plasmas, spark discharges and APGDs, although further research is required to fully explore their capabilities.

We believe that increasing the energy efficiency (and/or the CO_2 conversion, i.e., the performance in general) is possible by reducing the back reactions, i.e., recombination of CO with O/O_2, back into CO_2. This is possible, among others, by fast quenching, i.e., fast cooling of the gas immediately after the reactor, as demonstrated for instance by Hecimovic et al. [36], who reported a rise in conversion by a factor 7 upon adding a quenching nozzle after a MW plasma. Removing the O/O_2 from the gas mixture, to avoid these back reactions, has also been demonstrated already by means of scavenging materials or chemicals [43], membranes [44], chemical looping materials [50], by solid oxide electrolysis [45], or by means of a carbon bed [46–48], where the O atoms and O_2 molecules can react with the C atoms, to form extra CO, and at the same time they are consumed, hence not available anymore for recombination reactions with CO. This was demonstrated to increase both the CO_2 conversion and energy efficiency by a factor two. Finally, performance improvements are also possible by smart reactor design, affecting the gas flow dynamics inside the reactor, to make sure that most of the gas passes through the active plasma, and also to enhance mixing between hot and cold gas, for bringing energetic O atoms in contact with unconverted CO_2, thus producing more CO, instead of recombining with CO into CO_2, as demonstrated by van de Steeg et al. [12].

Another approach to suppress the back reactions is the combination with heterogeneous catalysts. Although forward and back reactions are promoted by non-thermal plasma, CO_2 conversion is higher than the thermal equilibrium (Fig. 10.7). In other words, the plasma promotion effect on the forward reaction is greater than on the back reaction. This is possible because the forward reaction is promoted by the Eley–Rideal pathway, or direct interaction of excited CO_2 and surface hydrogen, where the flux of those species to the catalysts is decoupled from the saturated surface coverage of CO_2, or the equilibrium limitation. Meantime, the back reaction would proceed by the Langmuir–Hinshelwood pathway where the surface coverage of product species would be governed by the equilibrium. The appropriate catalyst combination is the key issue that promotes the forward reaction selectively and leads to higher energy efficiency [128]. Deep insight into plasma–catalyst interaction is thus a critically important subject of research, enabling better designability of catalysts, including reactor design such as fluidized-bed plasma [119].

Note, however, that the efficiency target of 60% for DRM, as mentioned above, was defined in the case of syngas production. While this is indeed the major product of DRM, plasma can in principle also produce value-added compounds in one step, such as oxygenates or higher hydrocarbons, when using suitable catalysts. This will give plasma technology a clear advantage, drastically reducing the efficiency target to compete with other technologies, which need two steps for this. However, clearly more research is needed to design catalysts tailored to the plasma environment, in order to selectively produce these value-added compounds.

This chapter also discusses recent advances in plasma catalysis, not only for CO_2 splitting and DRM, but also for artificial photosynthesis (CO_2/H_2O) and CO_2 hydrogenation (CO_2/H_2), presenting some success stories, but also the current limitations and what is needed to overcome them. Plasma catalysis is an emerging technology for the production of fuels and chemicals, and research on plasma catalysis has been increasing in recent years. Plasma catalysis is considered a very promising electrification technology besides electrochemistry; it is expected to emerge as a technology contributing to zero-carbon chemical processes. However, commercial applications of plasma catalysis still face challenges that need to be addressed. To enhance the selectivity of targeted products from plasma-catalytic CO_2 conversion, as well as the energy efficiency, rational catalyst design becomes essential, considering the complex plasma–catalyst interactions. This will enable plasma-catalytic processes to become economically competitive compared to conventional processes. For example, by carefully designing catalysts that are compatible with plasma conditions, high CO_2 conversion and CH_4 selectivity can be achieved in plasma-catalyzed CO_2 methanation using a DBD reactor, resulting in over 60% energy efficiency. The development of advanced in situ catalyst characterization under plasma conditions and plasma catalysis modeling is crucial for gaining valuable insights into plasma-catalytic reaction mechanisms, especially plasma-assisted surface reactions. These insights, in turn, will contribute to the design and development of a more efficient plasma-catalytic process for CO_2 conversion into fuels and chemicals. In addition, another challenge with plasma technology is the lack of clear guidelines on the scale-up steps (with the exception of thermal plasmas). Nevertheless, industrial ozonizers appear to be a good example of a DBD that can provide guidelines for scaling up.

In general, plasma technology is highly promising for CO_2 conversion, especially when combined with renewable electricity, due to its flexibility, in terms of energy source (i.e., all types of renewable electricity), feed gas (not only pure CO_2 splitting, but also mixtures with any H-source), as well as operation (i.e., modular upscaling possible and turnkey process).

References

1. Bogaerts A, Neyts E, Gijbels R, Van der Mullen J (2002) Gas discharge plasmas and their applications. Spectrochim Acta Part B 57:609–658
2. Snoeckx R, Bogaerts A (2017) Plasma technology—a novel solution for CO_2 conversion? Chem Soc Rev 46:5805–5863
3. Kogelschatz U (2003) Dielectric-barrier discharges: their history, discharge physics, and industrial applications. Plasma Chem Plasma P 23:1–46
4. Van Laer K, Bogaerts A (2015) Fluid modelling of a packed bed dielectric barrier discharge plasma reactor. Plasma Sources Sci Technol 25:015002
5. Michielsen I et al (2017) CO_2 dissociation in a packed bed DBD reactor: first steps towards a better understanding of plasma catalysis. Chem Eng J 326:477–488
6. Bogaerts A, Neyts EC (2018) Plasma technology: an emerging technology for energy storage. ACS Energy Lett 3:1013–1027
7. Azizov RI et al. Akademiia Nauk SSSR Doklady, pp 94–98
8. Berthelot A, Bogaerts A (2017) Modeling of CO_2 splitting in a microwave plasma: how to improve the conversion and energy efficiency. J Phys Chem C 121:8236–8251
9. Wolf AJ, Righart TW, Peeters F, Bongers W, Van De Sanden M (2020) Implications of thermo-chemical instability on the contracted modes in CO_2 microwave plasmas. Plasma Sources Sci Technol 29:025005
10. den Harder N et al (2017) Homogeneous CO_2 conversion by microwave plasma: wave propagation and diagnostics. Plasma Processes Polym 14:1600120
11. D'Isa F, Carbone E, Hecimovic A, Fantz U (2020) Performance analysis of a 2.45 GHz microwave plasma torch for CO_2 decomposition in gas swirl configuration. Plasma Sources Sci Technol 29:105009
12. van de Steeg A et al (2021) Redefining the microwave plasma-mediated CO_2 reduction efficiency limit: the role of O-CO_2 association. ACS Energy Lett 6:2876–2881
13. van den Bekerom DC et al (2019) The importance of thermal dissociation in CO_2 microwave discharges investigated by power pulsing and rotational Raman scattering. Plasma Sources Sci Technol 28:055015
14. van de Steeg A et al (2022) The chemical origins of plasma contraction and thermalization in CO_2 microwave discharges. J Phys Chem Lett 13:1203–1208
15. Wolf A et al (2019) Characterization of CO_2 microwave plasma based on the phenomenon of skin-depth-limited contraction. Plasma Sources Sci Technol 28:115022
16. Nunnally T et al (2011) Dissociation of CO_2 in a low current gliding arc plasmatron. J Phys D Appl Phys 44:274009
17. Liu J-L, Wang X, Li X-S, Likozar B, Zhu A-M (2020) CO_2 conversion, utilisation and valorisation in gliding arc plasma reactors. J Phys D Appl Phys 53:253001
18. Ramakers M, Trenchev G, Heijkers S, Wang W, Bogaerts A (2017) Gliding arc plasmatron: providing an alternative method for carbon dioxide conversion. Chemsuschem 10:2642–2652
19. Cleiren E, Heijkers S, Ramakers M, Bogaerts A (2017) Dry reforming of methane in a gliding arc plasmatron: towards a better understanding of the plasma chemistry. Chemsuschem 10:4025–4036
20. Trenchev G, Kolev S, Wang W, Ramakers M, Bogaerts A (2017) CO2 conversion in a gliding arc plasmatron: multidimensional modeling for improved efficiency. J Phys Chem C 121:24470–24479
21. Montesano C, Faedda M, Martini LM, Dilecce G, Tosi P (2021) CH4 reforming with CO_2 in a nanosecond pulsed discharge. The importance of the pulse sequence. J CO_2 Util 49:101556
22. Jurković DL, Liu J-L, Pohar A, Likozar B (2021) Methane dry reforming over Ni/Al2O3 catalyst in spark plasma reactor: linking computational fluid dynamics (CFD) with reaction kinetic modelling. Catal Today 362:11–21
23. Babaeva NY, Naidis G (2021) On the efficiency of CO_2 conversion in corona and dielectric-barrier discharges. Plasma Sources Sci Technol 30:03LT03

24. Trenchev G, Nikiforov A, Wang W, Bogaerts A (2019) Atmospheric pressure glow discharge for CO_2 conversion: model-based exploration of the optimum reactor configuration. Chem Eng J 362:830–841
25. Wanten B et al (2022) Dry reforming of methane in an atmospheric pressure glow discharge: confining the plasma to expand the performance. J CO_2 Util 56:101869
26. Vertongen R, Bogaerts A (2023) How important is reactor design for CO2 conversion in warm plasmas? J CO_2 Util 72:102510
27. Ozkan A et al (2016) DBD in burst mode: solution for more efficient CO_2 conversion? Plasma Sources Sci Technol 25:055005
28. Van Laer K, Bogaerts A (2015) Improving the conversion and energy efficiency of carbon dioxide splitting in a zirconia-packed dielectric barrier discharge reactor. Energ Technol 3:1038–1044
29. Mei D et al (2016) Plasma-photocatalytic conversion of CO_2 at low temperatures: understanding the synergistic effect of plasma-catalysis. Appl Catal B 182:525–532
30. Li S, Gallucci F (2022) CO_2 capture and activation with a plasma-sorbent system. Chem Eng J 430:132979
31. Rusanov VD, Fridman A, Sholin GV (1981) The physics of a chemically active plasma with nonequilibrium vibrational excitation of molecules. Soviet Phys Uspekhi 24:447
32. Silva T, Britun N, Godfroid T, Snyders R (2014) Optical characterization of a microwave pulsed discharge used for dissociation of CO_2. Plasma Sources Sci Technol 23:025009
33. Bongers W et al (2017) Plasma-driven dissociation of CO_2 for fuel synthesis. Plasma Processes Polym 14:1600126
34. Wolf AJ, Peeters F, Groen P, Bongers W, Van De Sanden M (2020) CO2 conversion in nonuniform discharges: disentangling dissociation and recombination mechanisms. J Phys Chem C 124:16806–16819
35. Mercer ER et al (2023) Post-plasma quenching to improve conversion and energy efficiency in a CO_2 microwave plasma. Fuel 334:126734
36. Hecimovic A, Federico A, Carbone E, Fantz U (2022) Enhancement of CO_2 conversion in microwave plasmas using a nozzle in the effluent. J CO_2 Util 57:101870
37. Sun S, Wang H, Mei D, Tu X, Bogaerts A (2017) CO_2 conversion in a gliding arc plasma: performance improvement based on chemical reaction modeling. J CO_2 Util 17:220–234
38. Wang W, Mei D, Tu X, Bogaerts A (2017) Gliding arc plasma for CO_2 conversion: better insights by a combined experimental and modelling approach. Chem Eng J 330:11–25
39. Liu J-L, Park H-W, Chung W-J, Park D-W (2016) High-efficient conversion of CO 2 in AC-pulsed tornado gliding arc plasma. Plasma Chem Plasma P 36:437–449
40. Trenchev G, Bogaerts A (2020) Dual-vortex plasmatron: a novel plasma source for CO_2 conversion. J CO_2 Util 39:101152
41. Vermeiren V, Bogaerts A (2020) Plasma-based CO_2 conversion: to quench or not to quench? J Phys Chem C 124:18401–18415
42. Van Alphen S et al (2023) Modelling post-plasma quenching nozzles for improving the performance of CO_2 microwave plasmas. Chem Eng J 462:142217
43. Aerts R, Snoeckx R, Bogaerts A (2014) In-situ chemical trapping of oxygen in the splitting of carbon dioxide by plasma. Plasma Processes Polym 11:985–992
44. Chen G et al (2020) A novel plasma-assisted hollow fiber membrane concept for efficiently separating oxygen from CO in a CO_2 plasma. Chem Eng J 392:123699
45. Mori S, Matsuura N, Tun LL, Suzuki M (2016) Direct synthesis of carbon nanotubes from only CO_2 by a hybrid reactor of dielectric barrier discharge and solid oxide electrolyser cell. Plasma Chem Plasma P 36:231–239
46. Li Z et al (2020) Boudouard reaction driven by thermal plasma for efficient CO_2 conversion and energy storage. J Energy Chem 45:128–134
47. Huang J et al (2021) Enhanced conversion of CO_2 into O_2-free fuel gas via the Boudouard reaction with biochar in an atmospheric plasmatron. J CO_2 Util 45:101429
48. Girard-Sahun F, Biondo O, Trenchev G, van Rooij G, Bogaerts A (2022) Carbon bed post-plasma to enhance the CO_2 conversion and remove O_2 from the product stream. Chem Eng J 442:136268

49. Vertongen R, Trenchev G, Van Loenhout R, Bogaerts A (2022) Enhancing CO_2 conversion with plasma reactors in series and O_2 removal. J CO_2 Util 66:102252
50. Delikonstantis E et al (2022) Exceeding equilibrium CO_2 conversion by plasma-assisted chemical looping. ACS Energy Lett 7:1896–1902
51. Snoeckx R, Zeng Y, Tu X, Bogaerts A (2015) Plasma-based dry reforming: improving the conversion and energy efficiency in a dielectric barrier discharge. RSC Adv 5:29799–29808
52. Chun SM, Hong YC, Choi DH (2017) Reforming of methane to syngas in a microwave plasma torch at atmospheric pressure. J CO_2 Util 19:221–229
53. Sun H, Lee J, Bak MS (2021) Experiments and modeling of atmospheric pressure microwave plasma reforming of a methane-carbon dioxide mixture. J CO_2 Util 46:101464
54. Kelly S, Mercer ER, De Meyer R, Bals S, Bogaerts A Microwave plasma-based dry reforming of methane: a focus on carbon formation. Available at SSRN 4486414
55. Tu X, Whitehead JC (2014) Plasma dry reforming of methane in an atmospheric pressure AC gliding arc discharge: co-generation of syngas and carbon nanomaterials. Int J Hydrogen Energy 39:9658–9669
56. Martin-del-Campo J, Coulombe S, Kopyscinski J (2020) Influence of operating parameters on plasma-assisted dry reforming of methane in a rotating gliding arc reactor. Plasma Chem Plasma P 40:857–881
57. Li L et al (2019) Plasma-assisted CO_2 conversion in a gliding arc discharge: improving performance by optimizing the reactor design. J CO_2 Util 29:296–303
58. Van Alphen S et al (2021) Effect of N_2 on CO_2-CH_4 conversion in a gliding arc plasmatron: can this major component in industrial emissions improve the energy efficiency? J CO_2 Util 54:101767
59. Slaets J, Aghaei M, Ceulemans S, Van Alphen S, Bogaerts A (2020) CO_2 and CH_4 conversion in "real" gas mixtures in a gliding arc plasmatron: how do N_2 and O_2 affect the performance? Green Chem 22:1366–1377
60. Li K, Liu J-L, Li X-S, Zhu X, Zhu A-M (2016) Warm plasma catalytic reforming of biogas in a heat-insulated reactor: dramatic energy efficiency and catalyst auto-reduction. Chem Eng J 288:671–679
61. Li K et al (2018) Novel power-to-syngas concept for plasma catalytic reforming coupled with water electrolysis. Chem Eng J 353:297–304
62. Neyts E, Bogaerts A (2014) Understanding plasma catalysis through modelling and simulation—a review. J Phys D Appl Phys 47:224010
63. Bogaerts A et al (2020) The 2020 plasma catalysis roadmap. J Phys D Appl Phys 53:443001
64. Loenders B, Michiels R, Bogaerts A (2023) Is a catalyst always beneficial in plasma catalysis? Insights from the many physical and chemical interactions. J Energy Chem 85:501–533
65. Wang L, Yi Y, Guo H, Tu X (2018) Atmospheric pressure and room temperature synthesis of methanol through plasma-catalytic hydrogenation of CO_2. ACS Catal 8:90–100
66. Tu X, Whitehead JC (2012) Plasma-catalytic dry reforming of methane in an atmospheric dielectric barrier discharge: understanding the synergistic effect at low temperature. Appl Catal B 125:439–448
67. Sun Y et al (2022) Plasma-catalytic CO_2 hydrogenation over a Pd/ZnO catalyst: in situ probing of gas-phase and surface reactions. JACS Au 2:1800–1810
68. Chen G, Snyders R, Britun N (2021) CO_2 conversion using catalyst-free and catalyst-assisted plasma-processes: recent progress and understanding. J CO_2 Util 49:101557
69. Yu Q, Kong M, Liu T, Fei J, Zheng X (2012) Characteristics of the decomposition of CO_2 in a dielectric packed-bed plasma reactor. Plasma Chem Plasma P 32:153–163
70. Uytdenhouwen Y et al (2019) How process parameters and packing materials tune chemical equilibrium and kinetics in plasma-based CO_2 conversion. Chem Eng J 372:1253–1264
71. Chen G, Georgieva V, Godfroid T, Snyders R, Delplancke-Ogletree M-P (2016) Plasma assisted catalytic decomposition of CO_2. Appl Catal B 190:115–124
72. Ray D, Subrahmanyam C (2016) CO_2 decomposition in a packed DBD plasma reactor: influence of packing materials. RSC Adv 6:39492–39499

73. Mei D, Tu X (2017) Atmospheric pressure non-thermal plasma activation of CO_2 in a packed-bed dielectric barrier discharge reactor. Chem Phys Chem 18:3253–3259
74. Wang L et al (2020) Plasma-enhanced direct conversion of CO_2 to CO over oxygen-deficient Mo-doped CeO_2. Chem Commun 56:14801–14804
75. Ashford B, Wang Y, Poh C-K, Chen L, Tu X (2020) Plasma-catalytic conversion of CO_2 to CO over binary metal oxide catalysts at low temperatures. Appl Catal B 276:119110
76. Mei D, Zhu X, He Y-L, Yan JD, Tu X (2014) Plasma-assisted conversion of CO2 in a dielectric barrier discharge reactor: understanding the effect of packing materials. Plasma Sources Sci Technol 24:015011
77. Zhang H et al (2020) Plasma-enhanced catalytic activation of CO_2 in a modified gliding arc reactor. Waste Disposal Sustain Energy 2:139–150
78. Chen G et al (2017) An overview of CO_2 conversion in a microwave discharge: the role of plasma-catalysis. J Phys D Appl Phys 50:084001
79. Zhu F et al (2017) Plasma-catalytic reforming of CO_2-rich biogas over Ni/γ-Al_2O_3 catalysts in a rotating gliding arc reactor. Fuel 199:430–437
80. Long H, Shang S, Tao X, Yin Y, Dai X (2008) CO_2 reforming of CH_4 by combination of cold plasma jet and Ni/γ-Al_2O_3 catalyst. Int J Hydrogen Energy 33:5510–5515
81. George A et al (2021) A review of non-thermal plasma technology: a novel solution for CO_2 conversion and utilization. Renew Sustain Energy Rev 135:109702
82. Zeng Y, Zhu X, Mei D, Ashford B, Tu X (2015) Plasma-catalytic dry reforming of methane over γ-Al_2O_3 supported metal catalysts. Catal Today 256:80–87
83. Tu X, Gallon HJ, Twigg MV, Gorry PA, Whitehead JC (2011) Dry reforming of methane over a Ni/Al_2O_3 catalyst in a coaxial dielectric barrier discharge reactor. J Phys D Appl Phys 44:274007
84. Mei DH, Liu SY, Tu X (2017) CO_2 reforming with methane for syngas production using a dielectric barrier discharge plasma coupled with Ni/γ-Al_2O_3 catalysts: process optimization through response surface methodology. J CO_2 Util 21:314–326
85. Andersen JA, Christensen JM, Østberg M, Bogaerts A, Jensen AD (2020) Plasma-catalytic dry reforming of methane: screening of catalytic materials in a coaxial packed-bed DBD reactor. Chem Eng J 397:125519
86. Li Y et al (2023) CH_4 and CO_2 conversion over boron nitride-supported Ni catalysts with BO defects in DBD plasma. Fuel Process Technol 242:107655
87. Sheng Z, Kim H-H, Yao S, Nozaki T (2020) Plasma-chemical promotion of catalysis for CH_4 dry reforming: unveiling plasma-enabled reaction mechanisms. Phys Chem Chem Phys 22:19349–19358
88. Sheng Z, Watanabe Y, Kim H-H, Yao S, Nozaki T (2020) Plasma-enabled mode-selective activation of CH_4 for dry reforming: first touch on the kinetic analysis. Chem Eng J 399:125751
89. Sheng Z, Kameshima S, Yao S, Nozaki T (2018) Oxidation behavior of Ni/Al_2O_3 catalyst in nonthermal plasma-enabled catalysis. J Phys D Appl Phys 51:445205
90. Sheng Z et al (2019) Factors determining synergism in plasma catalysis of biogas at reduced pressure. J Phys D Appl Phys 52:414002
91. Wang J, Zhang K, Mertens M, Bogaerts A, Meynen V (2023) Plasma-based dry reforming of methane in a dielectric barrier discharge reactor: importance of uniform (sub) micron packings/catalysts to enhance the performance. Appl Catal B Environ 337:122977
92. Kameshima S, Mizukami R, Yamazaki T, Prananto LA, Nozaki T (2018) Interfacial reactions between DBD and porous catalyst in dry methane reforming. J Phys D Appl Phys 51:114006
93. Ray D, Reddy PMK, Subrahmanyam C (2018) Ni-Mn/γ-Al_2O_3 assisted plasma dry reforming of methane. Catal Today 309:212–218
94. Zhang A-J, Zhu A-M, Guo J, Xu Y, Shi C (2010) Conversion of greenhouse gases into syngas via combined effects of discharge activation and catalysis. Chem Eng J 156:601–606
95. Mei D et al (2023) Plasma-catalytic reforming of biogas into syngas over Ni-based bimetallic catalysts. Chem Eng J 462:142044
96. Zeng YX et al (2018) Low temperature reforming of biogas over K-, Mg- and Ce-promoted Ni/Al_2O_3 catalysts for the production of hydrogen rich syngas: understanding the plasma-catalytic synergy. Appl Catal B 224:469–478

97. Chen X et al (2021) CH4 dry reforming in fluidized-bed plasma reactor enabling enhanced plasma-catalyst coupling. J CO_2 Util 54:101771
98. Zeng Y et al (2023) Biogas reforming for hydrogen-rich syngas production over a Ni–K/Al_2O_3 catalyst using a temperature-controlled plasma reactor. Int J Hydrogen Energy 48:6192–6203
99. Zeng Y et al (2022) Plasma-catalytic biogas reforming for hydrogen production over K-promoted Ni/Al_2O_3 catalysts: effect of K-loading. J Energy Inst 104:12–21
100. Diao Y et al (2022) Plasma-assisted dry reforming of methane over Mo2C-Ni/Al_2O_3 catalysts: effects of β-Mo2C promoter. Appl Catal B 301:120779
101. Mei D, Ashford B, He YL, Tu X (2017) Plasma-catalytic reforming of biogas over supported Ni catalysts in a dielectric barrier discharge reactor: effect of catalyst supports. Plasma Processes Polym 14:1600076
102. Vakili R et al (2020) Plasma-assisted catalytic dry reforming of methane (DRM) over metal-organic frameworks (MOFs)-based catalysts. Appl Catal B 260:118195
103. Gallon HJ, Tu X, Whitehead JC (2012) Effects of reactor packing materials on H_2 production by CO_2 reforming of CH_4 in a dielectric barrier discharge. Plasma Processes Polym 9:90–97
104. Jiang T et al (2002) Plasma methane conversion using dielectric-barrier discharges with zeolite A. Catal Today 72:229–235
105. Zhang K, Eliasson B, Kogelschatz U (2002) Direct conversion of greenhouse gases to synthesis gas and C_4 hydrocarbons over zeolite HY promoted by a dielectric-barrier discharge. Ind Eng Chem Res 41:1462–1468
106. Zhang K, Kogelschatz U, Eliasson B (2001) Conversion of greenhouse gases to synthesis gas and higher hydrocarbons. Energy Fuels 15:395–402
107. Eliasson B, Liu C-J, Kogelschatz U (2000) Direct conversion of methane and carbon dioxide to higher hydrocarbons using catalytic dielectric-barrier discharges with zeolites. Ind Eng Chem Res 39:1221–1227
108. Michielsen I, Uytdenhouwen Y, Bogaerts A, Meynen V (2019) Altering conversion and product selectivity of dry reforming of methane in a dielectric barrier discharge by changing the dielectric packing material. Catalysts 9:51
109. Wang Y et al (2022) Catalyst-free single-step plasma reforming of CH_4 and CO_2 to higher value oxygenates under ambient conditions. Chem Eng J 450:137860
110. Li J et al (2023) One-step plasma reforming of CO_2CH_4 into hydrogen and liquid fuels: the roles of Cu and Fe sites on products distribution. Fuel Process Technol 242:107648
111. Li D et al (2023) Orientating the plasma-catalytic conversion of CO_2 and CH\ toward liquid products by using a composite catalytic bed. Appl Catal A 650:119015
112. Wang L et al (2023) Direct conversion of CH_4 and CO_2 to alcohols using plasma catalysis over Cu/Al$(OH)_3$ catalysts. Chem Eng J 466:143347
113. Dou L et al (2022) Disentangling metallic cobalt sites and oxygen vacancy effects in synergistic plasma-catalytic CO_2/CH_4 conversion into oxygenates. Appl Catal B 318:121830
114. Mei D et al (2023) Plasma-enabled catalytic dry reforming of CH4 into syngas, hydrocarbons and oxygenates: insight into the active metals of γ-Al_2O_3 supported catalysts. J CO_2 Util 67:102307
115. Wang Y et al (2022) Insight into the synthesis of alcohols and acids in plasma-driven conversion of CO_2 and CH_4 over copper-based catalysts. Appl Catal B 315:121583
116. Wang L, Yi Y, Wu C, Guo H, Tu X (2017) One-step reforming of CO_2 and CH_4 into high-value liquid chemicals and fuels at room temperature by plasma-driven catalysis. Angew Chem Int Ed 56:13679–13683
117. Zou J-J, Zhang Y-P, Liu C-J, Li Y, Eliasson B (2003) Starch-enhanced synthesis of oxygenates from methane and carbon dioxide using dielectric-barrier discharges. Plasma Chem Plasma P 23:69–82
118. Martin-del-Campo J, Uceda M, Coulombe S, Kopyscinski J (2021) Plasma-catalytic dry reforming of methane over Ni-supported catalysts in a rotating gliding arc—spouted bed reactor. J CO_2 Util 46:101474
119. Chen X, Kim HH, Nozaki T (2023) Plasma catalytic technology for CH_4 and CO_2 conversion: a review highlighting fluidized-bed plasma reactor. Plasma Processes Polym e2200207

120. Cui Z et al (2022) Plasma-catalytic methanol synthesis from CO_2 hydrogenation over a supported Cu cluster catalyst: insights into the reaction mechanism. ACS Catal 12:1326–1337
121. Porosoff MD, Yan B, Chen JG (2016) Catalytic reduction of CO_2 by H_2 for synthesis of CO, methanol and hydrocarbons: challenges and opportunities. Energy Environ Sci 9:62–73
122. Zeng Y, Tu X (2017) Plasma-catalytic hydrogenation of CO_2 for the cogeneration of CO and CH_4 in a dielectric barrier discharge reactor: effect of argon addition. J Phys D Appl Phys 50:184004
123. Sun Y et al (2020) Reverse water-gas shift in a packed bed DBD reactor: investigation of metal-support interface towards a better understanding of plasma catalysis. Appl Catal A 591:117407
124. Liu L et al (2020) Low temperature catalytic reverse water-gas shift reaction over perovskite catalysts in DBD plasma. Appl Catal B 265:118573
125. Liu L, Zhang Z, Das S, Xi S, Kawi S (2020) $LaNiO_3$ as a precursor of Ni/La2O3 for reverse water-gas shift in DBD plasma: effect of calcination temperature. Energy Convers Manage 206:112475
126. Liu L et al (2022) Plasma-catalytic carbon dioxide conversion by reverse water–gas shift over La0. 9Ce0. 1B0. 5B'0.5 O3-δ perovskite-derived bimetallic catalysts. Chem Eng J 431:134009
127. Li J et al (2019) Effect of plasma on catalytic conversion of CO_2 with hydrogen over Pd/ZnO in a dielectric barrier discharge reactor. J Phys D Appl Phys 52:244001
128. Kim D-Y et al (2022) Cooperative catalysis of vibrationally excited CO_2 and alloy catalyst breaks the thermodynamic equilibrium limitation. J Am Chem Soc 144:14140–14149
129. Zeng Y, Tu X (2015) Plasma-catalytic CO_2 hydrogenation at low temperatures. IEEE Trans Plasma Sci 44:405–411
130. Xu W, Zhang X, Dong M, Zhao J, Di L (2019) Plasma-assisted Ru/Zr-MOF catalyst for hydrogenation of CO_2 to methane. Plasma Sci Technol 21:044004
131. Chen H et al (2020) Effect of metal dispersion and support structure of Ni/silicalite-1 catalysts on non-thermal plasma (NTP) activated CO_2 hydrogenation. Appl Catal B 272:119013
132. Mikhail M et al (2021) Tailoring physicochemical and electrical properties of Ni/CeZrOx doped catalysts for high efficiency of plasma catalytic CO_2 methanation. Appl Catal B 294:120233
133. Nizio M et al (2016) Hybrid plasma-catalytic methanation of CO_2 at low temperature over ceria zirconia supported Ni catalysts. Int J Hydrogen Energy 41:11584–11592
134. Ahmad F et al (2020) Low-temperature CO_2 methanation: synergistic effects in plasma-Ni hybrid catalytic system. Acs Sustain Chem Eng 8:1888–1898
135. Chen H et al (2020) Nonthermal plasma (NTP) activated metal–organic frameworks (MOFs) catalyst for catalytic CO_2 hydrogenation. AIChE J 66:e16853
136. Chen H et al (2019) Coupling non-thermal plasma with Ni catalysts supported on BETA zeolite for catalytic CO_2 methanation. Catal Sci Technol 9:4135–4145
137. Mu Y et al (2020) Kinetic study of nonthermal plasma activated catalytic CO_2 hydrogenation over Ni supported on silica catalyst. Ind Eng Chem Res 59:9478–9487
138. Bacariza M et al (2018) DBD plasma-assisted CO_2 methanation using zeolite-based catalysts: structure composition-reactivity approach and effect of Ce as promoter. J CO_2 Util 26:202–211
139. Biset-Peiró M, Guilera J, Zhang T, Arbiol J, Andreu T (2019) On the role of ceria in Ni-Al_2O_3 catalyst for CO_2 plasma methanation. Appl Catal A 575:223–229
140. Wang B et al (2021) Improvement of the activity of CO_2 methanation in a hybrid plasma-catalytic process in varying catalyst particle size or under pressure. J CO_2 Util 46:101471
141. Wang B et al (2020) Coupling experiment and simulation analysis to investigate physical parameters of CO_2 methanation in a plasma-catalytic hybrid process. Plasma Processes Polym 17:1900261
142. Wierzbicki D et al (2020) Ni-Fe layered double hydroxide derived catalysts for non-plasma and DBD plasma-assisted CO_2 methanation. Int J Hydrogen Energy 45:10423–10432
143. Biset-Peiró M, Mey R, Guilera J, Andreu T (2020) Adiabatic plasma-catalytic reactor configuration: energy efficiency enhancement by plasma and thermal synergies on CO_2 methanation. Chem Eng J 393:124786

144. Mikhail M et al (2019) Plasma-catalytic hybrid process for CO_2 methanation: optimization of operation parameters. React Kinet Mech Catal 126:629–643
145. Lan L, Wang A, Wang Y (2019) CO_2 hydrogenation to lower hydrocarbons over ZSM-5-supported catalysts in a dielectric-barrier discharge plasma reactor. Catal Commun 130:105761
146. Parastaev A, Hoeben WF, van Heesch BE, Kosinov N, Hensen EJ (2018) Temperature-programmed plasma surface reaction: an approach to determine plasma-catalytic performance. Appl Catal B 239:168–177
147. Lee CJ, Lee DH, Kim T (2017) Enhancement of methanation of carbon dioxide using dielectric barrier discharge on a ruthenium catalyst at atmospheric conditions. Catal Today 293:97–104
148. Benrabbah R et al (2017) Plasma DBD activated ceria-zirconia-promoted Ni-catalysts for plasma catalytic CO_2 hydrogenation at low temperature. Catal Commun 89:73–76
149. Nizio M et al (2016) Low temperature hybrid plasma-catalytic methanation over Ni-Ce-Zr hydrotalcite-derived catalysts. Catal Commun 83:14–17
150. Jwa E, Lee S, Lee H, Mok Y (2013) Plasma-assisted catalytic methanation of CO and CO_2 over Ni–zeolite catalysts. Fuel Process Technol 108:89–93
151. Ronda-Lloret M et al (2020) CO_2 hydrogenation at atmospheric pressure and low temperature using plasma-enhanced catalysis over supported cobalt oxide catalysts. Acs Sustain Chem Eng 8:17397–17407
152. Jiang X, Nie X, Guo X, Song C, Chen JG (2020) Recent advances in carbon dioxide hydrogenation to methanol via heterogeneous catalysis. Chem Rev 120:7984–8034
153. Eliasson B, Kogelschatz U, Xue B, Zhou L-M (1998) Hydrogenation of carbon dioxide to methanol with a discharge-activated catalyst. Ind Eng Chem Res 37:3350–3357
154. Du J et al (2022) Numerical investigation on the heterogeneous pulsed dielectric barrier discharge plasma catalysis for CO_2 hydrogenation at atmospheric pressure: effects of Ni and Cu catalysts on the selectivity conversions to CH_4 and CH_3OH. Plasma Processes Polym 19:2100111
155. Joshi N, Loganathan S (2021) Methanol synthesis from CO_2 using Ni and Cu supported Fe catalytic system: understanding the role of nonthermal plasma surface discharge. Plasma Processes Polym 18:2000104
156. Zhang X et al (2023) Boosting methanol production via plasma catalytic CO_2 hydrogenation over a $MnOx/ZrO_2$ catalyst. Catal Sci Technol 13:2529–2539
157. Kong T, Jiang Y, Xiong Y (2020) Photocatalytic CO_2 conversion: what can we learn from conventional COx hydrogenation? Chem Soc Rev 49:6579–6591
158. Nitopi S et al (2019) Progress and perspectives of electrochemical CO_2 reduction on copper in aqueous electrolyte. Chem Rev 119:7610–7672
159. Snoeckx R, Ozkan A, Reniers F, Bogaerts A (2017) The quest for value-added products from carbon dioxide and water in a dielectric barrier discharge: a chemical kinetics study. Chemsuschem 10:409–424
160. Chen G et al (2015) Simultaneous dissociation of CO_2 and H_2O to syngas in a surface-wave microwave discharge. Int J Hydrogen Energy 40:3789–3796
161. Hoeben W, Van Heesch E, Beckers F, Boekhoven W, Pemen A (2015) Plasma-driven water assisted CO_2 methanation. IEEE Trans Plasma Sci 43:1954–1958
162. Ma X et al (2019) Plasma assisted catalytic conversion of CO_2 and H_2O over Ni/Al_2O_3 in a DBD reactor. Plasma Chem Plasma P 39:109–124
163. Mahammadunnisa S, Reddy EL, Ray D, Subrahmanyam C, Whitehead JC (2013) CO_2 reduction to syngas and carbon nanofibres by plasma-assisted in situ decomposition of water. Int J Greenhouse Gas Control 16:361–363
164. Yang S et al (2021) CO_2 Reduction to higher hydrocarbons by plasma discharge in carbonated water. ACS Energy Lett 6:3924–3930
165. Wang A et al (2019) Nonthermal plasma-catalytic conversion of biogas to liquid chemicals with low coke formation. Energy Convers Manage 191:93–101

Chapter 11
Transition Metal-Promoted Carboxylation of Unsaturated Compounds with CO_2

Man Zhao, An-Guo Wu, and Liang-Nian He

Abstract Carbon dioxide (CO_2), a resourceful, renewable, and non-toxic C_1 resource, is of great significance if applied to produce value-added chemicals. In recent years, CO_2 valorization has been well exploited as expected. Among them, functionalizing organic substrates with CO_2 to synthesize high value-added carboxylic acid compounds is extremely attractive. Since carboxylic acids and its derivatives are essential structural motifs in pharmaceuticals, pesticides, fine chemicals, commodities, and natural products. Moreover, dicarboxylic acids are very appealing monomers for the manufacture of functional plastics and rubbers. Therefore, considerable efforts have been devoted to developing synthetic methods of carboxylic acids and expanding their applications. Notably, carboxylation reactions of unsaturated organic compounds (e.g., alkenes, alkynes, aromatic hydrocarbons) with CO_2 catalyzed by transition metals have been regarded as an efficacious way to synthesize diverse functionalized carboxylic acids and its derivatives. In this chapter, we summarize the advancement of carboxylation reactions of unsaturated compounds such as (1) carboxylation of C–H bond, (2) heterocarboxylation, (3) hydrocarboxylation, and (4) dicarboxylation, promoted by transition metal complexes such as Ni, Cu, Pd, Au, Rh, and so on. We are focusing on their rationales behind and several kinds of important products which include aryl carboxylic acid, α, β-unsaturated carboxylic acid, functionalized carboxylic acid, and dicarboxylic acid.

11.1 Introduction

Confronting with the huge amount of CO_2, converting it into high value-added chemicals can not only effectively alleviate environmental issues but also deliver excellent economic benefits, which is one of the main breakthrough points toward minimizing carbon emissions. However, the percentage of CO_2 resources recovered and utilized

---

M. Zhao · A.-G. Wu · L.-N. He (✉)
State Key Laboratory and Institute of Elemento-Organic Chemistry, College of Chemistry, Nankai University, Weijin Rd. 94, Tianjin 300071, People's Republic of China
e-mail: heln@nankai.edu.cn

G. Zhang et al. (eds.), *Advances in CO_2 Utilization*, Green Chemistry and Sustainable Technology, https://doi.org/10.1007/978-981-99-8822-8_11

globally is still extremely weak, and the major constraint is that CO_2 is relatively stable and cost-effective to utilize. In organic synthesis, CO_2 is evolving as a renewable resource into a variety of integrated and valuable key products, making a unique contribution to carbon neutrality, as expected, reducing the dependence on fossil resources for chemical manufacturing and being of great importance to promoting the construction of a sustainable society with a green carbon cycle.

CO_2 represents the highest oxidation state of carbon and the lowest energy state which has a high standard enthalpy of formation (− 394 kJ mol^{-1}) and a high standard entropy (213.8 J k^{-1} mol^{-1}) [1]. In addition, CO_2 molecules, having a non-polar linear geometry, consist of two strong carbon–oxygen bonds. As mentioned above, CO_2 is thermodynamically stable and kinetically inert, which is the fundamental issue for its selective activation and conversion difficulties. As a result, a large energy input is required while using CO_2 as a raw material for the preparation of organic compounds. What motivates the high thermodynamic stability and kinetic inertness of CO_2 is the challenging task for efficient conversion of CO_2 into high value-added chemicals. Strategies currently enabling to surmount these constraints are as follows: (a) exploiting efficient catalyst schemes for activating CO_2 or motivating other substrates involving in to kinetically enhance CO_2 conversion; (b) coupling CO_2 conversion with thermodynamically favorable reactions to compensate for the energy required for CO_2 conversion; (c) reacting CO_2 with high free-energy substrates, such as epoxides, organometallics, etc.; (d) generating energetic compounds via photochemistry or electrochemistry which could be used to activate CO_2 [2, 3].

Considering that carboxylic acids and their derivatives are not only essential structural motifs for pharmaceutical [4] (Fig. 11.1) and pesticides, but also widely present in natural products, which occupies an important role in the chemical industry for the manufacture of plastics, rubbers, detergents, cosmetics, and so on [5, 6]. Therefore, the use of CO_2 as a carbon source in organic synthesis has been very appealing for the preparation of carboxylic acids through the construction of carbon–carbon bonds by incorporating CO_2 into the organic backbone [7]. Since CO_2 is a relatively stable molecule, intensive energy is necessary to break the molecular structure of the original molecule. To reduce the activation energy of the reaction, transition metal-catalyzed activation of CO_2 is a common choice. Among these reactions, unsaturated compounds are of great concern owing to the high reactivity and easy availability as one of the petrochemical species. In this context, transition metal-promoted carboxylation of unsaturated organic compounds (e.g., olefins, alkynes, aromatics) with CO_2 has been well documented, providing a valuable reference for the acquisition and further study of various functionalized carboxylic acids [8]. Wherein, the formation of a carboxylic acid is typically achieved by nucleophilic addition of carbon–metal bond intermediates of unsaturated compounds to CO_2 and can also be achieved by using low valent transition metals to form metal ring intermediates with unsaturated substrates [9]. Up to now, carboxylation reactions with CO_2 have been accomplished by several transition metals, i.e., Cu, Ag, Au; Rh, Pd, Ni, etc. [8, 10]. Among them, Ni and Cu have been most widely reported for the carboxylation of unsaturated substrates. This kind of reactions can be catalyzed by transition metals to generate new organometallic reagents in situ and then complete the carboxylation with CO_2.

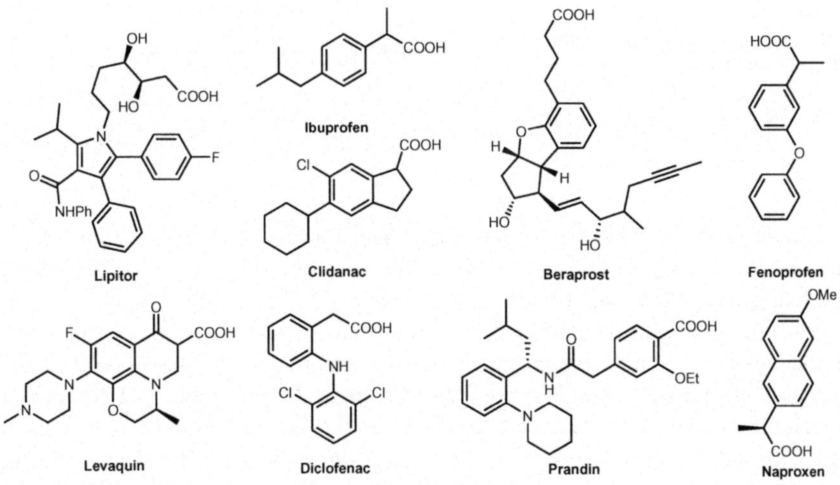

Fig. 11.1 Carboxylic acids in pharmaceuticals

Organometallic reagents (e.g., organoaluminum reagents, organozinc reagents, and Grignard reagents) are suitable for such reactions. In this chapter, we are focusing on the advancement of carboxylation reactions of unsaturated compounds promoted by transition metal complexes, which includes four parts, respectively, carboxylation of C–H bond, heterocarboxylation, hydrocarboxylation, and dicarboxylation.

11.2 Carboxylation of C–H Bond

11.2.1 Carboxylation of Alkenes

Acrylates as well as its downstream products are ubiquitous in daily life and are produced in millions of tons worldwide. Acrylates yielded from CO_2 and ethylene, one of a high value conversion strategy of CO_2 is regarded as a remarkably appealing approaches. To date, catalytic processes for directly catalyzing the functionalization of olefins with CO_2 into industrially relevant target molecules have not been established. A major breakthrough in this "dream reaction" has been achieved by proposing a convenient method for the catalytic formation of lithium acrylate from CO_2 and ethylene in combination with a homogeneous nickel catalyst and "hard" Lewis acid. The turnaround number is as high as 21 (Scheme 1) [11].

In this regard, a catalytic cycle for the Ni-catalyzed acrylate formation has been proposed as depicted in Scheme 2. The first step (**I**) involves the oxidative coupling of ethylene and CO_2 on a Ni(0) phosphine complex. Subsequently, Lewis acid (Li^+)

Scheme 1 Ni-catalyzed carboxylation of activated ethylene with CO_2

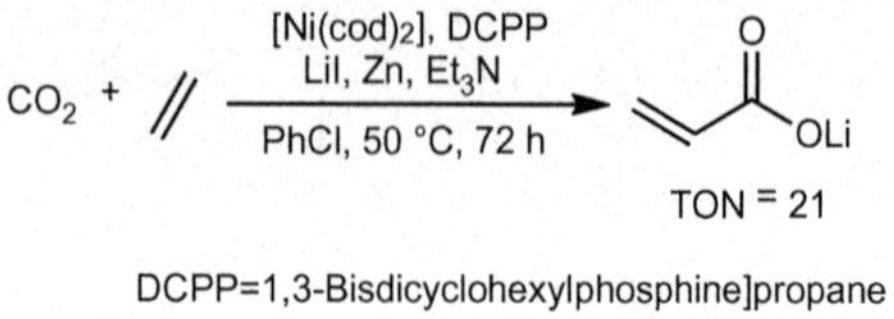

is partially coordinated to the oxygen of nickelalactone to facilitate Ni–O bond dissociation (**II**). Then, β-H elimination (**III**) occurs, followed by the release of lithium acrylate and regeneration of Ni(0) in the presence of a base (**IV**).

With the increasing understanding of the reaction, one targets the formation of acrylates for exergonic reactions rather than high internal energy acrylic acid ($\Delta G =$ 21 vs. $-$ 59 kJ mol^{-1}), which overcomes the kinetically infeasible β-H elimination from dtbpe-linked nickelalactone **1** ($\Delta G = 21$ vs. $-$ 59 kJ mol^{-1} in THF) (Scheme 3). The kinetically infeasible β-H elimination from dtbpe-linked nickelalactone **1** (ΔG $\neq$ 103 kJ mol^{-1}) is overcome by α-deprotonation with an alkoxide base [12].

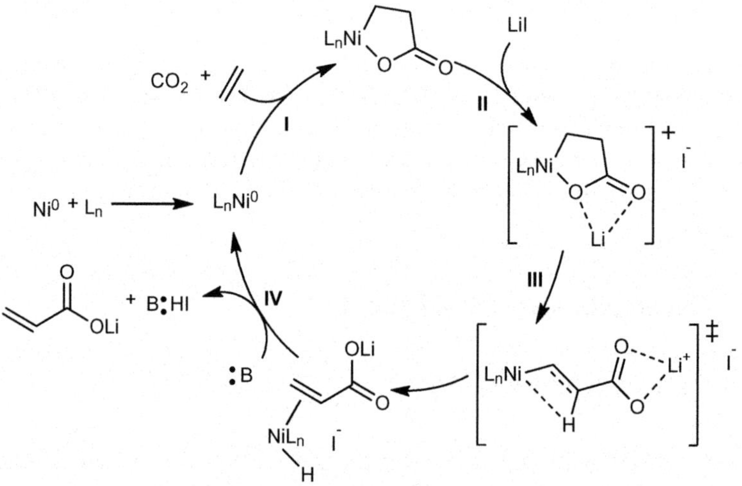

Scheme 2 Proposed mechanism for the nickel-catalyzed carboxylation of activated ethylene with CO_2

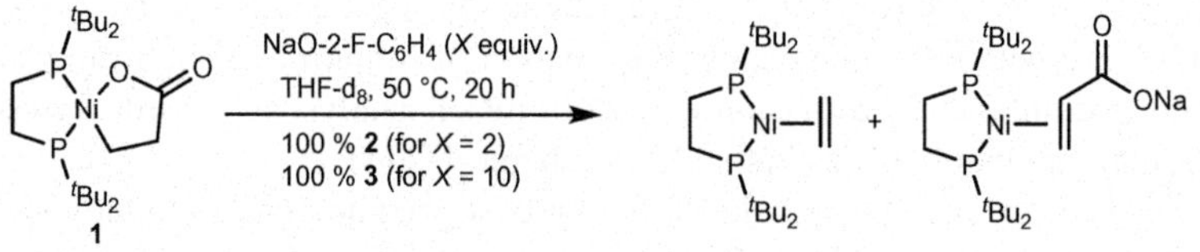

Scheme 3 Nickelalactone cleavage by sodium 2-fluorophenoxide

Scheme 4 Pd-catalyzed carboxylation of ethylene with CO_2

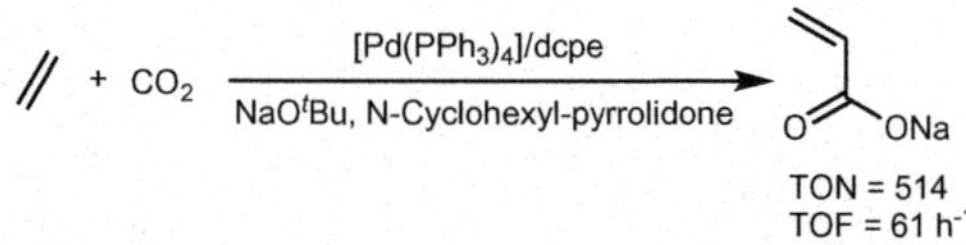

Considering that the turnover numbers of the above catalytic reactions are still low, a continuous process has been developed using amides as solvents (e.g., *N*-cyclohexylpyrrolidone or *N*, *N*-dibutylformamide) and $[Pd(PPh_3)_4]$ (PPh_3 = triphenylphosphine) combined with 1,2-bis(dicyclohexylphosphino)ethane (dcpe) as catalysts to avoid the use of solid reducing agents (e.g., Zn) and to improve the catalytic activity (Scheme 4). In this manner, the Pd-catalyzed synthesis of sodium acrylate from ethylene and CO_2 in the presence of alcoholate bases is significantly improved. A turnover number greater than 500 is achieved in a single run. What is even better is that the catalyst recovery of does not require any additional regeneration step owing to the use of amide solvents [13].

Besides, rhodium complexes containing guanidinium-based NCN pincer ligands have been designed and successfully applied to the synthesis of acrylates based on CO_2 and ethylene (Scheme 5). According to the theoretical comparison of the sidearm effect and experimental results, the repulsive force between the pπ electrons of guanidine and the filled dπ orbitals of the rhodium center increases the electron density of rhodium, which promotes the formation of favorable thermodynamic effects in rhodium lactones. This work bridges the gap in the lack of VIII B metal-based catalysts for acrylate synthesis and provides a new design approach for pincer-linked d_8 metal catalysts to promote desirable redox processes using $p\pi$–$d\pi$ interactions [14].

More intriguingly, the C–H carboxylation of undirected olefins has been achieved by an automated tandem copper-catalyzed process (Scheme 6). The resulting multifunctional catalyst enables the highly selective tandem functionalization of alkenyl C–H bonds by in situ conversion of copper hydride after β-hydride elimination. Fascinatingly, (a) the optimized reaction system accelerates the β-hydride elimination of the alkyl copper intermediate while blocking its interaction with CO_2, which allows the resulting copper hydride and borated alkenes to participate further reactions; (b) the transient reaction of the unstable copper hydride with CO_2 leads to the formation

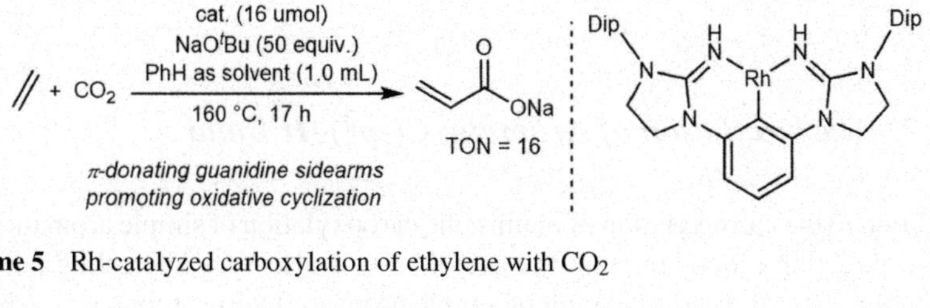

Scheme 5 Rh-catalyzed carboxylation of ethylene with CO_2

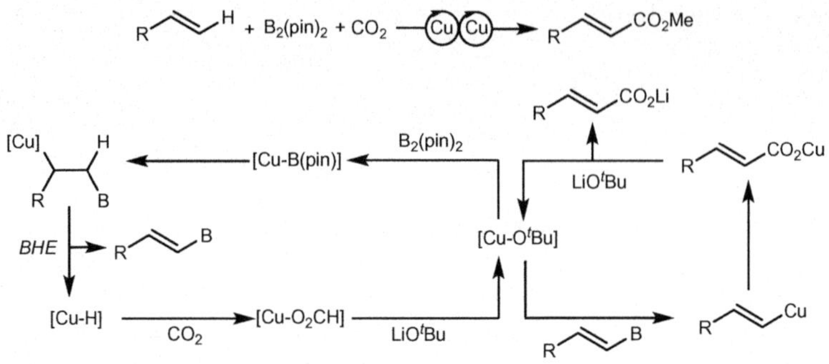

Scheme 6 Copper-catalyzed carboxylation of undirected alkenyl C–H bonds with CO_2

of the key active species, copper *tert*-butoxide, by circumventing the migration insertion after β-hydride elimination; (c) copper *tert*-butoxide achieves C–H carboxylation of non-directed olefins via a tandem catalytic cycle; (d) CO_2 as a multitasking carboxylation reagent enables not only the conversion of copper hydrogenation but also induces the selective carboxylation of streptenyl borate intermediates [15].

A possible reaction mechanism for the copper-catalyzed selective C–H carboxylation reaction of olefins with CO_2 is provided as illustrated in Scheme 7. The initial complexation between the copper chloride and LiO-^{t}Bu gives tert-butanol copper **A**, which upon reaction with $B_2(pin)_2$ produces boryl copper complex **B**. Subsequently, selective insertion of the C–C double bond of olefin substrate **1** into the Cu–B bond of **B** will give β-boryl copper species **C**. Although CO_2 can be inserted into the Cu–C bond to produce the boron carboxylation intermediate **J**, this catalytic system significantly accelerates the β-hydride elimination of **C** to give the copper hydrogenated **D** and alkenyl boronic acid ester **F**. Afterward, the instantaneous reaction of the **D** with CO_2 leads to the formation of copper formate **E**, the reaction with LiO–^{t}Bu can release lithium formate and regenerate the key active species **A**. Later, the reaction of chain alkenyl boronic acid ester **F** with Lewis basic intermediate **A** leads to the metallization of the chain alkenyl group to the copper center and produces chain alkenyl copper The reaction with CO_2 then yields the copper acrylate intermediate **H**. Finally, the complexation between LiO–^{t}Bu and **H** will yield lithium acrylate **I** and regenerate *tert*-butyl copper **A**. Treating lithium salt **I** with methyl iodide will yield the methyl acrylate product.

11.2.2 Carboxylation of Aromatic C(sp^2)–H Bond

In addition to the carboxylation of olefins, the carboxylation of simple aromatic rings without directing groups through the use of CO_2 is also very interesting but challenging. In 2014, the carboxylation of simple aromatic hydrocarbons was achieved

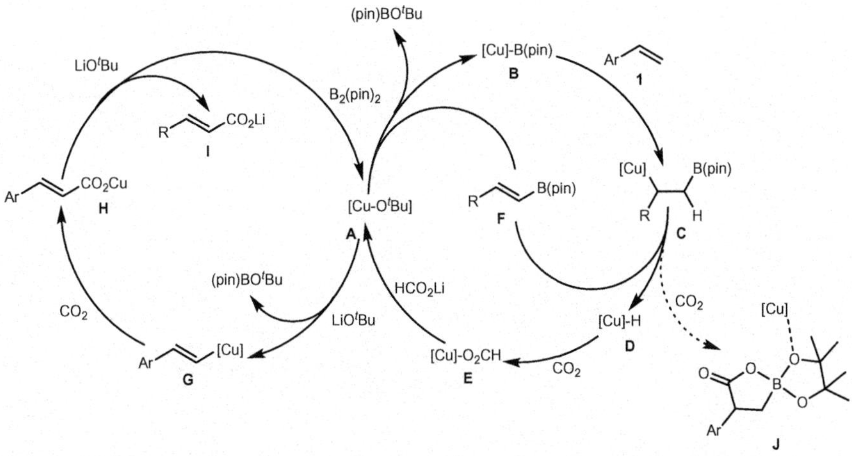

Scheme 7 Proposed mechanism for the copper-catalyzed carboxylation of undirected alkenyl C–H bond with CO_2

with the assistance of undirected groups by a rhodium-catalyzed C–H bond activation strategy under CO_2 at 1 atm (Scheme 8). The reaction displays broad versatility and interesting regioselectivity, which has not been achieved previously through the use of acid or base promotion [16].

For this reason, one possible reaction mechanism scheme is proposed by the authors as shown in Scheme 9. The reaction starts with the generation of methylrhodium(I) complex **A** and $AlMe_{1.5}(OEt)_{1.5}$ from **1c**, followed by oxidative addition of sp^2 C–H bonds of simple aromatics to **A** to give aryl(hydride)(methyl)rhodium(III) intermediate **B**. Reduction of eliminated methane from **B** gives highly reactive aryl rhodium(I) complex **C**. Nucleophilic addition of **C** to CO_2 gives rhodium(I) benzoate complex **D**, **D** obtains methylrhodium(I) **A** by transmetallation with $AlMe_{1.5}(OEt)_{1.5}$.

Meanwhile, a [(NHC)Au(I)]-based (NHC = *N*-heterocyclic carbene) catalytic system has been developed for the carboxylation of aromatic and heteroaromatic C–H bonds (Scheme 10). The high basic strength of the Au–OH species is the origination of the activation process, allowing simple functionalization of the C–H bond without the use of other organometallic reagents [17].

Moreover, a simple copper-mediated scheme is also developed in which the N–H or C–H bond can be directly functionalized using an easily

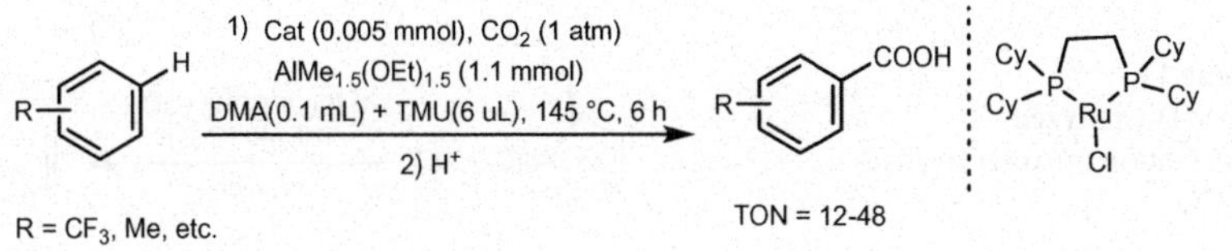

Scheme 8 Ru-catalyzed carboxylation of Ar–H bonds with CO_2

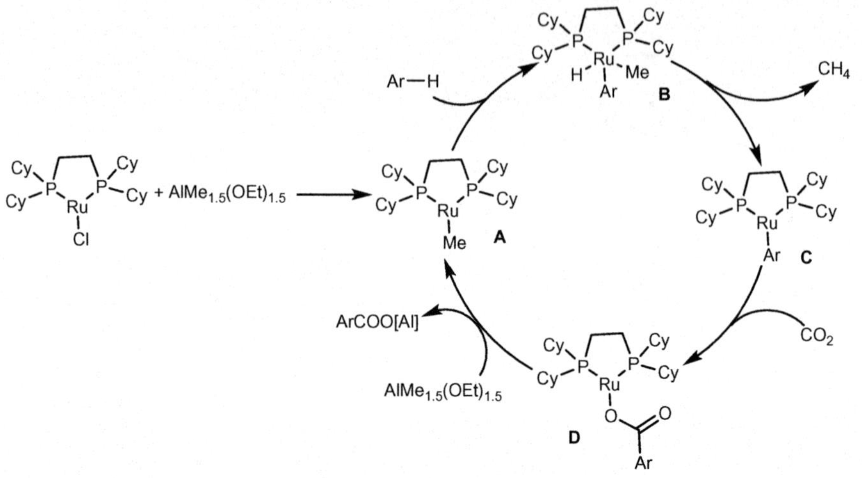

Scheme 9 Proposed mechanism for the Ru-catalyzed carboxylation of Ar–H bonds with CO_2

Scheme 10 NHC-Au-catalyzed carboxylation of Ar–H bonds with CO_2

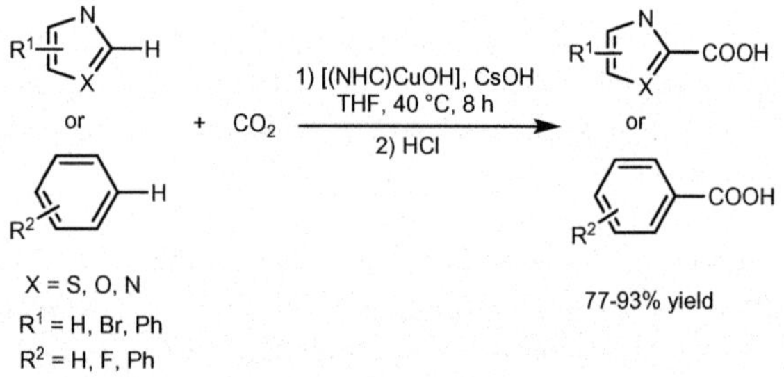

prepared catalyst (Scheme 11). The novel [1,3-bis(2,6-diisopropylphenyl)imidazol-2-ylidene]copper(I) hydroxide [Cu(IPr)(OH)] allows effortless activation and carboxylation of N–H and C–H bonds with pKa values less than 27.7 [18].

Mechanistic studies reveal that the catalytic cycle is very similar to that of the gold-catalyzed carboxylation reaction, in which 2-Oxazolidinone (**2a**) is formed by the primary decomposition of [Cu(IPr)(OH)](**1**) to produce copper(I)-oxazolidinone species **A** (Scheme 12). The cycle proceeds with the salt complex decomposition involving KOH to regenerate **1**, along with the precipitation of potassium 3(2H)-Oxazolecarboxylicacid.

Scheme 11 Copper(I)-catalyzed carboxylation of Ar–H bonds with CO_2

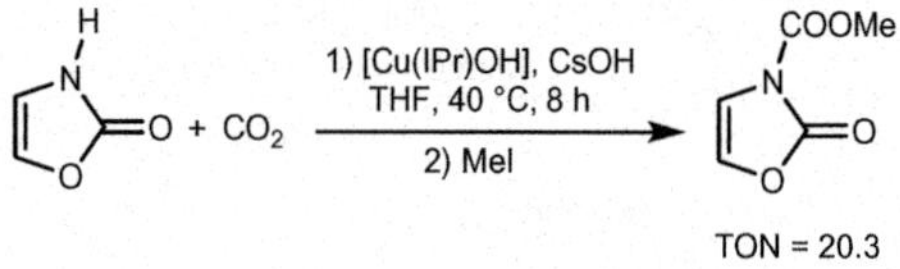

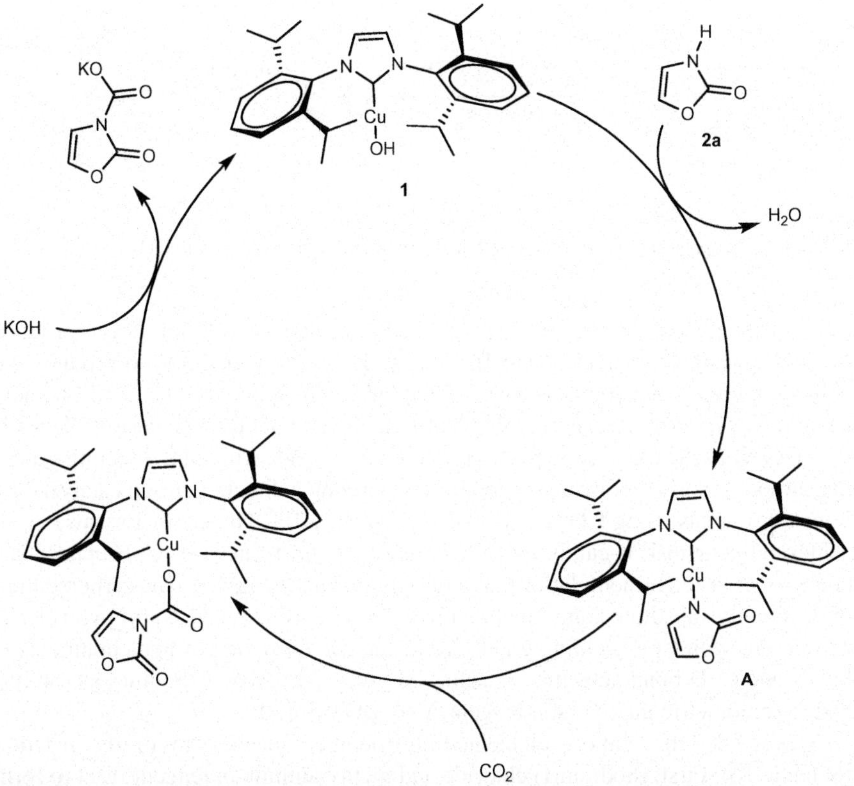

Scheme 12 Proposed mechanism for the copper(I)-catalyzed carboxylation of Ar–H bonds with CO_2

Despite the carboxylation reaction of aryl C–H bonds with CO_2 has shown high synthetic efficiency, there are still limitations such as narrow substrate range, high catalyst loading, harsh reaction conditions, poor reaction selectivity, and directed group dependence considering the inertness and selectivity challenges of C–H bonds. An efficient copper-catalyzed carboxylation reaction of arylthioanthracenes salts has been carried out with CO_2 under room temperature (Scheme 13). The reaction is performed using a low-loading cuprous chloride with good functional group tolerance at 1 atm CO_2, providing an effective site-selective C–H carboxylation method for the late-stage carboxylation modification of drug molecules. Through this approach, various aromatic carboxylic acids and several carboxylate modified drug molecules are easily accessible in combination with C–H thionation [19].

To this end, a rational reaction pathway is proposed. Initially, CuCl reacts with $ZnEt_2$ to form the CuEt intermediate (Scheme 14). Then, the catalytically active Cu(0) species is formed by the release of H_2 and C_2H_4. Next, C–S bond cleavage occurs via single electron transfer (SET) of Cu(0) to form aryl CuI intermediates **I**, $CuBF_4$ and thianthrene (TT). $CuBF_4$ can be reduced to Cu(0) by $ZnEt_2$ in the

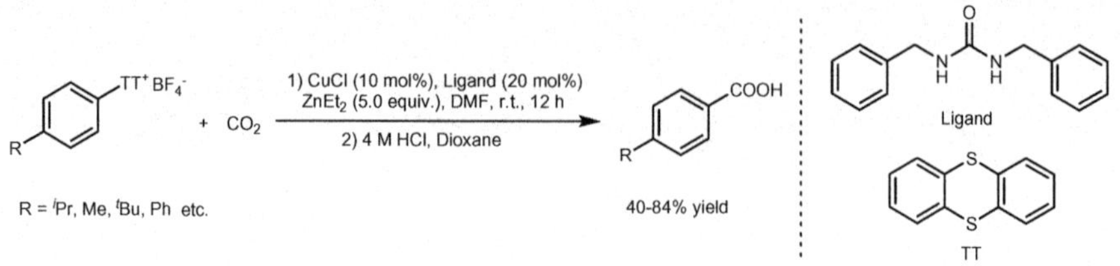

Scheme 13 Copper-catalyzed carboxylation of aryl thianthrenium salts with CO_2

same manner. Subsequently, CO_2 is inserted into the C–CuI bond of **I** to produce the carboxylate CuI intermediate **II**. Finally, **II** reacts with $ZnEt_2$ to produce the corresponding carboxylic acid zinc $ArCO_2ZnEt$, and regenerates CuEt for the next catalytic cycle. Protonation of $ArCO_2ZnEt$ yields the final product carboxylic acid.

Apart from the carboxylation of aromatic compounds without a directing group, the carboxylation of aromatic compounds containing a directing group catalyzed by transition metals using CO_2 is also developed. In 2011, Nobuharu Iwasawa et al. achieved the first Rh(I)-catalyzed carboxylation of unactivated aryl C–H bonds via a chelation-assisted o-methylation catalytic approach (Scheme 15). The carboxylation of differently substituted and functionalized 2-aryl pyridines and 1-aryl pyrazoles is carried out with CO_2 yielding good yields, and the formation of nucleophilic C–C bonds and C–H bond activation is achieved in the presence of rhodium as catalyst and stoichiometric methylation reagent $AlMe_2(OMe)$ [20].

Meanwhile, the authors elucidated the detailed mechanism of the reaction (Scheme 16). First, rhodium(I) chloride and methylaluminum reagent react to form methylrhodium(I) **A**. Then, the key aryl rhodium(I) **C** is obtained by chelation-assisted C–H bond activation (oxidative addition) of the generated aryl rhodium

Scheme 14 Proposed mechanism for the copper(I)-catalyzed carboxylation of aryl thianthrenium salts with CO_2

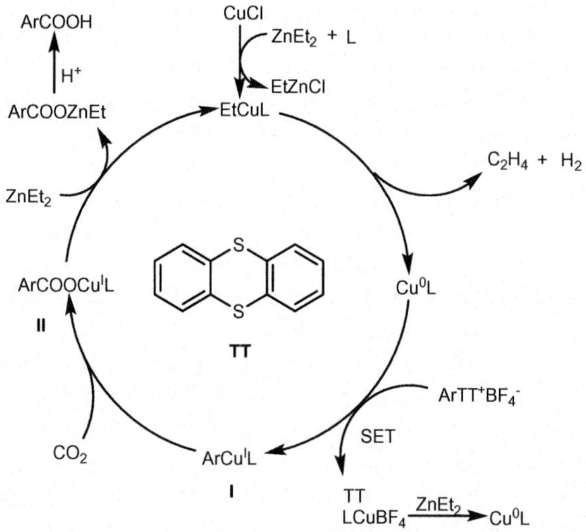

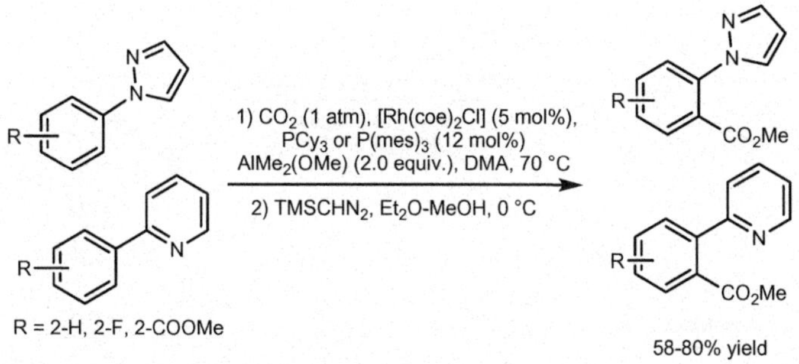

Scheme 15 Rhodium-catalyzed carboxylation of Ar–H bonds with CO_2

(III) species **B**, and then the methane is eliminated by reduction. Subsequently, the oxidative addition reaction of aryl rhodium(I) intermediate **C** with CO_2 will give rhodium (II) carboxylate **D** intermediate, which undergoes metal transfer with methylaluminum reagent to give aluminum carboxylate **E**, while methyl rhodium(I) **A** is regenerated.

Then, an efficient and versatile ligand has been developed to enable selective carboxylation of 2-aryl phenols with CO_2 under redox-neutral conditions via $Rh_2(OAc)_4$ catalysis (Scheme 17). This new approach is compatible with a wide scope of substrates, including challenging heteroaromatics, and yields a range of

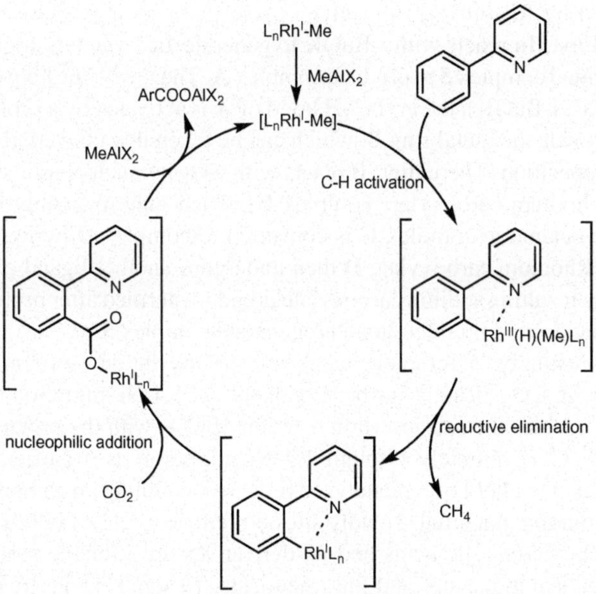

Scheme 16 Proposed mechanism for the Rhodium-catalyzed carboxylation of Ar–H bonds with CO_2

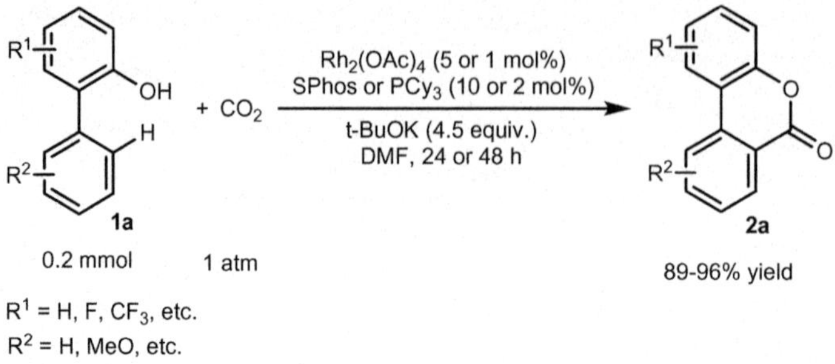

Scheme 17 Rh-promoted carboxylation of Ar–H bonds with CO_2

carboxylation products in high yields, making an important contribution to key structural motifs in biologically important molecules. Remarkably, the reaction occurs selectively on the less nucleophilic phenyl group. The use of a phosphine ligand in combination with $Rh_2(OAc)_4$ successfully overcomes the traditionally dominant Kolbe–Schmitt type reaction of phenol derivatives under alkaline conditions and effectively carboxylates the non-acidic C–H bonds in several important heterocycles, revealing the importance of the ligand in promoting the reactivity and site selectivity in the C–H carboxylation process. In addition, the reaction does not involve sensitive reducing or oxidizing agents, providing a relatively benign and moderate catalytic system for the direct carboxylation of aromatic hydrocarbons with CO_2 [21].

Based on the experimental results and previous studies on the C–H activation catalyzed by $Rh_2(OAc)_4$, a catalytic cycle pathway is proposed as shown in Scheme 18. First, **1a** reacts with *t*-BuOK to generate **1a′**, which is coordinated with the in situ formed complex **5** to produce complex **A**. Then, with the help of an external base ROK (e.g., *t*-BuOK), the aryl C–H bond of **A** is activated by a proton extraction process to provide the metal ring **B**, which can be reversibly converted back to **A** by proton decomposition. Thereafter, **B** reacts with CO_2 by nucleophilic carboxylation to form the rhodium carboxylate complex **C**, which may reversibly regenerate **B**. The unstable octameric complex **C** is converted to complex **D** by ligand exchange with KOAc. Rhodium carboxylate **D** then undergoes another ligand exchange with KOAc and the resulting sodium carboxylate is endo-esterified after proton hydrolysis to generate the desired product **2a** and regenerate complex **5**.

More interestingly, a remotely catalyzed carboxylation reaction of aromatic compounds with CO_2 (1 atm) has been realized via 1,4-Ni migration (Scheme 19). The reaction enables the site-insertion reaction of CO_2 with the previously unfunctionalized sp^2 C–H remotely through 1,4-Ni migration as a carrier, providing a novel prospect in the field of Ni-catalyzed reductive coupling and an unconventional platform for binding potential-friendly sites on remote sp^2 C–H bonds [22].

Heterocyclic carboxylic acids and its derivatives are valuable structural motifs for pharmaceutical molecules and fine chemicals. To some extent, transition metal-catalyzed carboxylation reactions using CO_2 in aromatic heterocycles have also been exploited. For instance, NHC–copper complexes have been used as excellent catalysts

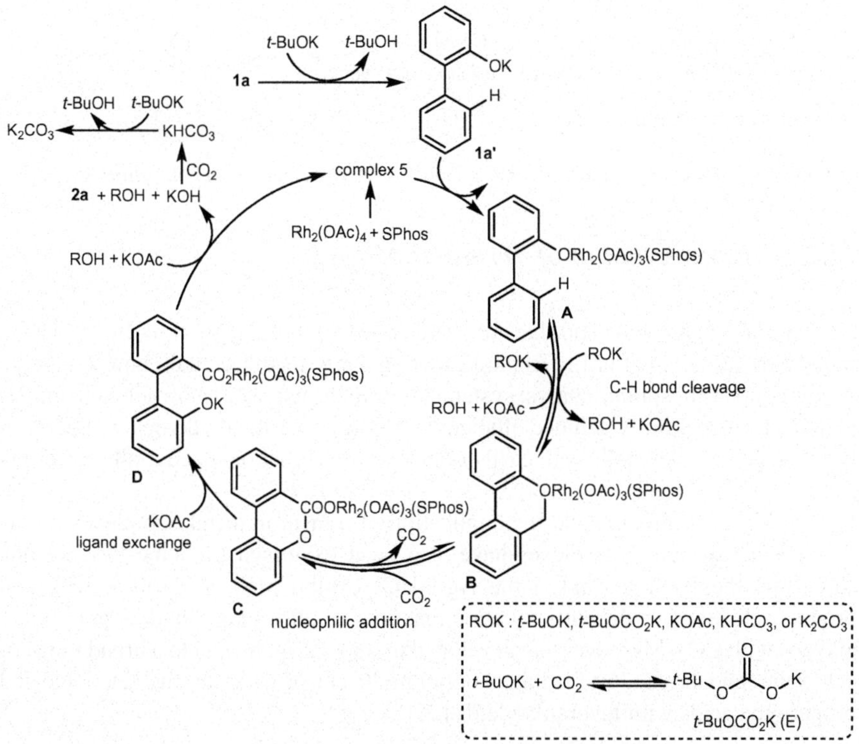

Scheme 18 Proposed mechanism for the Rh-promoted carboxylation of Ar–H bonds with CO_2

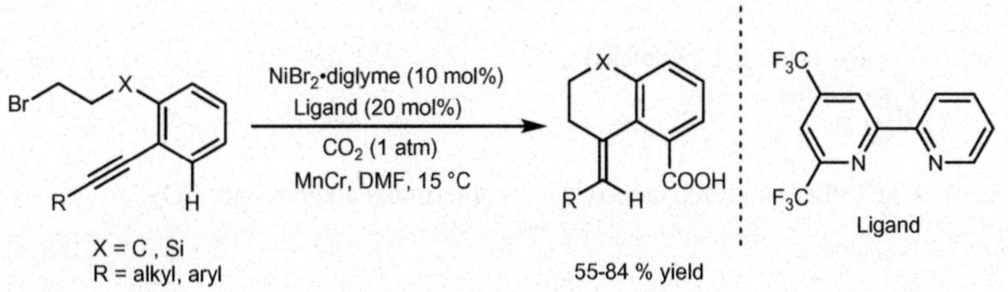

Scheme 19 Ni-mediated remote carboxylation of Ar–H bonds with CO_2

for the direct carboxylation of C–H bonds of aromatic heterocycles with CO_2 after alkyl iodide treatment, yielding the corresponding carboxylate esters (Scheme 20). This method has the unique advantages of adopting CO_2 as the C_1 source, simple heterocycles as substrates (no need to prepare pre-activated substrates) and relatively inexpensive copper complexes as catalysts. Thus, this method poses an economical and environmentally benign approach for the synthesis of heterocyclic carboxylic acid esters [23].

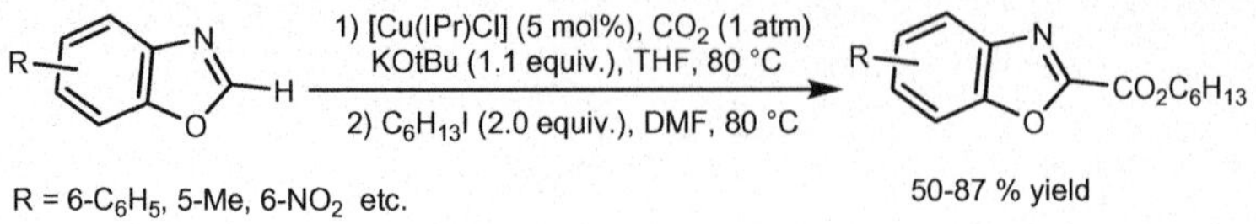

Scheme 20 NHC-copper-catalyzed carboxylation of heteroaromatic C–H bond with CO_2

11.2.3 Carboxylation of Terminal Alkynes

Alkyne carboxylic acids and their derivatives, which are widely found in bioactive molecules, are of great utility as multifunctional intermediates in organic synthesis. A new copper/phosphine catalyst system has been designed, as shown in Scheme 21, which can successfully achieve efficient coupling of terminal alkynes with CO_2 in the presence of alkyl halides to easily obtain various functionalized 2-alkynyl esters [24].

Besides, to further expand the application of transition metal complexes in this domain, silver-based complexes have also been examined (Scheme 22). Results show that these simple and stable Ag-NHCs exhibit high catalytic activity and product selectivity in the carboxylation reactions of alkynyl aryl and alkyl terminal alkynes with CO_2. Moreover, the developed method is applicable to a broad range of substrates, and the catalyst loading of only 1 mol% can provide various functionalized propargylic acids with favorable yields [25].

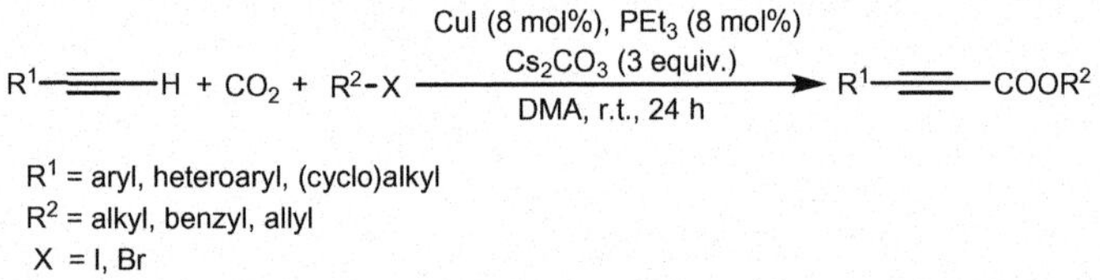

Scheme 21 CuCl/PEt$_3$-catalyzed carboxylation of terminal alkynes with CO_2

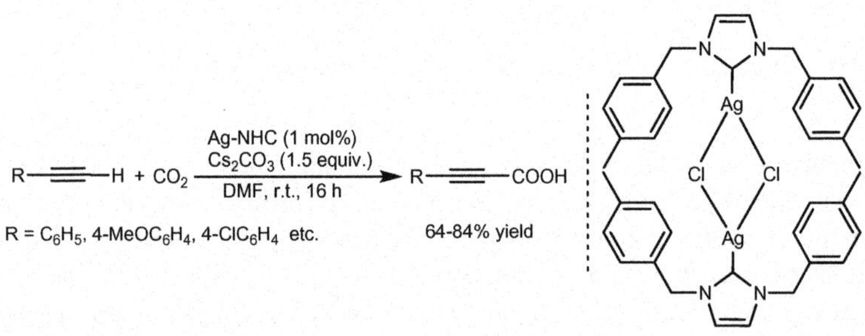

Scheme 22 Ag-NHC-catalyzed carboxylation of terminal alkynes with CO_2

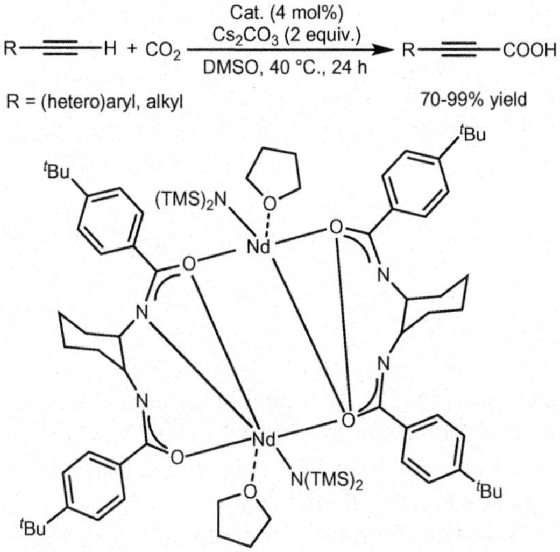

Scheme 23 Nd-catalyzed carboxylation of terminal alkynes with CO_2

In addition, bis(amidate) rare-earth metal amides $\{LRE[N(SiMe_3)_2]\cdot THF\}_2(H_2L = N,N'$-(cyclohexane-1,2-diyl)bis(4-tert-butylbenzamide); RE = Nd(2)), prepared in tetrahydrofuran with $RE[N(SiMe_3)_2]_3$ bridged to the amide-derivative H_2L, has also been found to be an effective catalyst for the direct carboxylation of terminal alkynes with CO_2 at atmospheric pressure (Scheme 23). Various propargylic acids with favorable functional group tolerance can be successfully synthesized under mild conditions with high to excellent yields [26].

11.3 Hydrocarboxylation

Most reaction pathways of hydrocarboxylation require that active metal hydride species is added to unsaturated bonds to form carbon–metal bond into which CO_2 is inserted (Scheme 24A). In addition, this type of reaction can also be achieved through the oxidation ring metallization of CO_2 and unsaturated hydrocarbons with low valent transition metals (Scheme 24B).

11.3.1 Hydrocarboxylation of Alkynes

In recent years, transition metal catalyzed hydrocarboxylation of alkynes with CO_2 has received considerable attention because it may directly produce α, β-unsaturated carboxylic acid, which is an important substructure sequence widely found in drugs and biologically active natural products. An efficient nickel-catalyzed three-component highly regio- and stereoselective *syn*-hydrocarboxylation

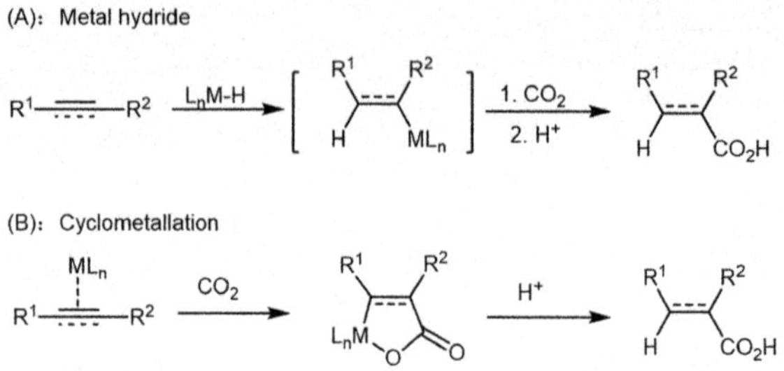

Scheme 24 Two reaction modes for hydrocarboxylation of unsaturated hydrocarbons with CO_2

of internal alkynes using CO_2 mediated by CsF was developed by Ma and coworkers (Scheme 25) [27]. This catalytic system is efficient with just 1–3 mol% of [Ni(cod)$_2$] needed for the smooth activation of CO_2, which greatly reduces the catalyst loading. When one side is an alkyl substituent, the reaction requires 3 mol% of catalyst, with yields ranging from 68 to 91%.

Based on control experiments, they proposed a possible mechanism (Scheme 26). The alkyne reacts with Ni(0)L and the resulting complex is transmetallized with diethylzinc, which explains the *syn* stereoselectivity, followed by β-hydrogen elimination and reduction elimination. The *cis* hydrozincization of the alkyne with diethylzinc results in a vinyl-zinc intermediate, which is then carboxylated with CO_2.

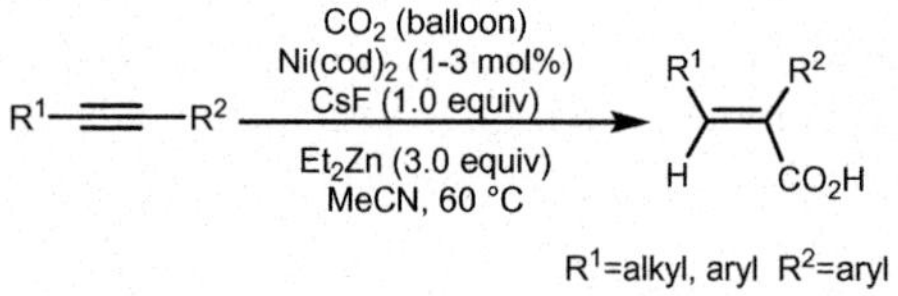

Scheme 25 Ni-catalyzed hydrocarboxylation of alkyne with CO_2

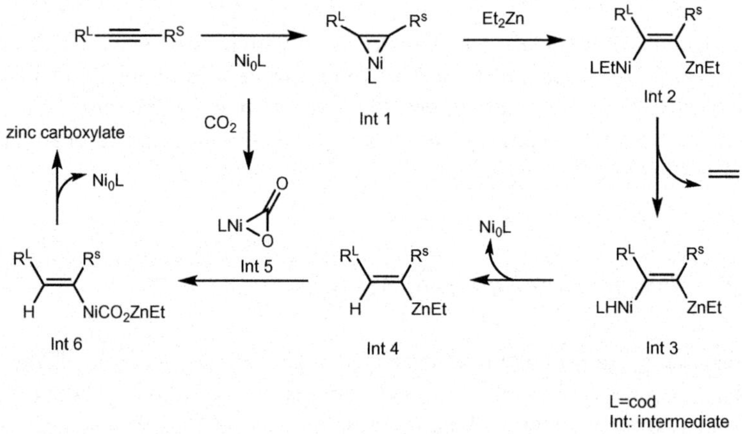

Scheme 26 The mechanism for the Ni-catalyzed hydrocarboxylation of alkyne with CO_2

In addition to organozinc, Grignard reagents can also be hydride source. RMgBr as hydride source, Fe-and Ti-catalyzed hydrocarboxylation of internal alkynes with CO_2 are reported separately (Scheme 27) [28, 29]. However, the yield of substrates with electron withdrawing groups on the benzene ring connected to the triple bond will be as low as 40%. In addition, when using unsymmetrical internal alkynes as substrates, the regioselectivity of some products is moderate especially using Fe catalysts.

Mechanisms of the two systems are similar (Scheme 28). Cp_2TiCl_2 is reduced to give Cp_2TiH by β-hydride elimination. The intermediate A undergoes transmetallation with the Grignard reagent to release hydromagnesiated product C that reacts with CO_2.

In cases above-mentioned, substrate scope is limited to internal alkynes, while terminal alkynes are not applicable. The hydroalumination or methylalumination of various alkynes is combined with the carboxylation of the resulting alkenylaluminum species with CO_2 to realize the hydrocarboxylation of terminal acetylenes (Scheme 29) [30]. The NHC–copper complex [CuCl(IPr)], acting as an excellent catalyst for the carboxylation reactions, matches well with various catalyst systems

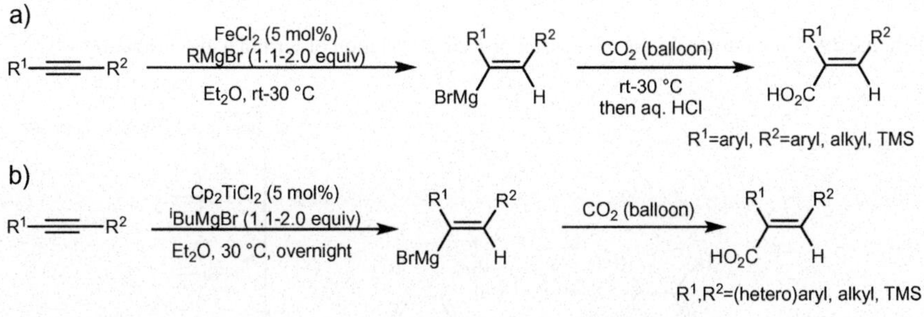

Scheme 27 Fe/Ti-catalyzed hydrocarboxylation of internal alkynes with CO_2

Scheme 28 The mechanism for the Fe/Ti-catalyzed hydrocarboxylation of internal alkynes with CO_2

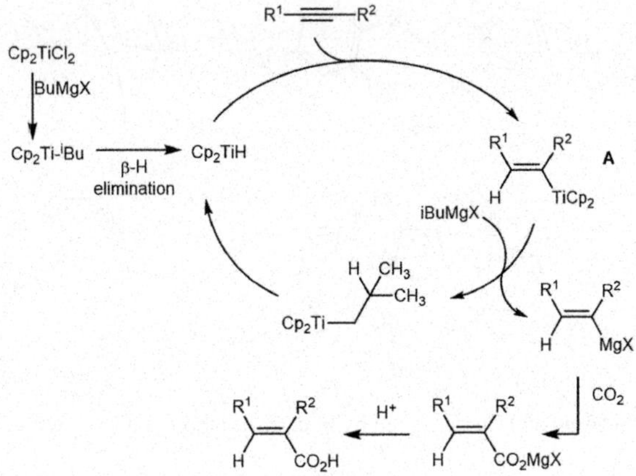

for the hydroalumination or methylalumination reactions. The regio- and stereoselectivity of the overall reaction is determined solely by the selection of ligands in methylalumination or hydroalumination process.

Metal–organic reagents are air-sensitive, so it is appealing that the use of mild and easy-to-handle hydrosilane as reducing agent realizes highly efficient hydrocarboxylation of alkynes (Scheme 30) [31]. A copper(I) hydride species (A) is generated in situ by the aid of the strong silicon–fluorine interaction. The regioselectivity is determined by the stability of the copper alkenyl intermediate (B). It is worth mentioning that when PMHS (polymethylhydrosiloxane) is used as the hydride source, the yield of target product can reach 80%. As a by-product of the silicon industry, PMHS has stable properties, low price, and potential for industrial application.

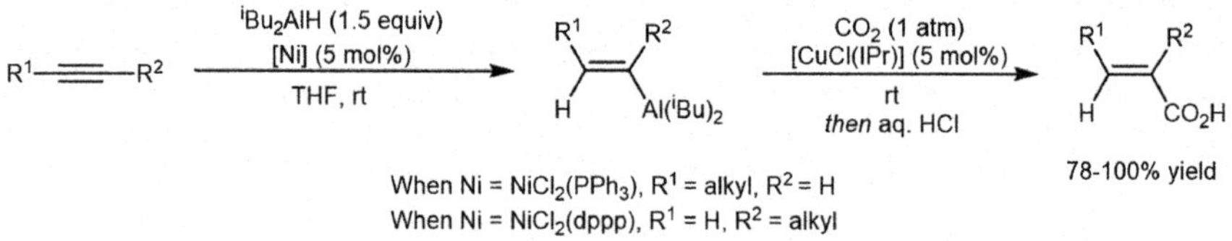

Scheme 29 Cu-catalyzed α-carboxylation or β-carboxylation of terminal alkynes with CO_2

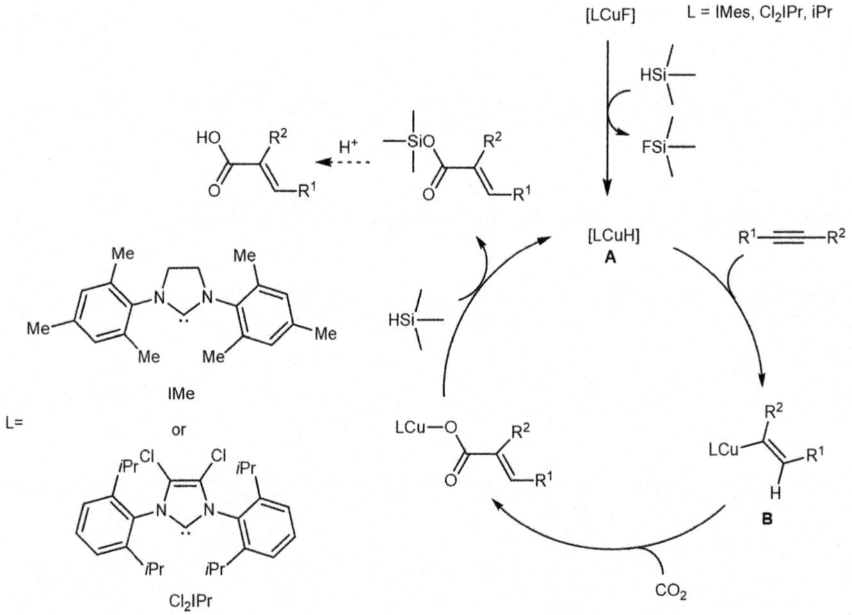

Scheme 30 Cu-catalyzed hydrocarboxylation of alkyne with CO_2

Recently, a highly regioselective Pd-catalyzed hydrocarboxylation of alkynes with CO_2 was developed (Scheme 31) [32]. This transformation offers an attractive route toward straightforward synthesis of various α-acrylic acids and a powerful tool for the late-stage functionalization of alkyne-containing complex molecules, including natural products and pharmaceuticals.

Plausible mechanistic pathways are tentatively proposed in Scheme 32. First, the active palladium catalyst **1** is generated in situ. Subsequently, the reaction might proceed via the oxidative cyclization process to afford five-membered palladiumalactone intermediate **2** (Scheme 32, path A), which could react with $PhSiH_3$ and undergo the reductive elimination to regenerate the Pd(0)L and afford intermediate **3** which released product **2a**. Additionally, the reaction also might proceed the hydropalladation that the active palladium species reacts with $PhSiH_3$ firstly as shown in Scheme 32 (path B) on the left.

Scheme 31 Pd-catalyzed hydrocarboxylation of alkyne with CO_2

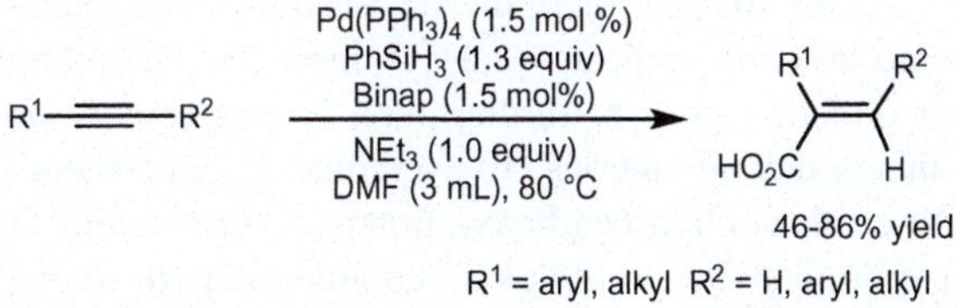

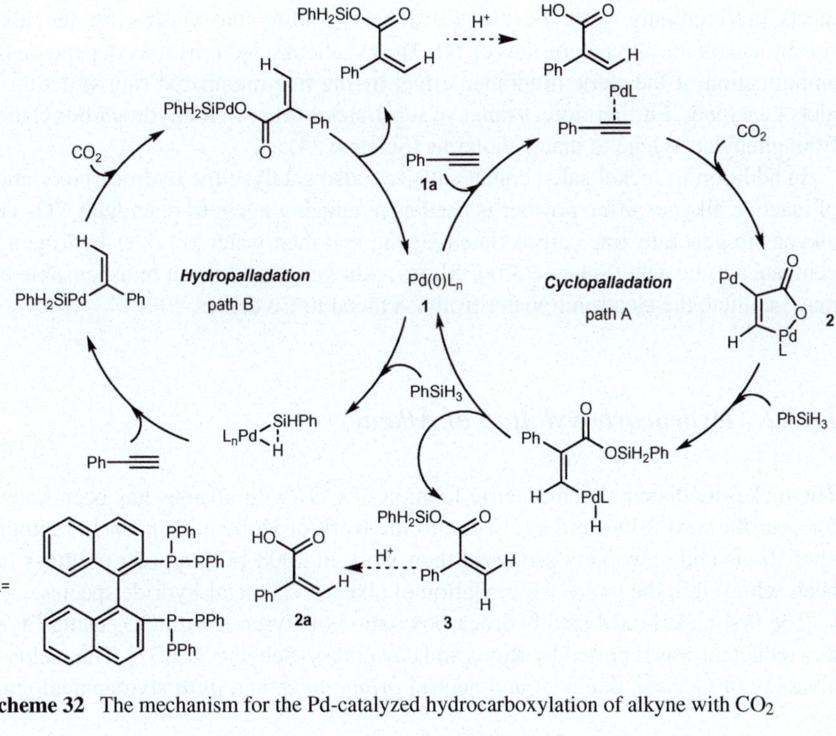

Scheme 32 The mechanism for the Pd-catalyzed hydrocarboxylation of alkyne with CO_2

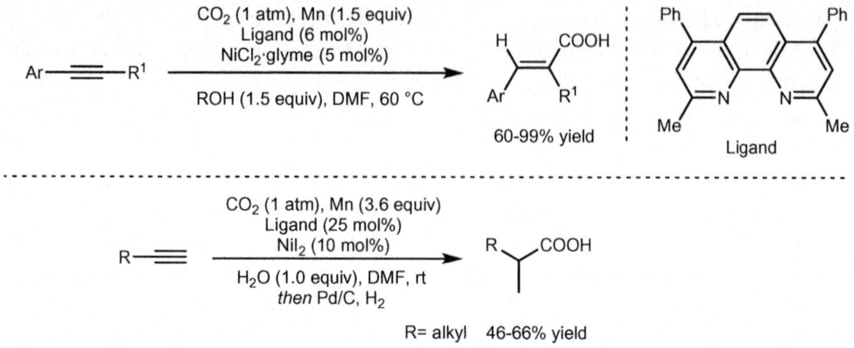

Scheme 33 Ni-catalyzed hydrocarboxylation of aryl alkynes with CO_2 and alcohols/H_2O

A close inspection of these procedures with metal–organic reagents and hydrosilane, however, indicates that CO_2 insertion preferentially occurs adjacent to aromatic or directing groups; furthermore, low selectivity profiles are found for sterically unbiased combinations. To overcome these problems, Martin group has described a Ni-catalyzed hydrocarboxylation of alkynes with CO_2 that occurs with an exquisite regioselectivity profile using commercially available alcohols and water as proton sources (Scheme 33) [33, 34].

The mechanistic scenario consists of the intermediacy nickelalactone **II**, which reacts preferentially with the alcohol donor avoiding the clash with the alkyl substituent on the alkyneterminus of **III**. The excellent selectivity may depend on the amplification of the steric hindrance effect by the five-membered ring structure of nickelalactones. Furthermore, a range of substrates work well for hydrocarboxylation from phenylacetylene to inactive olefins (Scheme 34).

In addition to nickel salts, cobalt salts can also catalyze the hydrocarboxylation of inactive alkynes. Zinc powder is used as a reducing agent to react with CO_2 and alkynes to generate zinc carboxylate reagent, and then water provides hydrogen to generate acrylic acid (Scheme 35) [35]. As additives, Et_4NI is not indispensable but may facilitate the electron transfer from Zn metal to Co center.

11.3.2 Hydrocarboxylation of Alkenes

The nickel-mediated stoichiometric fixation of CO_2 with alkenes has been known for over the next 20 years largely due to the work of Höberg [36]. Encouragingly, after Rovis and coworkers proposed their work in 2008 [37], a great progress has been achieved in the hydrocarboxylation of alkenes via metal-hydride species.

The first nickel-catalyzed hydrocarboxylation of styrenes with CO_2 using Et_2Zn as a reductant was reported by Rovis and coworkers (Scheme 36) [37]. The yields of a variety of electron deficient and neutral *ortho, meta,* and *para* styrene analogues

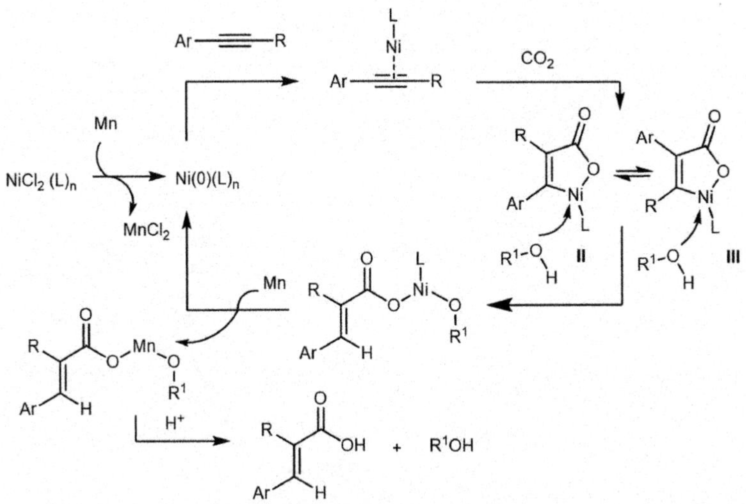

Scheme 34 The mechanism of Ni-catalyzed hydrocarboxylation of aryl alkynes with CO_2 and alcohols

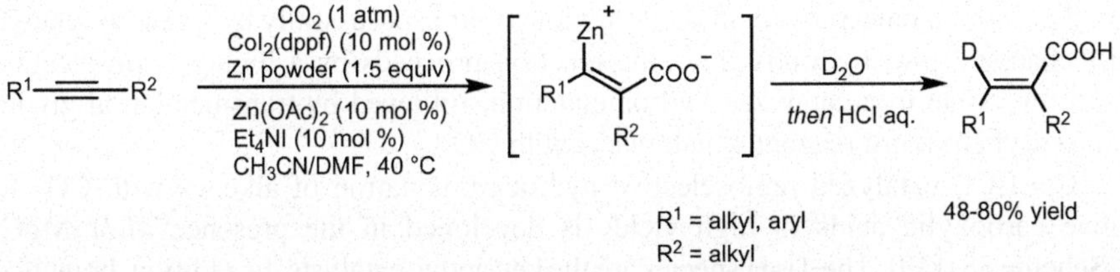

Scheme 35 Co-catalyzed hydrocarboxylation of aryl alkynes with CO_2 and H_2O

vary from 56 to 92%. This reaction represents the foundation of a methodology to incorporate CO_2 into more complex synthetic intermediates.

The catalytic cycle begins with insertion of styrene into the nickel-hydride bond providing benzyl nickel species **C**. A transmetalation generates the benzylic zinc species **D** and the product of net hydrozincation of the alkene, while also regenerating Et–Ni complex **A**. β-Hydride elimination and release of ethylene from **A** generates the presumed active catalyst **B**. A separate catalytic cycle involving transmetalation back to nickel (**D** to **C**) generates another benzylic nickel species which undergoes insertion of CO_2 prior to transmetalation with Et_2Zn, producing the hydrocarboxylation product **F** and regenerating precatalyst **A**.

Afterward, Grignard reagents are also used for the hydrogenation and carboxylation of olefins. The iron-catalyzed hydrocarboxylation of aryl alkenes has been developed using a highly active bench-stable iron(II) precatalyst and hydride source EtMgBr to give α-aryl carboxylic acids in excellent yields and regioselectivity (Scheme 37) [38]. Significantly, styrene derivatives bearing electron-donating

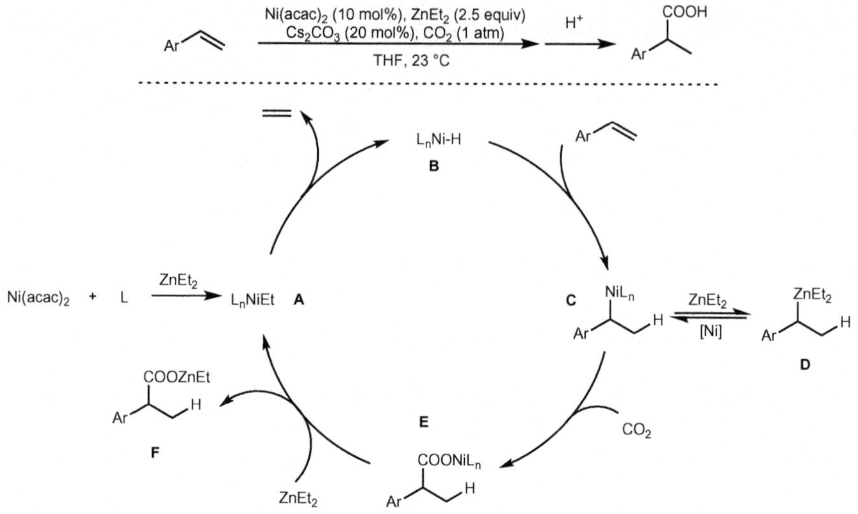

Scheme 36 Ni-catalyzed hydrocarboxylation of aryl alkenes with CO_2 and $ZnEt_2$

groups, which have proved difficult previously, are successfully hydrocarboxylated. Mechanistic investigations show the reaction proceeds by a highly regioselective and reversible iron-catalyzed hydrometalation, followed by transmetalation giving an α-aryl Grignard reagent which reacts with CO_2.

Cp_2TiCl_2-catalyzed regioselective hydrocarboxylation of alkenes with CO_2 to give carboxylic acids in high yields is developed in the presence of ^{i}PrMgCl (Scheme 38) [39]. The Ti–H species are the key active catalysts. In addition, branched and straight chain carboxylic acids can be selectively obtained from electron-rich styrene and aliphatic terminal alkene, respectively.

Martin group reported a catalytic protocol that controls the site-selective incorporation of CO_2 to a wide range of inactive olefins utilizing water as formal hydride source (Scheme 39) [34]. The author also tried the hydrocarboxylation of industrially relevant olefins. **2a** could be obtained as a single regioisomer from unrefined mixtures of olefins and propionic acid (**2b**) could be obtained in 20 turnover numbers (TON).

Promising and attractive substrates for the carboxylation reaction with CO_2 are alkylboron compounds given their great availability, functional group tolerance, and substrate scope. In 2011, Sawamura [40] and Hou [41] independently reported a combination of alkene hydroboration and carboxylation with CO_2 by a one-pot two-step sequence method (Scheme 40). They used copper catalysis and alkoxides as additives to form aliphatic, linear carboxylic acids. The 9-BBN dimer was exploited for the hydroboration step.

A catalytic cycle for the carboxylation of alkylboranes with CO_2 is proposed as shown in Scheme 41. The metathesis reaction between [(IPr)CuCl] and MeOLi should give the [(IPr)Cu(OMe)], which on reaction with an alkylborane gives the

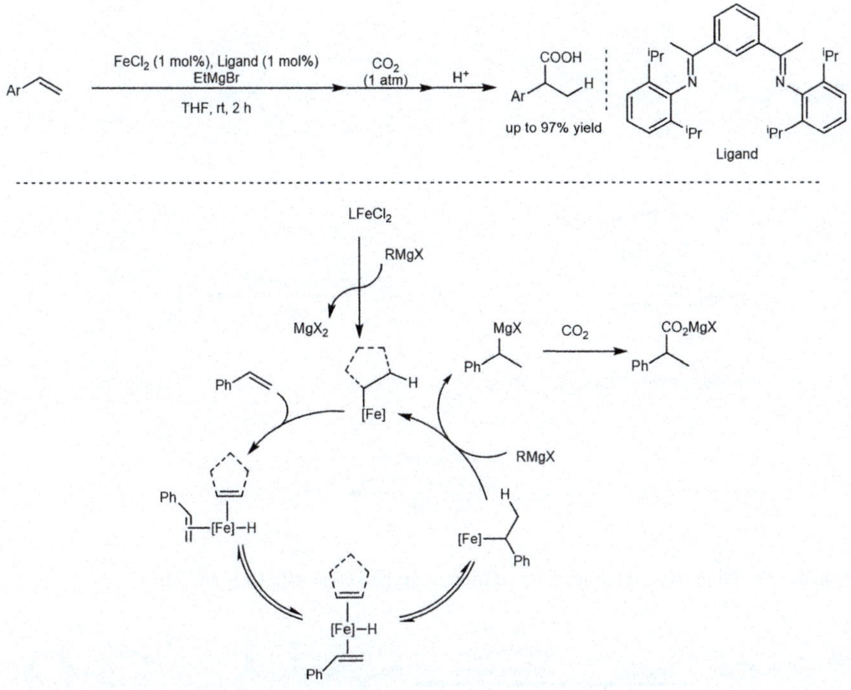

Scheme 37 Fe-catalyzed hydrocarboxylation of aryl alkenes with CO_2 and EtMgBr

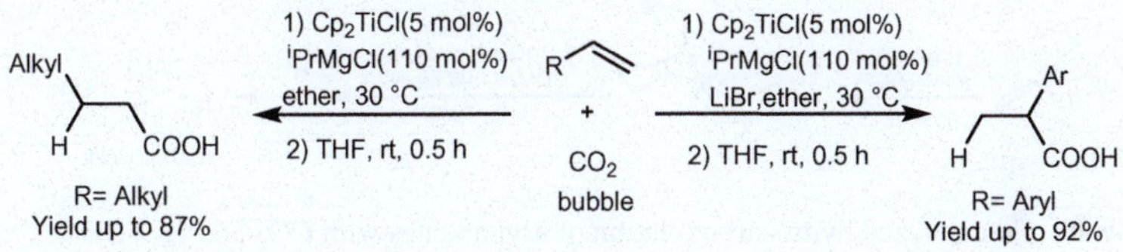

Scheme 38 Ti-catalyzed hydrocarboxylation of aryl alkenes with CO_2 and EtMgBr

adduct **A**. The subsequent transmetalation of the alkyl group to the copper atom would generate [(IPr)CuR]. Nucleophilic addition of the alkyl group to CO_2 should yield the carboxylate [(IPr)Cu(OCOR)], which upon metathesis with MeOLi could regenerate [(IPr)Cu(OMe)] and release the lithium carboxylate RCO_2Li.

Although the reaction could be well applied to unactivated alkenes, unfortunately internal alkenes could not be employed, which might arise from the inherent low reactivity of secondary alkylboranes. This problem was successfully solved by replacing alkoxides with CsF and developed an efficient methodology for accessing secondary carboxylic acids through a hydroboration–carboxylation sequence of internal olefins (Scheme 42) [42].

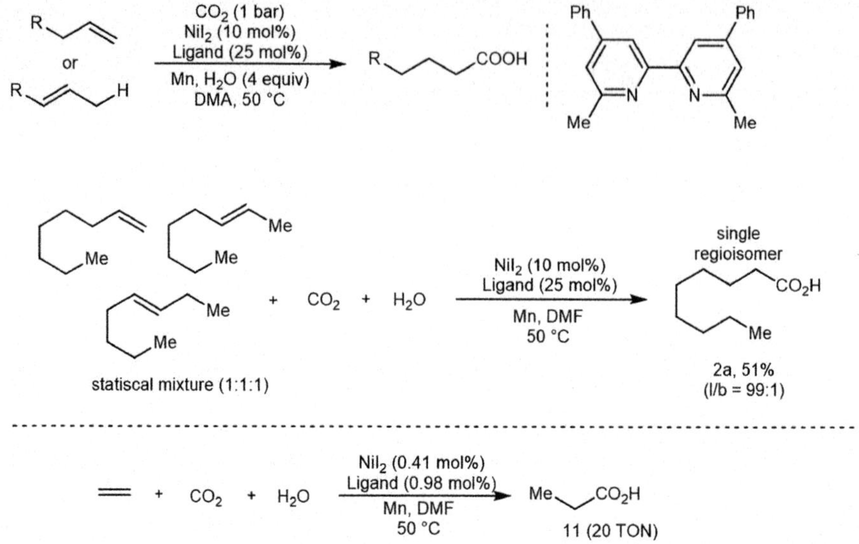

Scheme 39 Ni-catalyzed hydrocarboxylation of aryl alkenes with CO_2 and H_2O

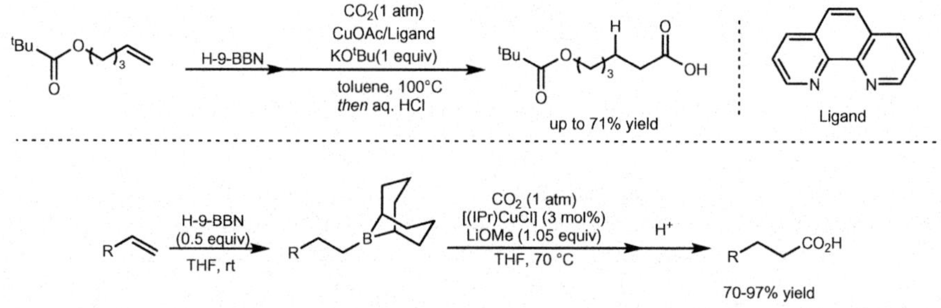

Scheme 40 Cu-catalyzed hydrocarboxylation of aryl alkenes with CO_2 and H-9-BBN

Scheme 41 The mechanism for the Cu-catalyzed hydrocarboxylation of aryl alkenes with CO_2 and H-9-BBN

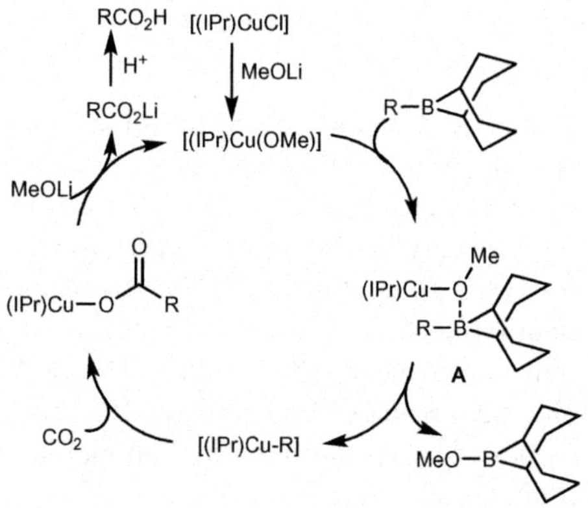

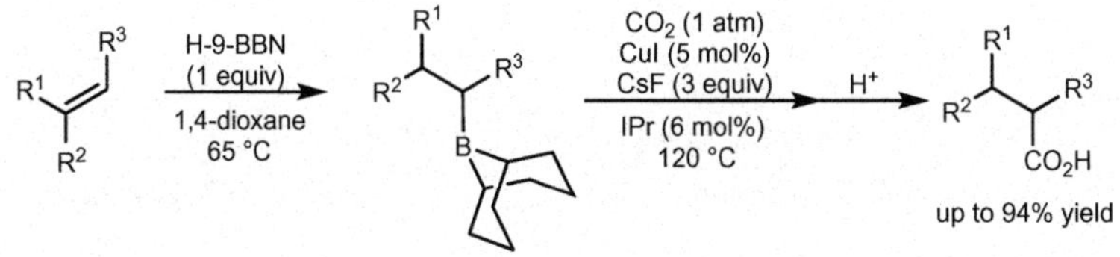

Scheme 42 Cu-catalyzed hydrocarboxylation of aryl alkenes with CO_2 and H-9-BBN using CsF

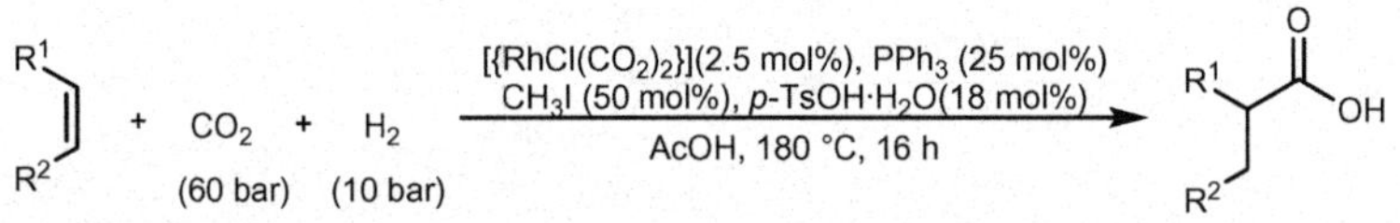

Scheme 43 Rh-catalyzed hydrocarboxylation of aryl alkenes with CO_2 and H_2

Hydrogen, which is cheap and easy to store and transport, can be produced by photocatalysis and electrocatalysis, is a preferred reagent for industrial production. An effective catalytic system is identified for the direct formation of carboxylic acids from non-activated olefins, CO_2, and H_2 (Scheme 43) [43]. Internal and terminal olefins are converted with regioselectivities similar to those observed in hydroformylation. Other substrates that are able to form rhodium–alkyl intermediates under the reaction conditions may also undergo this transformation, as shown in particular for primary and secondary alcohols. Mechanistic studies suggest that the formal hydrocarboxylation results from a combination of rhodium-catalyzed rWGSR and hydroxycarbonylation cycles.

11.4 Heterocarboxylation of Unsaturated Substrates with CO_2

11.4.1 Silacarboxylation of Unsaturated Substrates

The silacarboxylation of the first unsaturated substrate was described by Fleming's team in 1980 (Scheme 44A) [44]. In this work, the silyl-copper reagent reacted with hex-1-yne to give the intermediate vinyl-copper reagent, which reacted stereospecifically with CO_2.

In 2002, it was reported that silylcupration reactions of 1,3-dienes with a cyanocuprate reagent $PhMe_2SiCuCNLi$ produce a (4-silyl-2-alken-1-yl)–copper complex, which is trapped by electrophiles, on which the selectivity of the electrophilic trapping step is very dependent, and the electrophilic cleavage can occur in either an S_E2 or an S_E2' manner (Scheme 44B) [45]. CO_2 as an electrophile react exclusively via an S_E2' pathway, resulting in highly regioselective 1, 2-addtion. In this

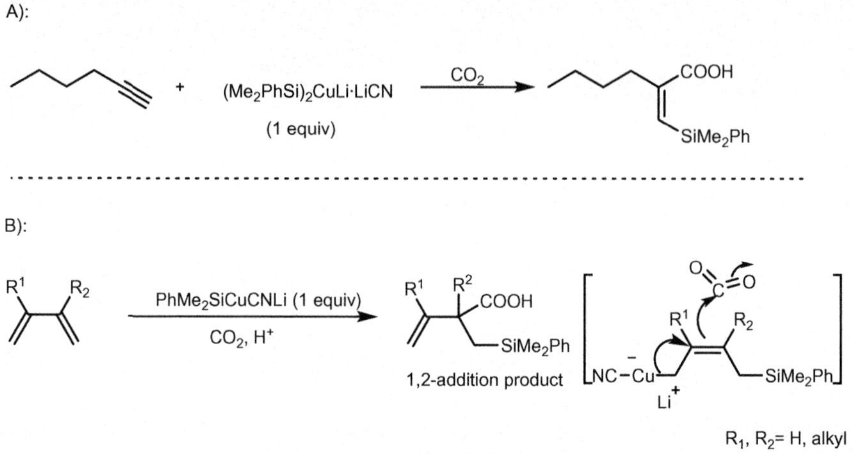

Scheme 44 Stoichiometric reaction for the silacarboxylation of unsaturated substrates

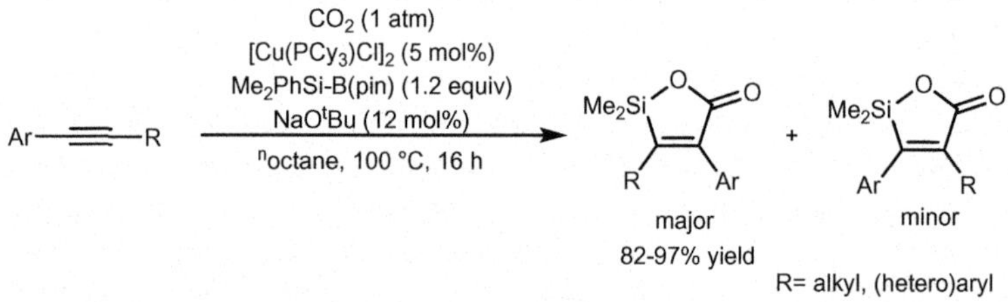

Scheme 45 Cu-catalyzed silacarboxylation of internal alkynes

context, Tsuji group reported the first catalytic silacarboxylation of internal alkynes employing CO_2 and silylborane in the presence of a copper catalyst (Scheme 45) [46]. The reaction afforded silalactone products regioselectively in good to high yields.

In the possible mechanism, *syn* addition of **A** to the alkyne affords the β-silylalkenylcopper intermediate **B**. Intramolecular cyclization via **C′** affords the phenylcopper species **D** by extrusion of the silalactone **2**. Significantly, the phenyl moiety on the silicon atom is indispensable for efficient Si–C bond cleavage. Finally, σ-bond metathesis of **D** with silylborane provides PhB(pin), and the silylcopper species **A** is regenerated (Scheme 46). Moreover, the silalactones are good substrates for the Hiyama cross-coupling reaction whose yield can reach 76% shown in Scheme 47.

Among unsaturated substrates for silylcupration, allene derivatives are very versatile reaction partners. A regiodivergent silacarboxylation of allenes using $PhMe_2Si$–B(pin) under a CO_2 atmosphere in the presence of a copper catalyst was developed(Scheme 48) [47]. The regioselectivity is highly controlled by the proper choice of ligands, and both carboxylated vinylsilanes and allylsilanes can be synthesized regiodivergently from a single substrate. The regioselectivity reversal might be

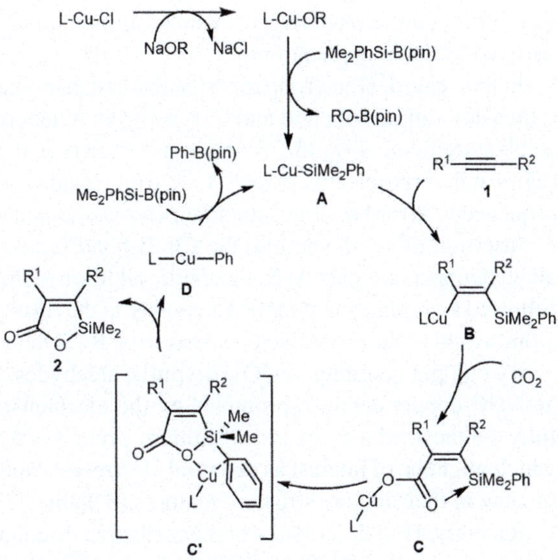

Scheme 46 The mechanism for the Cu-catalyzed silacarboxylation of internal alkynes

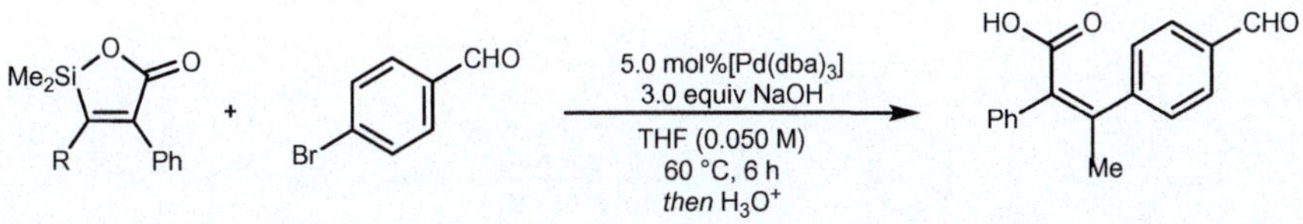

Scheme 47 The Hiyama cross-coulping reaction of silalactones with *p*-bromo benzaldehyde in THF

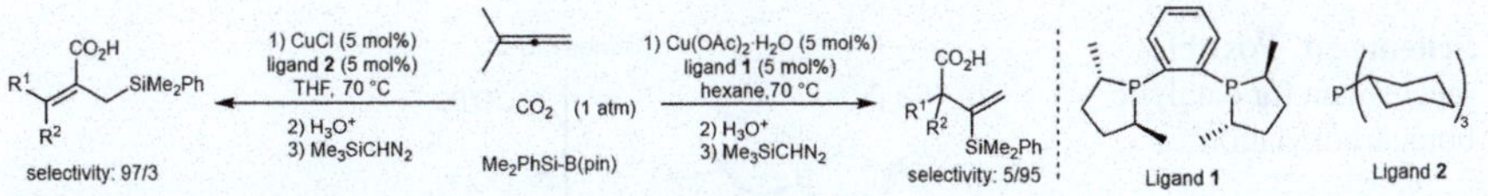

Scheme 48 Cu-catalyzed silacarboxylation of allenes with CO_2

attributed to the difference in relative steric bulk of the CuL (L = Me-DuPhos or PCy_3) and $SiMe_2Ph$ moieties.

11.4.2 Boracarboxyaltion of Unsaturated Substrates

The heterocarboxylation of alkynes (such as boracarboxylation) with CO_2, which would lead to the addition of both a heteroatom unit (e.g., boryl) and a carboxylate group to the C–C triple bond could serve as a useful method for the synthesis of multifunctionalized alkenes, which are potentially versatile building blocks for various synthetic applications.

In this regard, Hou group reported the first catalytic boracarboxylation of alkynes with a diborane compound and CO_2 using an *N*-heterocyclic carbene (NHC) copper catalyst (Scheme 49) [48]. Both a boryl moiety and a carboxylate group are added regio- and stereoselectively to the C–C triple bond via a borylcupration/carboxylation sequence to afford α, β-unsaturated β-boralactone derivatives.

Insertion of an alkyne into the Cu–B bond in a syn fashion would give β-boryl alkenylcopper complex **A**. Nucleophilic addition of the alkenylcopper species to CO_2 followed by replacement of the Cu moiety in the resulting carboxylate with the boron atom would yield β-boralactone derivative **B** (Scheme 50).

By one-pot coupling of CO_2, $B_2(pin)_2$, aldehydes, and LiOtBu in the presence of an NHC-copper catalyst, prompted by the mechanism above, Hou further successfully synthesized a new class of lithium cyclic boracarbonate ion pair compounds, which might be of interest as potential electrolyte candidates for lithium ion batteries in view of their unique structure features (Scheme 51) [49].

Recently, Hou described a highly selective difunctionalization of imines through incorporation of activation of CO_2 by intramolecular N/B Lewis pairs into a copper catalytic cycle (Scheme 52) [50]. A new family of lithium cyclic boracarbamate

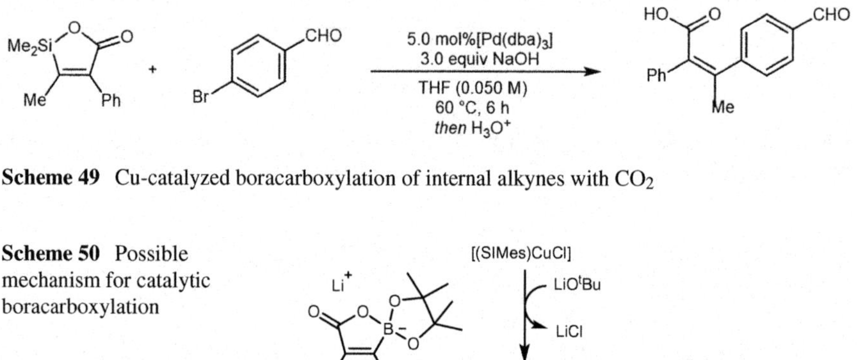

Scheme 49 Cu-catalyzed boracarboxylation of internal alkynes with CO_2

Scheme 50 Possible mechanism for catalytic boracarboxylation

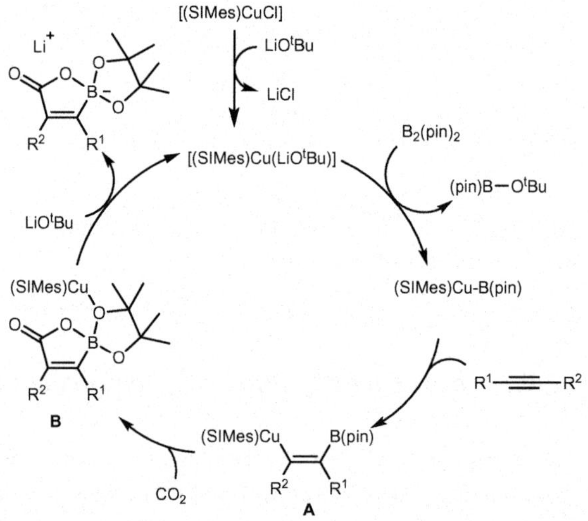

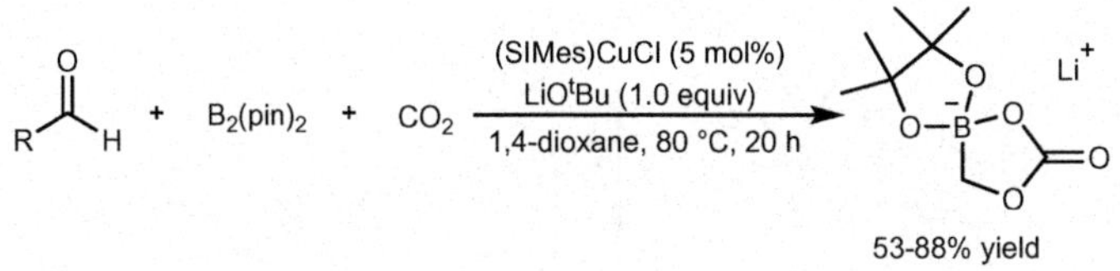

Scheme 51 Cu-catalyzed boracarboxylation of aldehydes with CO_2

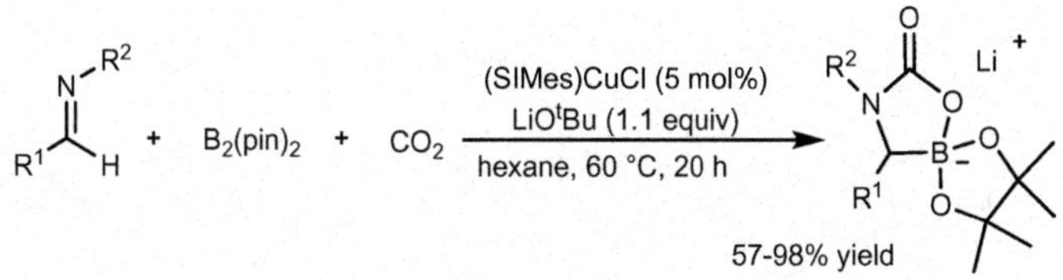

Scheme 52 Cu-catalyzed boracarboxylation of imines with CO_2

salts were obtained from this reaction, which can be easily converted into valuable N-carboxylated α-amino boronates derivatives.

Experimental and computational studies on the mechanistic aspect reveal that the traditional CO_2 insertion into the Cu–N bond of **B** is less favorable in the current catalytic reaction. The connection of CO_2 to a nitrogen atom and boron atom together with the migration of the (SIMes)Cu unit to a pinacolate oxygen atom leads to the generation of cyclic boracarbamate derivative **C** (Scheme 53).

Hetero(element)carboxylations above remain limited to alkynyl and allenyl substrates. In 2016, boracarboxylation was extended to less reactive olefinic substrates, which would be highly valuable to the synthetic community (Scheme 54) [51]. These novel boracarboxylated products can be converted into tropic acid, alkyl fluoride and the novel difluoroboralactone anion. Significantly, the two-step boracarboxylation/fluorination of olefins represents a new difunctionalization method that could have medicinal chemistry applications.

11.4.3 Azacarboxylation of Unsaturated Substrates

Ma and coworkers developed an efficient $Cu(OAc)_2$-catalyzed synthesis of indole-3-carboxylic acid and indolodihydropyran-2-one from the readily available 2-alkynylanilines with CO_2 in the presence of $ZnMe_2$ and CsF (Scheme 55) [52]. In 2016, Ma presented a transition metal-free methodology for the efficient construction of different indole-3-carboxylic acids with $ZnEt_2$ under a balloon atmosphere of CO_2 (Scheme 56) [53]. Functional groups including ester, cyano, nitro, acetyl, -OH, -NHTs, trifluoromethyl, etc. were smoothly tolerated. In the proposed mechanism, reaction may occur via a zinc-mediated cyclization reaction to form Int 1. $ZnEt_2$

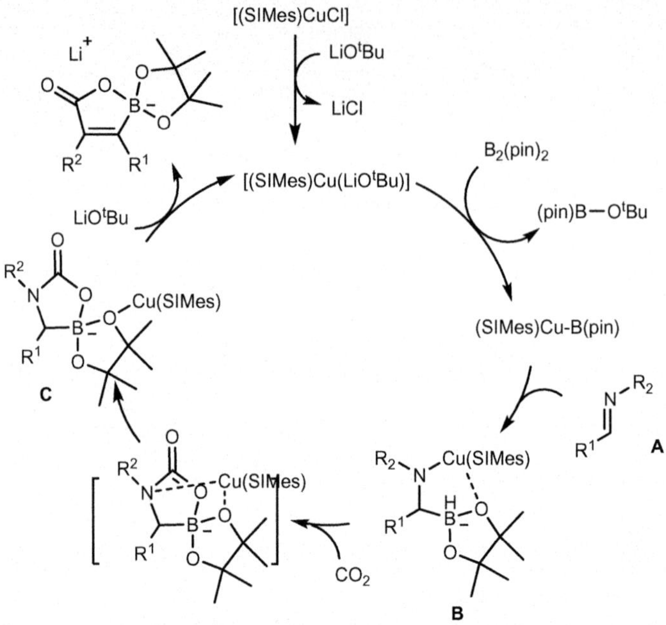

Scheme 53 Possible mechanism for catalytic boracarboxylation of imines with CO_2

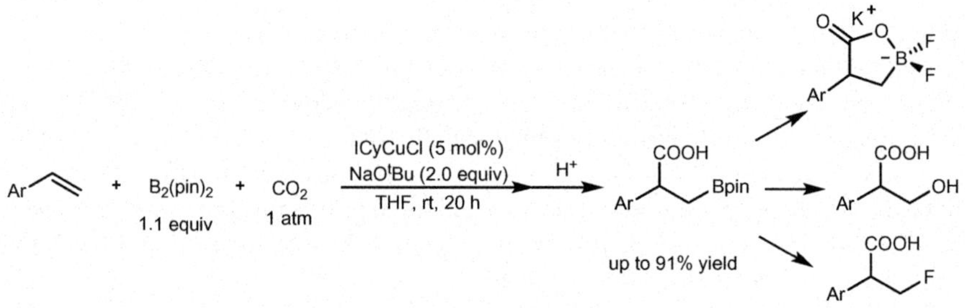

Scheme 54 Cu-catalyzed boracarboxylation of styrenes with CO_2

acted not only as a base but also as a Lewis acid. The final indole-3-carboxylic acid **2** is afforded upon protonation.

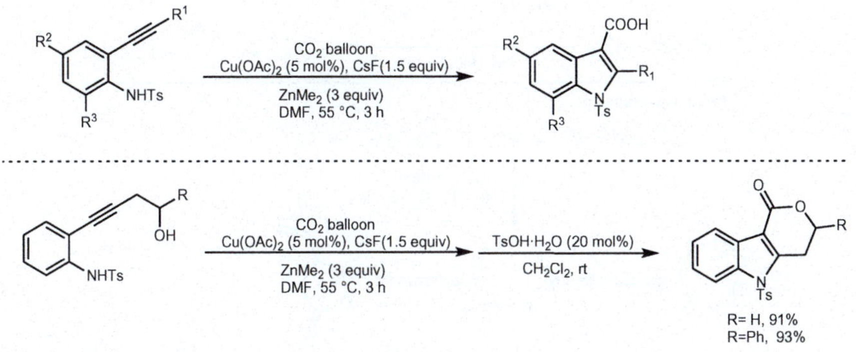

Scheme 55 Cu-catalyzed cyclic *anti*-nucleometallation-carboxylation of 2-alkynylanilines with CO_2

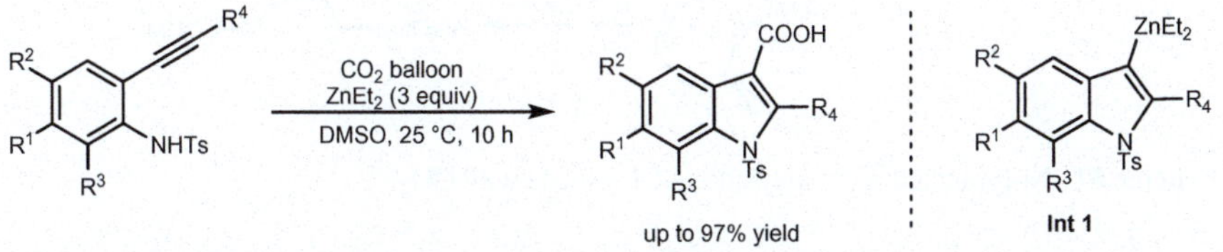

Scheme 56 Direct *anti*-azacarboxylation of 2-alkynylanilines with CO_2 mediated by $ZnEt_2$

11.5 Dicarboxylation

11.5.1 Dicarboxylation of Alkenes/Allenes with CO_2

Dicarboxylic acid compounds, which are bulky chemicals in industry, synthesized by the reaction of unsaturated compounds with CO_2 under reducing conditions have been extensively studied in the early stages. For example, the reaction of CO_2 with 1,3-butadiene at – 15 °C catalyzed by Ni(TMEDA) (TMEDA = *N,N′*-tetramethylenediamine) affords a nickel dicarboxylate complex, which subsequently reacted with CH_3OH/HCl to generate a *cis*-dicarboxylate in high yield (Scheme 57) [54]. In this reaction, oxidative metal cyclization of the Ni complex with 1,3-butadiene and CO_2 will form the η-3 nickel cyclic species, which then undergoes insertion of another molecule of CO_2 to form the desired dicarboxylate. Such work provides a solid foundation for this area.

In addition, 1,8-diazabicyclo[5.4.0]undec-7-ene (DBU) as a Ni-catalyzed oxidative coupling of 1,3-dienes with CO_2 in the presence of Me_2Zn has been found to be a superior ligand (Scheme 58) [55]. In the presence of DBU (2 equivalents of Ni) and $Ni(cod)_2$ (1 equivalent), 1,3-diene (1.1 equivalents) can be readily reacted with CO_2 (1 atm) under mild conditions (0 °C, 4 h) to give (*Z*)-1,4-dicarboxylated

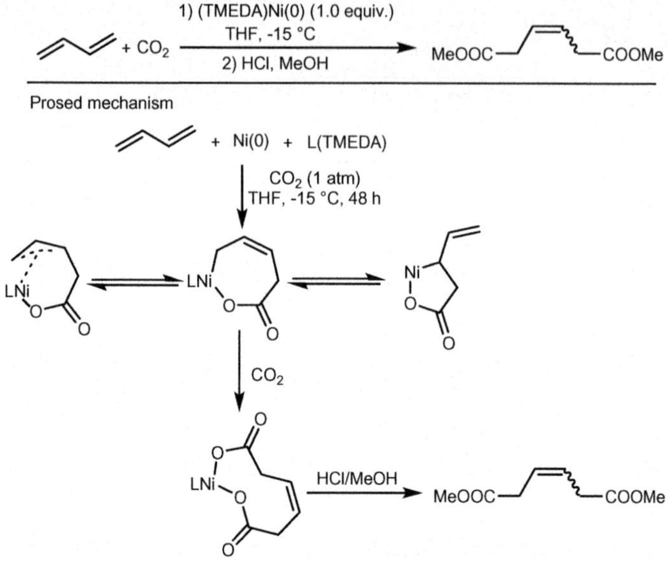

Scheme 57 Ni-promoted dicarboxylation of 1,3-diene with CO_2

products in a moderate to high yield. The reaction proceeds under very benign conditions, in a very short time and with a very simple procedures. It's interesting to note that the trans-selective dicarboxylation of cyclohexadiene suggests that the nickel cycle produced by migrating insertion and transmetallation may go through a back-side attack process rather than a 9-membered bulk cycle or reductive elimination, thereby dictating the precise morphology of the final product. However, when Ph_2Zn instead of Me_2Zn, arylated carboxylation appears to be the predominant pathway, indicating that spatial site resistance determines whether back-side attack or reductive elimination occurs first, with larger bulk barriers tending to undergo reductive elimination.

Although several progresses have been made in this field, the use of stoichiometric transition metal complexes has been frustrating. Until 2018, dicarboxylic acid products with excellent regioselectivity and a wide range of substrates under the conditions of $NiBr_4(TBA)_2$ as catalyst, *o*-phenanthroline derivatives as ligands and Mn as reducing agent are obtained at 50 °C (Scheme 59) [56]. Although the reaction mechanism is not well described, the *Z/E* selectivity problem of aliphatic dicarboxylic acids has been overcome by this tactic.

Fewer studies have been done on dicarboxylic acids and their derivatives catalyzed by transition metal complexes, exploiting ethylene with CO_2. In 1987, Heinz Hoberg first realized the preparation of bicarboxylic acid compounds from ethylene and CO_2 by reducing $FeCl_2$ with Mg in the presence of ethylene and PEt_3 to form bis(ethylene)bis(triethylphosphine)iron(0) complexes (Scheme 60) [57]. The salient

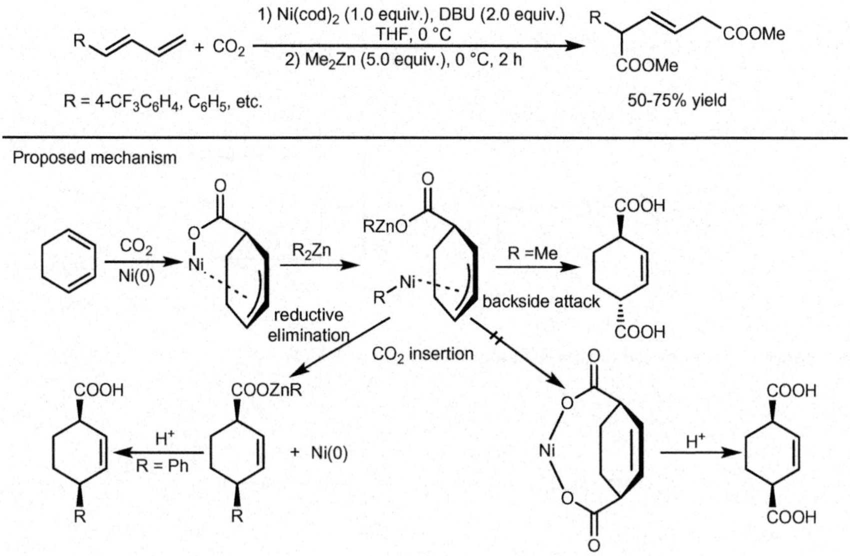

Scheme 58 The Ni-mediated dicarboxylation of 1,3-diene with CO_2

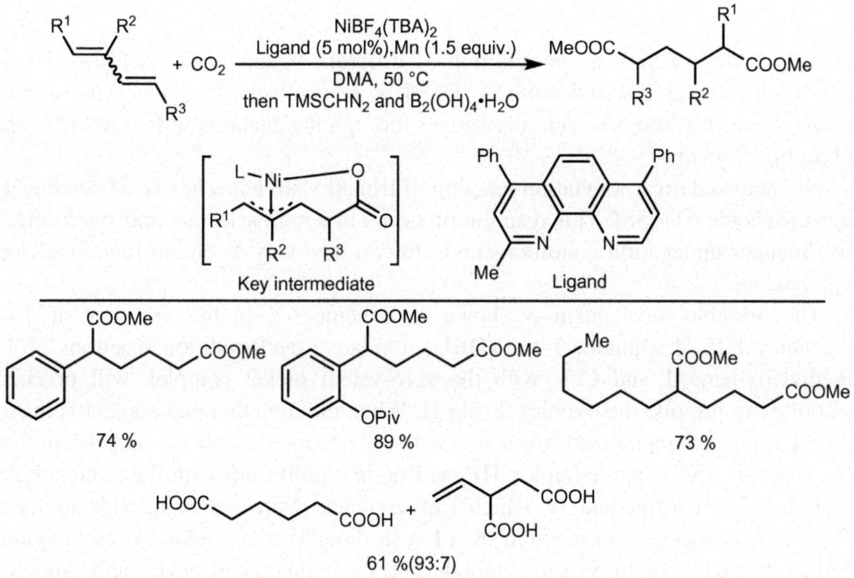

Scheme 59 Ni-catalyzed dicarboxylation of 1,3-dienes with CO_2

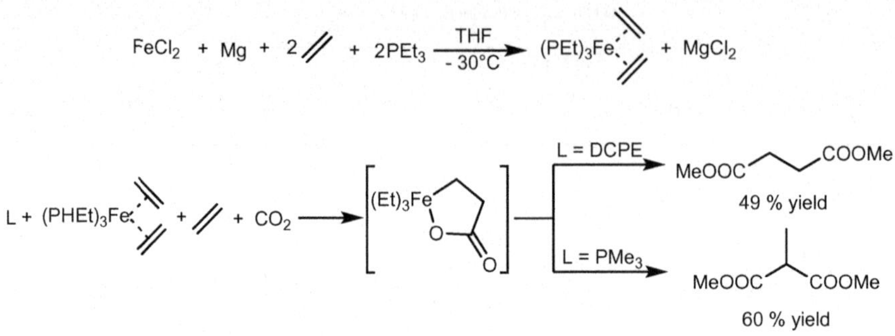

Scheme 60 Fe-promoted dicarboxylation of ethylene with CO_2

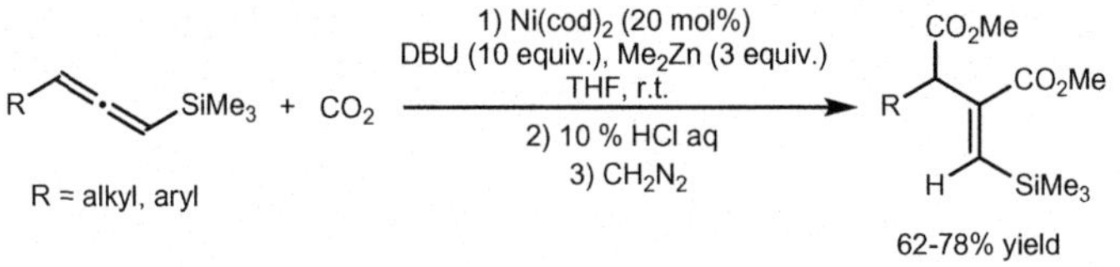

Scheme 61 Ni-promoted dicarboxylation of silylallenes with CO_2

feature of the reaction is the addition of different ligands such as PMe_3, P, P′-ethylenebis(dicyclohexylphosphine) (DCPE)s, respectively, to achieve the preparation of two bicarboxylic acid derivatives through the chelating effect of bidentate phosphine ligands.

Ni-catalyzed dicarboxylation reaction of trimethylsilylenes has also been developed (Scheme 61) [58]. This reaction proceeds in a regioselective and stereoselective manner under mild conditions and offers a new way to obtain functionalized vinylsilanes.

The possible mechanism is shown in Scheme 62. In the presence of 1,8-diazabicyclo[5. 4. 0]undec-7-ene (DBU), oxidative cycloaddition reactions of 1-methylsilylbene **1** and CO_2 with the zero-valent nickel complex will proceed smoothly to provide the oxonickel ring **II**. When the dimethylzinc reagent is introduced, II through metallization reaction with the zinc reagent to form nickel complex III′. The Si–allyl nickel complex **III′** will be in equilibrium with the nucleophilic *p*-allyl nickel intermediate **V** which further reacts with carbon dioxide to form **VI**. The transmetallation reaction of **VI** with dimethylzinc yields the carboxylate **VIII**, followed by hydrolysis to obtain the corresponding dicarboxylic acid. Another product of the transmetallation step is the dimethylnickel complex, which undergoes reductive elimination to regenerate the zero-valent nickel catalyst.

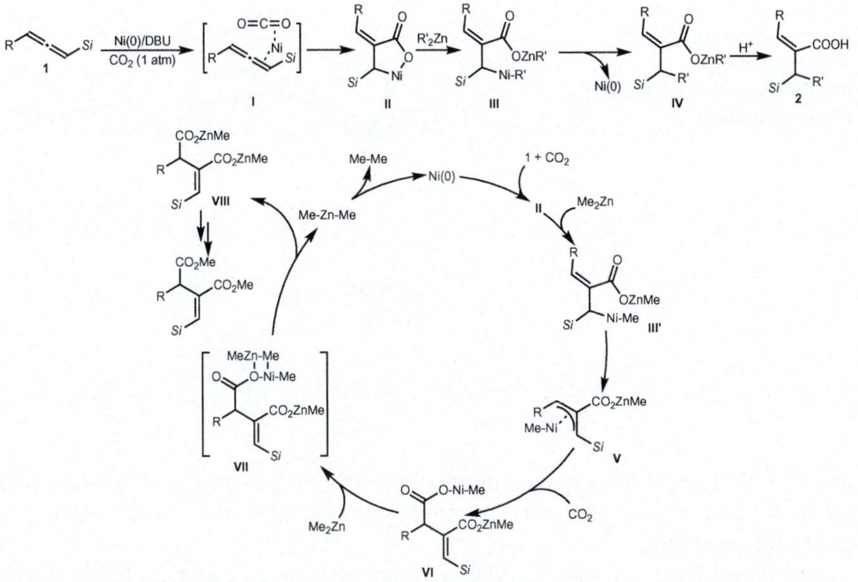

Scheme 62 Proposed mechanism for the Ni-promoted dicarboxylation of silylallenes with CO_2

11.5.2 Dicarboxylation of Alkynes with CO_2

High yields of malonic acid have been obtained by NHC–Cu (I) as a catalyst in the dicarboxylation of terminal alkynes, experiments using $(9\text{-BBN})_2$ to in situ generate diboron reagents and reacting with carbon dioxide in the presence of the additive CsF (Scheme 63) [59]. Since the electron rich ligand NHC provides strong nucleophilicity to the organic intermediate Cu(I) complexes enabling transmetallation reactions to occur, a high yield of malonic acid is achieved through a sequential process and hydrolysis. Notably, the key step to obtain the key intermediate 1,1-diboron reagent is the *anti*-Marshall hydrogenation reaction of alkynes.

Meanwhile, a proposed mechanistic cycle for the hydroboration–carboxylation approach has been offered (Scheme 64). An active IPr-linked copper fluoride species **A** is generated in situ from CuI, IPr and cesium fluoride. **A** then undergoes a σ-bond complexation reaction with a pre-formed alkylborane species to form an alkyl–copper

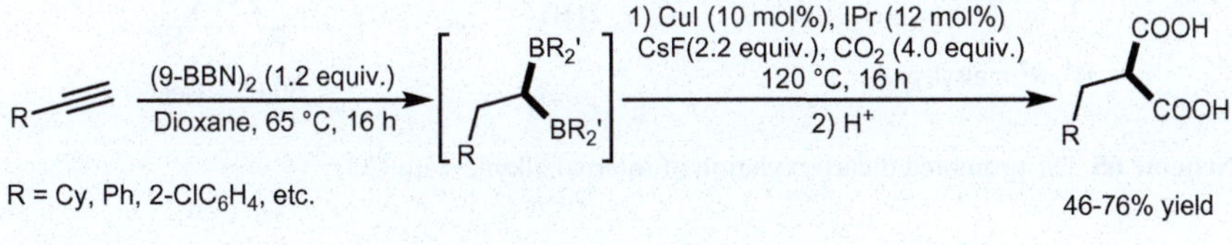

Scheme 63 Copper-promoted dicarboxylation of terminal alkynes with CO_2

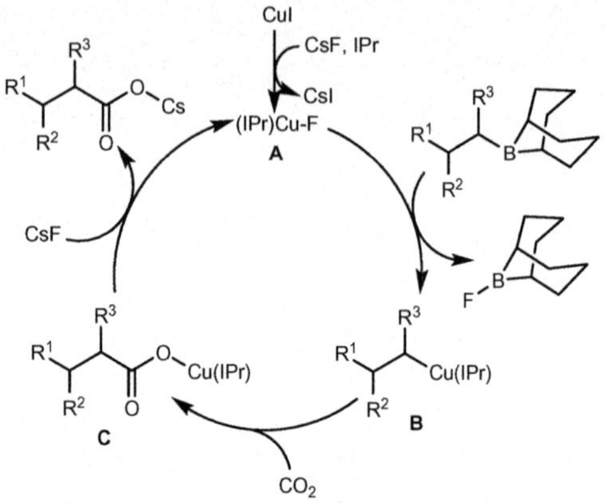

Scheme 64 Proposed mechanism for the copper-promoted dicarboxylation of terminal alkynes with CO_2

complex **B**. The complex is then inserted into CO_2 to generate complex **C**, which releases **A** and cesium carboxylate, eventually producing the desired carboxylic acid after acid quenching.

The nickel-catalyzed dicarboxylation of internal alkyne with CO_2 has also been discussed (Scheme 65) [60]. The reaction is conducted at room temperature in the presence of nickel as catalyst, zinc powder as reducing reagent and $MgBr_2$ as additive, by which various internal alkynes can be converted to the corresponding maleic anhydride derivatives in high to excellent yields. It should be noted that the DFT calculations reveal that $MgBr_2$ plays an indispensable role in the single-electron reduction and the second CO_2 insertion into the Ni(I)–C bond, so the key of the reaction is the generation of the Ni(I) metal ring intermediate and the second carboxylation promoted by $MgBr_2$.

For this purpose, a rational catalytic cycle is proposed (Scheme 66). Activated catalyst Ni(0)(bpy) is formed through the reduction of Ni(II) catalyst precursors by Zn (step a). Then, Oxidation cyclization of Ni(0)(bpy) with **1** and CO_2 yields Ni(II) metal cycle **4** (step b). Reduction of **4** with Zn and $MgBr_2$ gives the key intermediate of this dicarboxylation, Ni(I) species **b** (step c). Finally, reduction of **C** yields the dicarboxylation product **D** and regenerates the reactive Ni(0) catalyst species (step e).

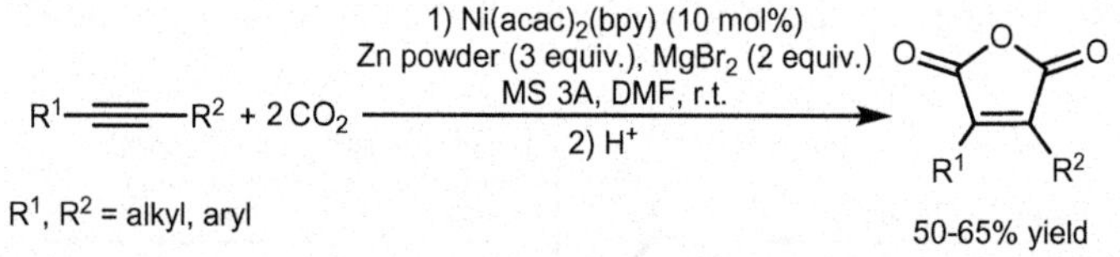

Scheme 65 Ni-promoted dicarboxylation of internal alkyne with CO_2

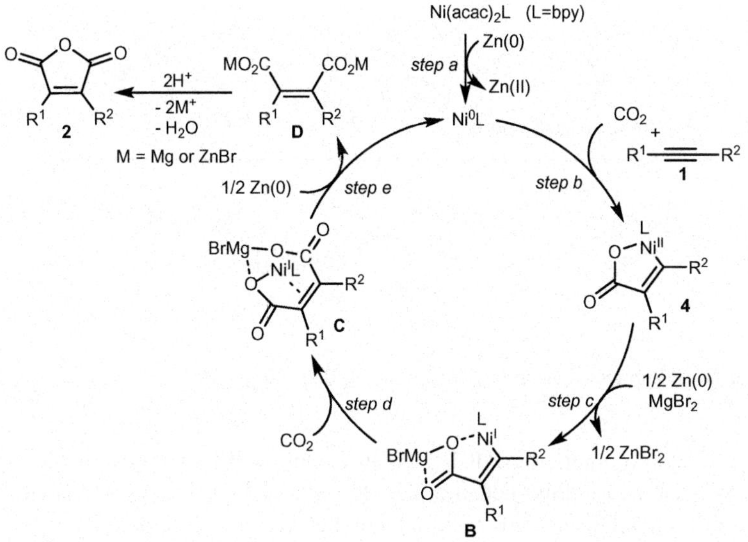

Scheme 66 Proposed mechanism for the Ni-promoted dicarboxylation of internal alkyne with CO_2

11.5.3 Dicarboxylation of Arene with CO_2

Considering the strong inertness of C–H bonds on unactivated aromatic compounds, the dicarboxylation reaction of stable aromatic compounds is extremely challenging with CO_2. In 2010, a catalyst design was performed by potentiometric titration of C–H bonds of aromatic compounds with pKa as a criterion (Scheme 67) [18]. The results showed that the novel [1,3-bis(2,6-diisopropylphenyl)imidazol-2-ylidene]copper(I) hydroxide [Cu(IPr)(OH)] allows easy activation and carboxylation of N–H and C–H bonds with pKa values less than 27.7. Therefore, the *N*-heterocyclic copper(I) complex as a catalyst is adopted for the carboxylation reaction of N–H/C–H bonds of aromatic compounds using CO_2 as the carboxylation reagent. It is noteworthy that the volume of base plays a vital role in the control of selectivity. Monocarboxylic acids are obtained when 1.05 equivalents of base are used, and dicarboxylic acids are given when 2.0 equivalents of base are employed.

In addition, Pd-catalyzed dearylation of indole derivatives with CO_2 and dicarboxylation reactions have also been undertaken (Scheme 68) [61]. Interestingly, this

Scheme 67 NHC-Cu-promoted dicarboxylation of arene with CO_2

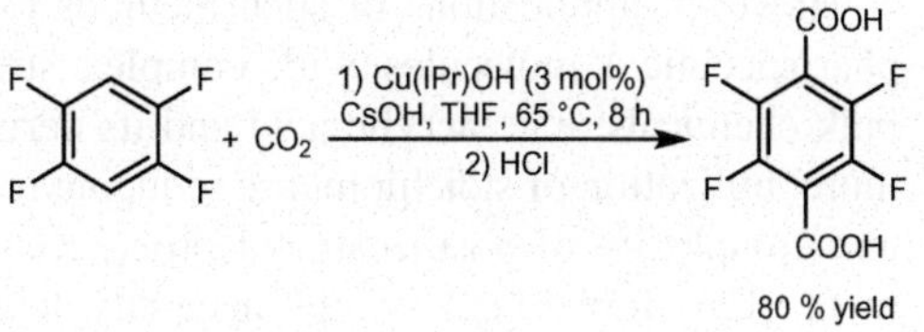

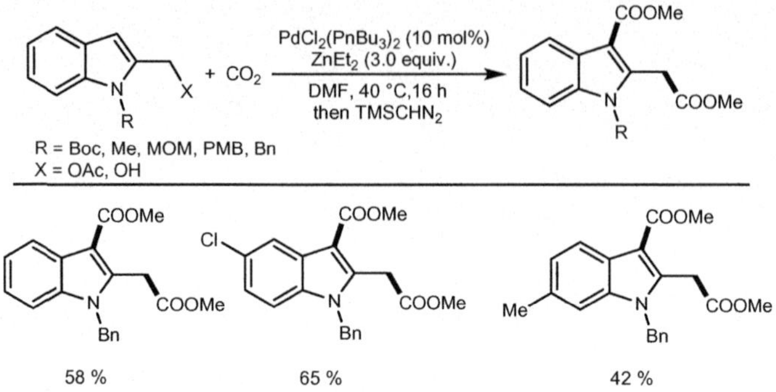

Scheme 68 Pd-promoted dicarboxylation of indole derivatives with CO_2

catalyst system comprising $PdCl_2[P(n\text{-}Bu)_3]_2$ and $ZnEt_2$ works well for carboxylation of various 3-indolylmethanol derivatives with CO_2 (1 atm) via dearomatization of the indole nucleus, affording 3-methyleneindoline-2-carboxylates. It is worth mentioning a competitive relationship between monocarboxylation and β-hydride elimination, resulting in the dicarboxylation product obtained in moderate yields.

11.6 Summary and Perspectives

As mentioned above, a wide range of transition metal catalytic systems have been developed for the carboxylation reactions of different types of unsaturated compounds with CO_2 and have shown promising achievements. The catalytic systems that have been designed with good functional group tolerance, high selectivity, and mild reaction conditions provide more diverse synthetic pathways for green organic synthesis. For the carboxylation of C–H bonds, the catalytic systems of Cu, Ag, and Au are widely used, which mostly relay on the reaction mechanism of hydrogen extraction under the synergistic effect of bases. When the C–H bond acidity of the substrate is low, catalytic systems represented by Rh and Pd are required. These catalysts systems can undergo electrophilic metallization with the C–H bond to form C–M bonds, which then react with CO_2 to form corresponding carboxylic acid compounds. For the hydrocarboxylation, the selection of hydride sources has gradually shifted from metal organic reagents to more environmentally friendly reagents such as water and hydrogen.

However, applications of such methods to the carboxylation of biological and pharmaceutical molecules with complex structures and scale up production of bulk chemicals, e.g., acrylic acid and its derivatives are still challenging. Furthermore, utilization of stoichiometric reducing reagents or additives, e.g., strong bases and complexity of post-treatment processes may also be harmful to the environment. Therefore, there is still necessity to develop greener and more efficient catalytic systems for the carboxylation reactions with CO_2. In addition, the catalytic systems developed so far are mainly suitable for the functionalization of unsaturated compounds with active sites, and efforts are still expected to develop high-efficiency and precise catalytic systems for the reaction of unactivated unsaturated compounds with CO_2.

Acknowledgements We are grateful to the Natural Science Foundation of Tianjin Municipal Science and Technology Commission (21JCZDJC00100), Nankai Cangzhou Bohai New Area Green Chemical Research Co. LTD (Grant No 20220130), and the Fundamental Research Funds for the Central Universities of Nankai University for financial support.

References

1. Aresta M, Dibenedetto A (2007) Utilisation of CO_2 as a chemical feedstock: opportunities and challenges. Dalton Trans 2975–2992
2. Sakakura T, Choi J-C, Yasuda H (2007) Transformation of carbon dioxide. Chem Rev 107:2365–2387
3. Davies J, Lyonnet JR, Zimin DP, Martin R (2021) The road to industrialization of fine chemical carboxylation reactions. Chem 7:2927–2942
4. Maag H (2007) In: Stella VJ et al (eds) Prodrugs: challenges and rewards part 1. Springer New York, pp 703–729
5. Correa A, Martín R (2009) Metal-catalyzed carboxylation of organometallic reagents with carbon dioxide. Angew Chem Int Ed 48:6201–6204
6. Gooßen LJ, Rodríguez N, Gooßen K (2008) Carboxylic acids as substrates in homogeneous catalysis. Angew Chem Int Ed 47:3100–3120
7. Börjesson M, Moragas T, Gallego D, Martin R (2016) Metal-catalyzed carboxylation of organic (pseudo)halides with CO_2. ACS Catal 6:6739–6749
8. Yu D, Teong SP, Zhang Y (2015) Transition metal complex catalyzed carboxylation reactions with CO_2. Coordin Chem Rev 293–294:279–291
9. Zhang L, Hou Z (2017) Transition metal promoted carboxylation of unsaturated substrates with carbon dioxide. Curr Opin Green Sustain Chem 3:17–21
10. Ackermann L (2011) Transition-metal-catalyzed carboxylation of C-H bonds. Angew Chem Int Ed 50:3842–3844
11. Hendriksen C, Pidko EA, Yang G, Schäffner B, Vogt D (2014) Catalytic formation of acrylate from carbon dioxide and ethene. Chem Eur J 20:12037–12040
12. Huguet N et al (2014) Nickel-catalyzed direct carboxylation of olefins with CO_2: one-pot synthesis of α, β-unsaturated carboxylic acid salts. Chem Eur J 20:16858–16862
13. Manzini S et al (2017) Enhanced activity and recyclability of palladium complexes in the catalytic synthesis of sodium acrylate from carbon dioxide and ethylene. Chem Cat Chem 9:2269–2274
14. Takegasa S, Lee MM, Tokuhiro K, Nakano R, Yamashita M (2022) Rhodium-catalyzed acrylate synthesis from carbon dioxide and ethylene by using a guanidine-based pincer ligand: perturbing occupied *d*-orbitals by $p\pi$-$d\pi$ repulsion makes a difference. Chem Eur J 28:e202201870
15. Sahoo H, Zhang L, Cheng J, Nishiura M, Hou Z (2022) Auto-tandem copper-catalyzed carboxylation of undirected alkenyl C-H bonds with CO_2 by harnessing β-hydride elimination. J Am Chem Soc 144:23585–23594
16. Suga T, Mizuno H, Takaya J, Iwasawa N (2014) Direct carboxylation of simple arenes with CO_2 through a rhodium-catalyzed C-H bond activation. Chem Commun 50:14360–14363
17. Boogaerts IIF, Nolan SP (2010) Carboxylation of C-H bonds using *N*-heterocyclic carbene gold(I) complexes. J Am Chem Soc 132:8858–8859

18. Boogaerts IIF, Fortman GC, Furst MRL, Cazin CSJ, Nolan SP (2010) Carboxylation of N-H/C-H bonds using *N*-heterocyclic carbene copper(I) complexes. Angew Chem Int Ed 49:8674–8677
19. Tang S, Zhao X, Yang L, Li B, Wang B (2022) Copper-catalyzed carboxylation of aryl thianthrenium salts with CO_2. Angew Chem Int Ed 61:e202212975
20. Mizuno H, Takaya J, Iwasawa N (2011) Rhodium(I)-catalyzed direct carboxylation of arenes with CO_2 via chelation-assisted C-H bond activation. J Am Chem Soc 133:1251–1253
21. Fu L et al (2018) Ligand-enabled site-selectivity in a versatile rhodium(ii)-catalysed aryl C-H carboxylation with CO_2. Nat Catal 1:469–478
22. Börjesson M et al (2020) Remote sp^2 C-H carboxylation via catalytic 1,4-Ni migration with CO_2. J Am Chem Soc 142:16234–16239
23. Zhang L, Cheng J, Ohishi T, Hou Z (2010) Copper-catalyzed direct carboxylation of C-H bonds with carbon dioxide. Angew Chem Int Ed 49:8670–8673
24. Inamoto K, Asano N, Kobayashi K, Yonemoto M, Kondo Y (2012) A copper-based catalytic system for carboxylation of terminal alkynes: synthesis of alkyl 2-alkynoates. Org Biomol Chem 10:1514–1516
25. Li S et al (2016) Carboxylation of terminal alkynes with CO_2 using novel silver *N*-heterocyclic carbene complexes. Dalton Trans 45:10577–10584
26. Cheng H, Zhao B, Yao Y, Lu C (2015) Carboxylation of terminal alkynes with CO_2 catalyzed by bis(amidate) rare-earth metal amides. Green Chem 17:1675–1682
27. Li S, Yuan W, Ma S (2011) Highly regio- and stereoselective three-component nickel-catalyzed syn-hydrocarboxylation of alkynes with diethyl zinc and carbon dioxide. Angew Chem Int Ed 50:2578–2582
28. Santhoshkumar R et al (2016) Synthesis of vinyl carboxylic acids using carbon dioxide as a carbon source by iron-catalyzed hydromagnesiation. Chem Cat Chem 8:2210–2213
29. Shao P, Wang S, Du G, Xi C (2017) Cp_2TiCl_2-catalyzed hydrocarboxylation of alkynes with CO_2: formation of α, β-unsaturated carboxylic acids. RSC Adv 7:3534–3539
30. Takimoto M, Hou Z (2013) Cu-catalyzed formal methylative and hydrogenative carboxylation of alkynes with carbon dioxide: efficient synthesis of alpha, beta-unsaturated carboxylic acids. Chemistry 19:11439–11445
31. Fujihara T, Xu T, Semba K, Terao J, Tsuji Y (2011) Copper-catalyzed hydrocarboxylation of alkynes using carbon dioxide and hydrosilanes. Angew Chem Int Ed 50:523–527
32. Xiong W et al (2020) Palladium-catalyzed highly regioselective hydrocarboxylation of alkynes with carbon dioxide. ACS Catal 10:7968–7978
33. Wang X, Nakajima M, Martin R (2015) Ni-catalyzed regioselective hydrocarboxylation of alkynes with CO_2 by using simple alcohols as proton sources. J Am Chem Soc 137:8924–8927
34. Gaydou M, Moragas T, Julia-Hernandez F, Martin R (2017) Site-selective catalytic carboxylation of unsaturated hydrocarbons with CO(2) and water. J Am Chem Soc 139:12161–12164
35. Nogi K, Fujihara T, Terao J, Tsuji Y (2016) Carboxyzincation employing carbon dioxide and zinc powder: cobalt-catalyzed multicomponent coupling reactions with alkynes. J Am Chem Soc 138:5547–5550
36. Hoberg H, Peres Y, Krüger C, Tsay Y-H (1987) A 1-Oxa-2-nickela-5-cyclopentanone from ethene and carbon dioxide: preparation, structure, and reactivity. Angew Chem Int Ed 26:771–773
37. Williams CM, Johnson JB, Rovis T (2008) Nickel-catalyzed reductive carboxylation of styrenes using CO_2. J Am Chem Soc 130:14936–14937
38. Greenhalgh MD, Thomas SP (2012) Iron-catalyzed, highly regioselective synthesis of alpha-aryl carboxylic acids from styrene derivatives and CO_2. J Am Chem Soc 134:11900–11903
39. Shao P, Wang S, Chen C, Xi C (2016) Cp_2TiCl_2-catalyzed regioselective hydrocarboxylation of alkenes with CO_2. Org Lett 18:2050–2053
40. Boogaerts II, Nolan SP (2011) Direct C-H carboxylation with complexes of the coinage metals. Chem Commun (Camb) 47:3021–3024
41. Ohishi T, Zhang L, Nishiura M, Hou Z (2011) Carboxylation of alkylboranes by *N*-heterocyclic carbene copper catalysts: synthesis of carboxylic acids from terminal alkenes and carbon dioxide. Angew Chem Int Ed 50:8114–8117

42. Juhl M et al (2017) Copper-catalyzed carboxylation of hydroborated disubstituted alkenes and terminal alkynes with cesium fluoride. ACS Catal 7:1392–1396
43. Ostapowicz TG, Schmitz M, Krystof M, Klankermayer J, Leitner W (2013) Carbon dioxide as a C(1) building block for the formation of carboxylic acids by formal catalytic hydrocarboxylation. Angew Chem Int Ed 52:12119–12123
44. Fleming I, Roessler F (1980) Reaction of a silyl-copper reagent with acetylenes: a new synthesis of vinylsilanes. J Chem Soc Chem Commun 276–277
45. Liepins V, Bäckvall J-E (2002) Regioselective 1,4-silylcupration of 1,3-dienes characterization and electrophilic trapping of the intermediate (σ-Allyl)copper complex. Eur J Org Chem 2002:3527–3535
46. Fujihara T, Tani Y, Semba K, Terao J, Tsuji Y (2012) Copper-catalyzed silacarboxylation of internal alkynes by employing carbon dioxide and silylboranes. Angew Chem Int Ed 51:11487–11490
47. Tani Y, Fujihara T, Terao J, Tsuji Y (2014) Copper-catalyzed regiodivergent silacarboxylation of allenes with carbon dioxide and a silylborane. J Am Chem Soc 136:17706–17709
48. Zhang L, Cheng J, Carry B, Hou Z (2012) Catalytic boracarboxylation of alkynes with diborane and carbon dioxide by an *N*-heterocyclic carbene copper catalyst. J Am Chem Soc 134:14314–14317
49. Carry B, Zhang L, Nishiura M, Hou Z (2016) Synthesis of lithium boracarbonate ion pairs by copper-catalyzed multi-component coupling of carbon dioxide, diboron, and aldehydes. Angew Chem Int Ed 55:6257–6260
50. Li Z et al (2020) $CO_{(2)}$ activation by lewis pairs generated under copper catalysis enables difunctionalization of imines. J Am Chem Soc 142:1966–1974
51. Butcher TW et al (2016) Regioselective copper-catalyzed boracarboxylation of vinyl arenes. Org Lett 18:6428–6431
52. Li S, Ma S (2012) Quadri-synergetic effect for highly effective carbon dioxide fixation and its application to indoloquinolinone. Adv Synth Catal 354:2387–2394
53. Miao B, Li S, Li G, Ma S (2016) Cyclic *anti*-azacarboxylation of 2-Alkynylanilines with carbon dioxide. Org Lett 18:2556–2559
54. Hoberg H, Apotecher B (1984) α, ω-Disäuren aus butadien und kohlendioxid an nickel(0). J Organomet Chem 270:c15–c17
55. Takimoto M, Mori M (2001) Cross-coupling reaction of oxo-π-allylnickel complex generated from 1,3-diene under an atmosphere of carbon dioxide. J Am Chem Soc 123:2895–2896
56. Tortajada A, Ninokata R, Martin R (2018) Ni-catalyzed site-selective dicarboxylation of 1,3-dienes with CO_2. J Am Chem Soc 140:2050–2053
57. Hoberg H, Jenni K, Angermund K, Krüger C (1987) C-C Linkages of ethene with CO_2 on an iron(0) complex-synthesis and crystal structure analysis of $[(PEt_3)_2Fe(C_2H_4)_2]$. Angew Chem Int Ed 26:153–155
58. Takimoto M, Kawamura M, Mori M, Sato Y (2005) Nickel-catalyzed regio- and stereoselective double carboxylation of trimethylsilylallene under an atmosphere of carbon dioxide and its application to the synthesis of chaetomellic acid a anhydride. Synlett 2005:2019–2022
59. Juhl M et al (2017) Correction to copper-catalyzed carboxylation of hydroborated disubstituted alkenes and terminal alkynes with cesium fluoride. ACS Catal 7:2247–2247
60. Fujihara T et al (2014) Nickel-catalyzed double carboxylation of alkynes employing carbon dioxide. Org Lett 16:4960–4963
61. Mita T, Ishii S, Higuchi Y, Sato Y (2018) Pd-catalyzed dearomative carboxylation of indolylmethanol derivatives. Org Lett 20:7603–7606

Chapter 12
Large-Scale Recycling of Carbon: Carbon Dioxide as Source of Carbon

Stepping Away from Fossil Carbon by Implementing the Carbon Cyclic Economy

Michele Aresta

Abstract Carbon dioxide can be used as building block for chemicals or source of carbon for energy products (fuels). The latter have a market that is over fifteen times that of chemicals and are, thus, of greater interest for the large-scale implementation of the CCU strategy (capture and utilization of CO_2). The drawback is that while the chemical use of CO_2 may not require neither an energy input nor hydrogen, the conversion of CO_2 into fuels requires both energy and hydrogen that cannot be derived from fossil-C. The solution is the use of C-free primary energy sources for powering the reduction process and water as source of hydrogen or, even better, of protons and electrons. Noteworthy, the reduction of CO_2 may either use H_2 or, alternatively, may be implemented avoiding the production and use of molecular hydrogen, mimicking nature that uses the *proton-coupled to electron transfer-PCET* strategy. In this chapter, the putative technologies that may promote a carbon cyclic economy (CCE) are discussed together with their readiness (on the short- to long-term) and degree of innovation. Technical barriers to the scale-up of electrolyzers for H_2 production are discussed, considering the existing options: AWE, PME, HTWE. Routes to avoiding H_2 production (either the electrolysis of CO_2 in water or the photochemical- or photo-electro-chemical co-processing of H_2O and CO_2) are analyzed. The direct use of solar energy is considered both at ambient temperature (co-processing of CO_2 and H_2O under solar irradiation), or at high temperature (processes based on the use of solar power concentrators-SPC). The analysis of the TRL of the various technologies concludes the discussion. A deep technological innovation that may improve the efficiency in the production and use of e-fuels or solar fuels is necessary for a real reduction of the impact of the chemical and energy sectors on climate change. Whether the substitution of H_2 for C-based fuels is a real benefit for climate is questioned.

M. Aresta (✉)
Innovative Catalysis for Carbon Recycling-IC2R, Via Camillo Rosalba 49, 70124 Bari, Italy
e-mail: michele.aresta@ic2r.com

G. Zhang et al. (eds.), *Advances in CO_2 Utilization*, Green Chemistry and Sustainable Technology, https://doi.org/10.1007/978-981-99-8822-8_12

12.1 Introduction

The CO_2 conversion into chemicals, materials, and fuels encompasses a variety of reactions that can be categorized as demanding "*low-*" (Class ***A***), "*average-*" (Class ***B***), and "*high-energy*" (Class ***C***) to occur. Moreover, they may or may not require hydrogen.

In *Class **A*** reactions, the carbon atom of CO_2 reacts with an Element-X (such as: N, O, halogens, Fig. 12.1) more electronegative than itself so that its oxidation state in the product remains + 4. The products are inorganic carbonates (an exergonic reaction), carbamates and polyurethanes, cyclic and linear organic carbonates and polycarbonates, or even ureas.

*Class **B*** includes the formations of C–C bonds (the oxidation state of C is lowered to + 3) such as in the direct synthesis of organic carboxylic acids from hydrocarbons, a reaction not easy to carry out because of the difficulty in C–H activation (Fig. 12.2) [1], or in esters, lactones, and pyrones, produced by reaction with electron-rich olefins, dienes, and alkynes [2].

Both *Class **A*** and ***B*** reactions have extensively been investigated since the late 1970s as do not require neither an high-energy input nor hydrogen, both still today derived from fossil-C at the rate of 82 and 95%, respectively. Using fossil-C for converting CO_2, it requires skill to avoid that more CO_2 is generated than transformed.

Finally, in *Class **C*** reactions, the oxidation state of C is lowered to + 2 (CO) and further down to − 4 (CH_4) (Fig. 12.3). These reactions in addition to energy require hydrogen and thus have a double barrier to overcome for large-scale deployment: the use of a non-fossil-C-based energy source and non-fossil-H_2.

The bulk chemicals and fuels sectors have quite different markets and emit different volumes of CO_2. The former in 2022 had a 7972 MUS$ global market

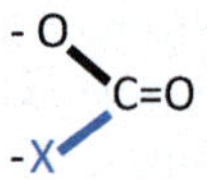

Fig. 12.1 Class ***A*** *reactions* and the nature of the CO_2 derivatives: X=O, inorganic and organic carbonates and polycarbonates; X=N, carbamates and polyurethanes

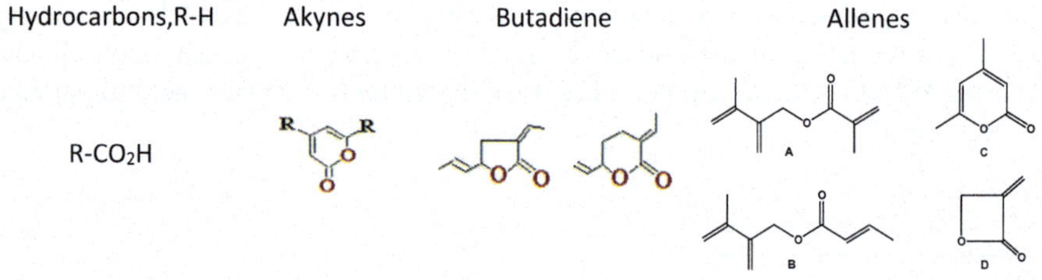

Fig. 12.2 Class ***B*** *reactions*: incorporation of CO_2 into an organic substrate with C–C bond formation

$$nCO_2 + mH_2 \rightarrow CO, CH_3OH, CH_4, C_nH_{2n+2}, C_nH_{2n}, C_nH_{2n+1}OH$$

Fig. 12.3 Class ***C***: energy products from CO_2

with a forecast to grow to 11,597 by 2029 with a 5.5% CAGR [3]. The emission of CO_2 was equal to 925 Mt in 2021 [4] for the Chemical Industry, while the energy sector (fuels) was responsible for the emission of 14.6 $Gt/_{CO2}$ [4]. As for today, 230 Mt_{CO2}/y are used by the chemical industry in the manufacture of chemical [2] recycling, thus, 24.8% of the emitted-C. Conversely, the production of fuels from CO_2 is almost equal to zero, due to the fact that, as mentioned above, most of energy and hydrogen are generated from fossil-C. A deep change in use of primary energy sources and in the production of hydrogen is necessary for a substantial recycling of carbon might be possible even for the production of fuels.

This paper will consider the latter class of reactions, namely the conversion of CO_2 into energy-rich products or fuels, a process that is a non-sense if framed into an energy-and- hydrogen supply strategy based on fossil-C. The options that make feasible the use of CO_2 as source of carbon for fuels will be discussed and the benefits therein.

12.1.1 Carbon Availability on Earth

Life has originated from CO_2 (and other key molecules containing hydrogen and nitrogen) that provided carbon for constructing organisms and life on our planet. Atmospheric CO_2 is the most abundant form of *easily accessible* carbon that we can dispose of (Table 12.1).

The atmosphere is, thus, the most abundant source of immediately available CO_2, but it is even the most diluted (415 ppm as for 2022). Subsurface CO_2 (up to 99% pure) is usually located close to volcanic areas and available all over the planet. It has been used so far for various purposes (from food packaging to enhanced oil recovery) [2], but its continued extraction is not recommended as it would cause transfer of carbon to the atmosphere, aggravating the impact on climate. Carbonate rocks amount at ca. 10^9 Gt_C that varies over very long time (see Table 12.1) because of the natural cycle of carbon (Fig. 12.4) that turns carbon (ca. 210 Gt_C) from the atmosphere or surface rocks into plants, microorganisms, or animals and finally into subsurface deposits.

The biotic part of the cycle is powered by sun that provides the necessary energy while water acts as a source of protons and electrons. Interestingly, nature does not make molecular hydrogen, if not in particular environments and usually at a very low rate: in CO_2 reduction processes, it uses a clever equivalent of elemental hydrogen, namely the "*proton coupled to electron transfer*" (PCET), "$H^+ + e^-$".

Nature offers the solution for large volume CO_2 conversion into fuels, namely to use: (i) water (or even organic waste) as source of protons and electrons (the reducing

Table 12.1 Distribution of carbon among different environments on Earth (as Gt_C)

Abiotic environment	References	Biotic environment	Reference
Atmospheric CO_2 780		*Global biosphere 550–560*	[7]
CO_2 deposits (degas rate 0.1 GtC/y) 30,000	[2]	Plants 450–460	
Fossil carbon: coal 607; oil 167; natural gas 167		Animals 2 (humans 0.06)	
Carbonate rocks[a] (sediments) 1.8×10^9	[5]	Bacteria 70; archaea 7; fungi 12; protists 4; viruses 0.2	
Ocean deep floor 37,000	[6]		
Above surface 43,500 (ca. 10–15% of all rocks)			

[a] Carbonate rocks were deposited essentially during the Phanerozoic eon, that includes the Paleozoic–mesozoic- and cenozoic era, spanning from 541 My to present. 2100×10^{15} t of carbonate rocks were deposited with a mass cycle of 8.6×10^{14} t_{CaCO3}/My and a decay constant of 0.0025 My^{-1}

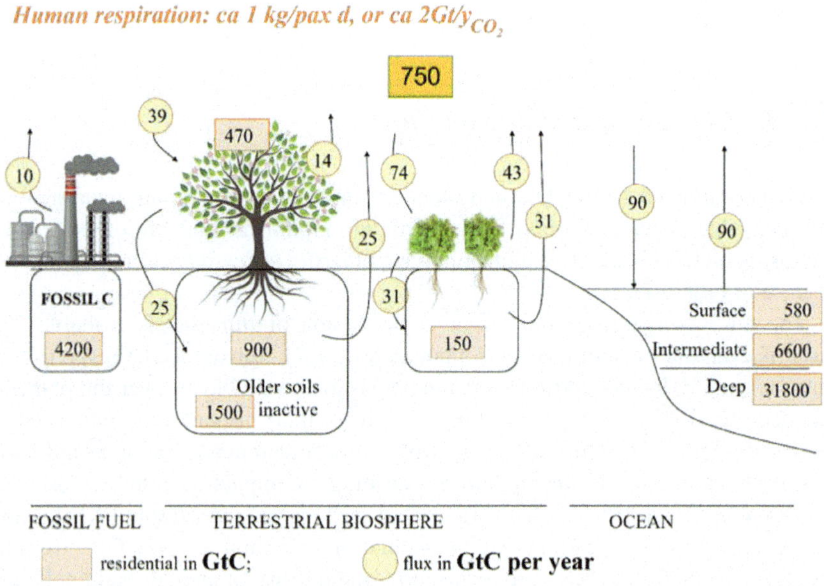

Fig. 12.4 The carbon cycle in nature

agent of CO_2) and (ii) perennial sources (solar-, wind-, hydro-, geo-power-SWHG) as primary energy. Worth to note, while solar energy can be used in the form of heat, radiations, and electrons (via PV), wind, hydro, and geo power can mostly provide non-fossil electrons. Therefore, we have two directions to follow: one based on the production of hydrogen from water and its use in the reduction of CO_2, the other based on the direct co-processing of CO_2 and water using electrochemistry, photochemistry, photo-electrochemistry, photo-bio-electro chemistry, or even cold-catalytic plasma.

12.2 Hydrogen from Water Electrolysis Using SWHG Energy and Chemothermal Reduction of CO_2: The Short-Term Option

The use of green-H_2 links the CO_2 reduction to technologies on stream, such as the synthesis of methanol or methane and the Fisher–Tropsch (FT) processes. The novelty with respect to processes on stream would be the use of non-fossil-H_2 and the substitution of CO_2 for CO. Hydrogen production via PV-water electrolysis (Eq. 12.1) is, thus, the very short-term technology for large-scale CO_2 reduction. The serious issue is to scale-up and reach the requested level of production of H_2 and the necessary continuity.

$$H_2O \rightarrow H_2 + 1/2O_2 \tag{12.1}$$

Water electrolysis is operated at industrial level since decades. The amount of hydrogen produced through such route is 1–2% of the over 90 Mt of H_2 produced in 2020. By 2030, over 220 Mt of H_2 are expected to be produced, 80 Mt of which by water electrolysis using wind or solar [8] energy. Such huge change requires a tremendous effort for building electrolyzers and installing wind (or solar) power. As an example, EU plan to install by 2030 65 GW electrolyzers (REPower2030 Programme), a very challenging plan considering the scarce EU availability of materials. The cost of production of electric energy is the barrier to the deployment of photovoltaic (PV)-H_2 production that has a cost 2–3 times (depending on the location) higher than H_2 produced via methane steam reforming-MSR. The reduction of the cost of both PV materials and electrochemical cells is mandatory for PV-H_2 may become competitive and be produced at large scale. A substantial change may occur after 2040 when the PV installed power will reach > 4500 GW [2] with respect to actual ca. 400 GW. By 2040, it is expected that new photomaterials will be operating (organics instead of Si-based) that will almost double the solar to electrons efficiency (from actual 20 to 40%) and reduce the PV-electrons cost. Supposed that enough PV cheap electrons will be available, the second issue is the availability of electrolyzers for water electrolysis. Three main technologies are on stream today: (i) alkaline water electrolysis (AWE), (ii) proton exchange membrane electrolysis (PEM), and (iii) high temperature water electrolysis (HTWE). In all cases, an electric current causes the water splitting with evolution of gaseous H_2 at the cathode (hydrogen evolution reaction-HER), and O_2 evolution at the anode (oxygen evolution reaction-OER). The thermodynamic potential of water splitting is 1.23 V, but in practice a higher value is needed for the reaction to occur due to cathodic and anodic overpotentials, that depend on the materials that constitute the electrodes and the operating conditions. Electrolyzers demand innovation for what concerns the electrodes (often made of precious costly materials), their life (if cheaper electrodes are used), the efficiency (how close to the thermodynamic potential the water-splitting is carried out), and the scale (how much H_2 can be produced per hour or day or year), targeting low capital (CAPEX) and operative (OPEX) costs. Although a rich literature exists [9–12], innovation mostly remains confidential to companies operating on the market. Table 12.2 summarizes the key properties of the electrolyzers on stream (AWE, PEM, HTWE) and relevant needs for improvement. In *AWE cells*, the two compartments for HER and OER host electrodes made of perforated sheets of porous metal immersed in a KOH solution that ensures optimal conductivity and pH for reducing the overpotential at minimum. Electrodes are electrically connected and separated by an OH-permeable diaphragm to avoid H_2 and O_2 mixing. The advantages of working at about 353 K and under ambient pressure are challenged by intrinsic drawbacks such as electrical losses, due to physical distance between electrodes, low purity of produced H_2, and thus, the need to spend energy for further purifying and compressing the gaseous product for its storage, transport, and use in chemical processes. On the other hand, the widespread application of this specific cell technology over the others is based on the good efficiency (approximately 80%) and quite long service life of the whole electrolyzer, properties which in turn rely on use of long-time optimized components in terms of materials and design.

The most important structural element of AWE resides in the diaphragm membrane, for which long-term stability, low gas permeability, and high ion (OH^-) conductivity are required; white asbestos, $Mg_3Si_2O_5(OH)_4$, has been the best and long-time used candidate. Health risks and fast corrosion by the hot caustic environment from the liquid electrolyte, thus reducing single component life as well, led to

Table 12.2 Working conditions of main electrolytic technologies and needs for improvement

	AWE	PEM	HTWE
Cathode	Steel covered with Ni	Platinum-black	Ni on ceramics
Anode	Ni-based mixed oxides	RuO_2, IrO_2, $Ir_xRu_{1-x}O_2$	Mixed oxides, perovskite-like
Operative conditions	*P*: ambient $T = 353$ K	5 MPa	973–1273 K
Compartment separation	OH-permeable membrane	A solid electrolytic membrane separates cathode and anode	
Strong points	Mature technology	Efficiency > 80% Very high H_2 purity > 99%	Production of H_2 under high-*T*/high-*P* with potential direct feeding to chemical processes
Weak points	Low H_2-purity. Electrodes corrosion. Long switch-off switch-on delay	High CAPEX and OPEX. Short life of the solid electrolytic-membrane	Poor lifetime of components due to the harsh operative conditions

development of efficient substitutes that were found in composite materials obtained by combination of porous polymers, as polytetrafluoroethene (PTFE) analogs or polysulfone, and ceramic materials such as zirconia, ZrO_2, or potassium titanate, K_2TiO_3. As for electrodes, while the "state of the art" in cathode materials is represented by perforated steel or iron plates with a surface Ni treatment, even by using Nickel Raney, mixed oxides containing Ni are preferred materials for the anode, where O_2 permeation is needed. Nickel is the elected metal for electrodes since it is resistant to alkaline media and has a low cost, thus nickel plating is now considered the best choice for the specific purpose. It is worth noting that nickel alloyed with metals such as Zn, Fe, Co, and V is widely employed to protect electrodes from deactivation, especially at the cathode side, due to nickel hydroxide, $Ni(OH)_2$, and hydride, NiH_x, formation in on/off cycles. The intrinsic long reactivation time can reveal as a severe drawback for coupling such cells with intermittent renewable energy sources.

Nowadays, the competitor to AWE technology is the *PEM cell*. While its development is strictly linked to the analogous fuel cells technology, it is now finding its own space as a mature technology even if with a still limited market. The advantages are represented by H_2 being produced at high pressure (ca. 5.0 MPa, that allows a direct feeding in most chemical plants without further pressurization) and high purity (99.99%, almost meeting fuel cells requirements) at 80% efficiency, and by fast activation times, that makes these cells suitable for intermittent use. Disadvantages are the high specification requirement for materials, meaning high CAPEX and OPEX. In PEM, a solid membrane capable of efficient H^+ ions transport is the electrolyte: It represents the most fragile part (thermal, mechanical, and chemical stability) and cost-rising in the system because of thinning due to degradation. Membranes are made of perfluorosulfonic acid polymers, among which Nafion® shows superior performance. Electrocatalysts for HER and OER are deposited onto the two membrane sides (membrane electrode assembly, MEA). Platinum-black (even supported on graphite nanofibers and multi-walled carbon nanotubes doped with N or B) is used as the standard element for cathodes. Open challenges are the carbon-support corrosion and the reduction of the Pt loading.

The anode is made of RuO_2 or the more stable IrO_2, or even mixtures of them, $Ir_xRu_{1-x}O_2$. Addition of a third oxide such as Ta_2O_5 or MoO_3, results in better OER performance, but lower stability. The high cost (as for 2022) of Pt (33.5 €/g), Ru (12.9 €/g), and Ir (34 €/g) and the quite expensive components, current collectors, and bipolar plates are a serious limitation to scale these cells. Metallic titanium is used at the anode- and carbon-based porous graphite at the cathode as collectors, the latter being susceptible to embrittlement and fracturing due to the high-pressure hydrogen environment.

High Temperature (973–1273 K) *Water Electrolysis-HTWE* (even known as Solid Oxide Electrolysis-SOE) is the third option, characterized by high reaction kinetics since WE is an endothermic process. HTWE would represent a best option whenever coupled to industrial processes that release high temperature heat that would be used to vaporize water to be fed to the solid electrolyte made of metal oxides. Key challenges in such cells are the fast on/off cycling and long-term stability of key components. Conductive oxide electrolyte is made of yttria-stabilized zirconia-YSZ, Y_2O_3-ZrO_2, often added with Sc_2O_3. Anodes are made of composite ceramic materials, perovskite-like structures, nickel modified rare earth metal-oxides, such as follows: $La_xSr_{1-x}MnO_3$, $La_xSr_{1-x}Co_yFe_{1-y}O_3$, $Ce_xY_{1-x}O_{1.95}$, $La_xNd_yPr_{2-x-y}NiO_{4+z}$. The cathode is made of ceramics such as YSZ or $Ce_{0.9}Gd_{0.1}O_{1.95}$ coupled with metallic nickel. Structural and electrical connections require materials with appropriate resistance (expansion coefficient) to high temperature and temperature swing and are usually made of steel or chromium enriched alloys.

As noted earlier, the cost of non-fossil electricity and hydrogen derived from it is a key parameter for its permeation into the industrial sector, transport and common life, so that fossil-C substitution may be scaled-up. The cost of H_2 generated from SWHG is predicted for a continuous decrease. Advances in materials and technology for WE are very key issues that need a rapid response for an effective deployment of CO_2 hydrogenation [9–12]. As a matter of fact, the availability of renewable (or green)-hydrogen at a cost comparable with that of MSR-H_2 would allow the effective large-scale conversion of CO_2, even using technologies on stream. The most immediately performing would be to apply reverse water gas shift (RWGS) for the conversion of CO_2 into CO (Eq. 12.2) and use CO and H_2 (Syngas) for making a variety of gaseous or liquid products (Eq. 12.3).

$$CO_2 + H_2 \rightarrow CO + H_2O \text{ (RWGS)} \tag{12.2}$$

$$xCO + yH_2 \rightarrow CH_{4,g},\ CH_3OH_l,\ C_{2-4,g},\ C_{5-n,l},\ \text{olefins, alcohols, aromatics} \tag{12.3}$$

An effective approach would, thus, be coupling a RWGS catalyst with a FT catalyst so that CO_2 conversion to CO and Syngas conversion into hydrocarbons could occur into a single reactor. As noted above, the benefit with respect to the state of the art would be avoiding the extraction of fossil-C, reducing the transfer of carbon from underground to the atmosphere, while continuing to use existing infrastructures.

On the other hand, the entire chain of production of energy vectors and use of chemical energy demands a profound revision for increasing the efficiency that today averages 33%. The efficiency issue (vide infra) stands whether either hydrogen will be directly burned or will be used for making fuels by reaction with captured-CO_2: In any case, the emission to the atmosphere of water vapor and heat will continue. At what extent this will perturb the delicate equilibrium of the atmosphere and continue to impact on climate? Such accelerated transfer of water from ground to the atmosphere and the relevant impact need to be assessed as water is a strong greenhouse gas (GHG), stronger than CO_2, and its accumulation into the atmosphere would cause violent downpours of water/snow/ice, more extreme than those we experience today [13, 14]. Besides the issue relevant to the effective impact on climate, even the economic cost of the production, storage, distribution, and use of H_2, that require special materials and new infrastructures, need a careful assessment.

12.3 Co-electrolysis of CO_2 and Water

To avoid the production of hydrogen and cut the CAPEX and OPEX costs relevant to its storage–transport–distribution, the co-electrolysis of CO_2 and water (CO2ER) can be exploited. In this case, electrons and protons produced in water oxidation will be transferred to CO_2 that will be reduced to C_1-C_2-C_n (Eqs. 12.4a and 12.4b).

$$H_2O \rightarrow 1/2O_2 + 2H^+ + 2e^- \tag{12.4a}$$

$$mCO_2 + n\left(2H^+ + 2e^-\right) \rightarrow C_mH_{2m+2},\ C_mH_{2m},\ C_mH_{2m+1}OH + 2mH_2O \tag{12.4b}$$

CO_2 electrochemical utilization/valorization is known since the late 1970s early 1980s [15–17].

The carboxylation of halogenated organic substrates (R-X, Eq. 12.5) and the reduction of CO_2 to $C_{1\text{-}2}$ species (Eq. 12.6) were both successfully attempted. The concern about the use of halogenated organics did not play in favor of further developing reaction 12.5, in which even the consumption of the cathode represented a drawback.

$$\text{R-X} + CO_2 + 2e^- \rightarrow \text{R-}CO_2^- + X^- \tag{12.5}$$

$$CO_2 + 8H^+ + 8e^- \rightarrow CH_4 + 2H_2O \tag{12.6}$$

The electrochemical synthesis of ethene, $H_2C = CH_2$, from CO_2 and water [18] (Eq. 12.7) is gaining momentum due to the large market and diversified uses of the organic compound: to label organics and polymers as "non-fossil" will most likely produce a large market acceptance.

The carboxylation of olefins [19] has been since long time another topic of great interest, as the route to mono or dicarboxylic acids, the latter useful co-monomers for polymers (Eq. 12.8).

$$2CO_2 + 12e^- + 12H^+ \rightarrow {}_2HC{=}CH_2 + 4H_2O \tag{12.7}$$

$$CH_2{=}CH_2 + 2CO_2 + 2e^- \rightarrow {}_2^-OC{-}CH_2{-}CH_2{-}CO_2^- \tag{12.8}$$

These days, the electrochemical reduction of CO_2 in water targets C_2-C_n species. Several Cu nanostructures, as nano-cubes [18], have been reported to be active but copper oxides (Cu_2O and CuO) produced in a controlled way with a high number of coordinatively unsaturated active sites seem to be the best option. Jung [20] observed that nanosized (a few nm) Cu_2O particles produce C_2H_4, and Lee [21] reported that a Cu foil activated with production of $Cu(OH)_2$ nanowire arrays, and anodized to produce surface Cu(I)-Cu(II) oxides is an active and stable electrocatalyst for preferential C_2H_4 production over CH_4. Engineering the electrode surface is a key issue: In fact, much less selectivity has been recorded [22] when the hydroxide precursor was annealed in air and the CuO produced was electrochemically reduced to obtain a nanoporous (100–200 nm) electrode surface. Activity and selectivity can be finely tuned by controlling the morphology: dendritic systems can stabilize Cu(I) and afford selectively ethene over methane (200:1) [23] and doping with light atoms such as B or N also enhances the activity and selectivity [24], via intermediates stabilization and porosity [25]. Combination of Cu atoms with other metals seems to produce advantages: Tandem catalysts with strict control on domains such as the Ag–Cu bimetallic generated by seed mediated methods [26] stabilize CO intermediate that is efficiently reduced to ethanol and ethene. High market demand for higher alcohols, especially ethanol and butanol, makes CO2ER intriguing.

One of the most performing processes reported in the literature is the reduction of CO_2 to ethene with 70+% faradic efficiency at – 0.55 V at engineered Cu electrode [27]. Scheme 12.1 [28] shows the putative reaction mechanism for CO_2-water electrochemical co-processing.

Reduction of CO_2 to CO is commonly reported as the first step of the reaction path. The production of ethene and other C_2 oxygenated species ($HOCH_2CH_2OH$,

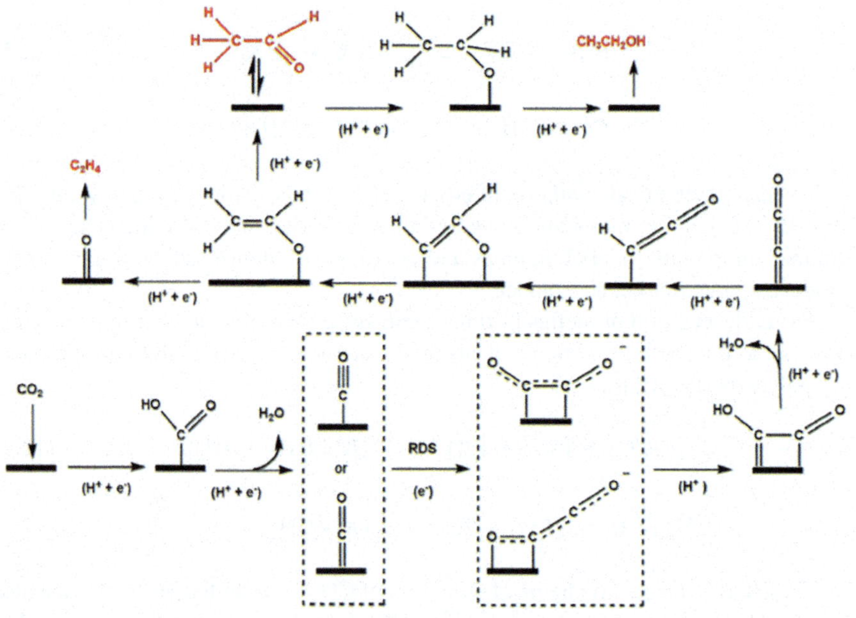

Scheme 12.1 Putative reaction path for the electrochemical co-processing of CO_2 and water [28]

$HCCH_2OH$, OHCCHO, HO_2CCH_2OH, HO_2CCHO, HO_2CCO_2H) is particularly interesting for the value of such species, their industrial use and the difficulty of making them via thermal catalysis in a selective way. High selectivity can be reached by careful engineering the electrodes, decorating the surface with specific nanoparticles in order to generate adjacent catalytic sites that can induce a C–C bond formation. Electrochemistry can also induce the formation of C_n species, but this is at a very low technology readiness level (TRL) now.

Co-electrolysis of CO_2 and water can, thus, bring to the production of added value chemicals and fuels, supposed that the right electrodes are built to promote C–C coupling.

As earlier said, the most likely reaction path (Scheme 12.1) proposes CO_2 reduction to CO as the first step, with the latter undergoing dimerization with C–C coupling and formation of metal bonded *CO-CO [29], that must be stabilized for undergoing hydrogenation to afford *CO-CHO or *CO-COH species stabilized on different Cu-facets. The preliminary hydrogenation of bonded *CO to bonded *CHO would represent an alternative to afford *CHO-CHO that is easier hydrogenated to deoxygenated species [30–35] *CO is the most abundant species and it is, thus, the most likely moiety to be dimerized. Nevertheless, *CO and *CHO coupling has a lower energy barrier on Cu{211} facets on which it is the favored reaction, while *CO-COH seems to be favored on Cu{100}. Therefore, the kinetics of dimerization of *CO is driven by Cu facets and surface morphology of the electrodes must be controlled for a faster process toward C_2^+ [33]. Routes that use the formate *OCHO or the carboxylate *CO_2H intermediates have been proposed but such species do not seem to have a great role in the formation of C_{2+} species, while are key intermediates in the formation of C_1 species.

R&I is necessary for upscaling technologies to manufacturing level that may produce benefit to both the chemical and energy industry, reducing costs, and guaranteeing higher safety with respect to water splitting and use of hydrogen. As a matter of fact, CO2ER to ethene can provide non-fossil-sourced raw material to the polymer industry (polyethene market of 160 Mt/y and polypropene market of 70 Mt/y), resulting in polymer production from recycled carbon with reduction of use of fossil-C and of its environmental impact. As for today, issues for large-scale CO2ER to C2+ products are as follows: high faradaic efficiencies toward target products, stable operation under high current density that may assure a continuous and constant (for its quality) production in the long term, facets operativity in selecting products, dynamic evolution of catalysts surfaces during electrolysis. Real time in situ detection techniques already known since decades (Raman, FTIR, IRAS, UV–VIS) or more recently developed (XRD, XPS, XAS, TEM, ESR) can assist in a better understanding of the surface modification under operation and its influence on rate of formation and stability of intermediates and selectivity toward end products [34, 35].

There is a key difference between the CO2ER and the thermo-chemo-catalytic hydrogenation mechanism of CO_2. In the latter, H_2 must be dissociated onto the catalyst surface and hydrogen atoms migrate to CO_2 and its reduced forms, in the former it is a "$H^+ + e^-$" transfer that provides the reduction.

12.4 Co-processing of CO_2 and Water Under Solar Irradiation

Co-processing of CO_2 and water under solar irradiation is studied since the 1980s but still is in its infancy as for a period of *ca.* thirty years, it has been investigated with a low profile. It can occur along one of the following routes: (i) *direct photochemical conversion*; (ii) *use of solar power concentrators-SPCs*; (iii) *use of photo-electrochemical cells*; (iv) *use of photo-bioreactors*. In (i) and (ii), solar radiations are directly used for promoting the reactions under chemocatalysis, but in quite different temperature conditions and following different mechanisms. In (iii) and (iv), solar power is converted into electrons used for CO_2 reduction using either chemical electro-catalysts or electro-biosystems. A large variety of products can be obtained, but the technologies need a boost in their scale-up. The moment being, they stay in the 2–4 TRL range.

Direct photochemical co-processing of CO_2 and water (or photochemical CO_2 reduction, PCO2R) is based on the use of semiconductors that can adsorb solar radiations and promote an electron excitation to higher energy levels (conduction band) leaving behind a "*hole*" in the valence band. Excited electrons can be used for the catalytic reduction of CO_2 in presence of protons generated in the water oxidation at the "*hole*" (Fig. 12.5).

As for today, the production of reduced forms of CO_2 is of the order of 2 $mmol_{CH4}/g_{cat}$ h [36] using an Au-Cu nanoalloy. Nanosized CuI has also been reported to be effective in CO_2 reduction with a mixed organic (isopropanol/glycerol)-water H^+-donor [37] with an interesting selectivity toward formic acid, accompanied by H_2 and CO. Interestingly, formic acid stays in the condensed phase, while H_2 and CO go to the gas phase where can be collected and used as Syngas. A major issue in PCO2R is represented by the multitude of multielectron–multiproton reduction processes that may occur in a narrow range of potential, as shown in Table 12.3.

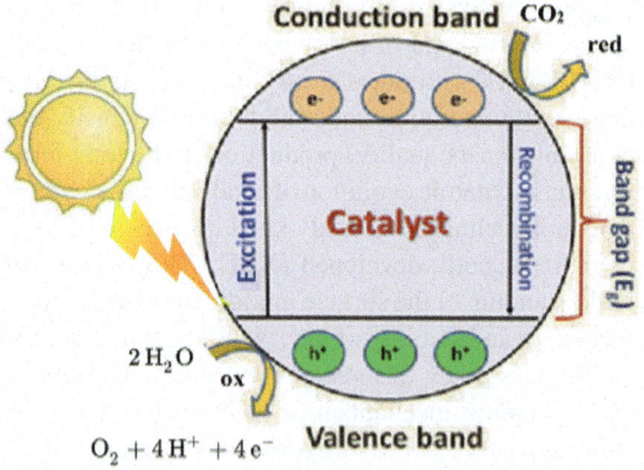

Fig. 12.5 Semiconductor at work

Table 12.3 Potential of single electron–multielectron reduction of CO_2 and of proton reduction

	Reaction	$E°$ [V] versus SHE *at pH 7 in water*
1	$CO_2 + e^- \rightarrow CO_2^-$	– 1.9 (– 2.1 in organic solvents)
2	$CO_2 + 2H^+ + 2e^- \rightarrow HCOOH$	– 0.61
3	$CO_2 + 2H^+ + 2e^- \rightarrow CO + H_2O$	– 0.52
4	$2CO_2 + 12H^+ + 12e^- \rightarrow C_2H_4 + 4H_2O$	– 0.34
5	$CO_2 + 4H^+ + 4e^- \rightarrow HCHO + H_2O$	– 0.51
6	$CO_2 + 6H^+ + 6e^- \rightarrow CH_3OH + H_2O$	– 0.38
7	$CO_2 + 8H^+ + 8e^- \rightarrow CH_4 + 2H_2O$	– 0.24
8	$2H^+ + 2e^- \rightarrow H_2$	– 0.42

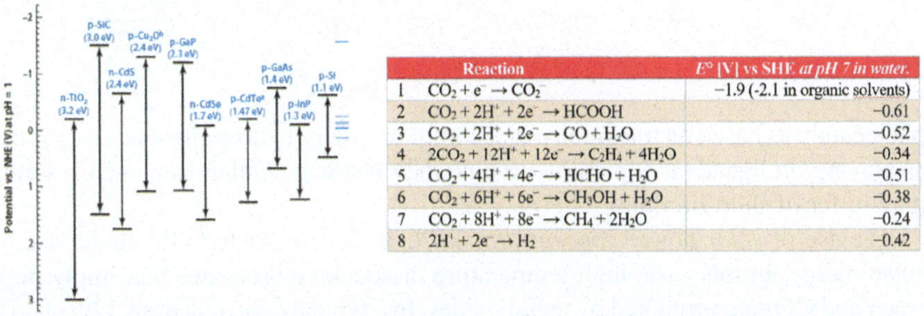

Fig. 12.6 Band gap of *p,n*-semiconductors and their relation to CO2R-potentials

Unless the reduction potential is very finely tuned, with the exclusion of the single electron CO_2 reduction that occurs at very negative potential, several processes may occur almost simultaneously, including water reduction to H_2. Tuning the reduction potential is not that easy when using solid semiconductors as photocatalysts because it requires a fine tuning of their band-gap, a complex task. Figure 12.6 shows the band-gap for a few photomaterials and how they correlate to the CO2R potentials.

Several challenges are still open in this option, among which the following are of paramount importance for technology scale-up: stabilization of charge separation, charge transfer kinetics, stability of catalytic centers and their ability to promote C–C bond formation, separation of reduction and oxidation reactions, light efficiency. The most performing photocatalysts are based on *p,n*-junctions, i.e., on coupling *p*- and *n*-semiconductors with a Z-scheme (Fig. 12.7) that allows the excited electron of a semiconductor to move to the valence band of a second one, preventing charge recombination. *p*-Type semiconductors (Cu oxides like) will more likely result in excess of electrons for reduction reactions, while *n*-type (TiO_2 like) semiconductors usually present excess of holes to drive oxidation reactions. Finding the right couple of semiconductors and their optimal structural fitting is a major research topic. The great advantage of this technology is that reactions occur at room temperature or close to it: rarely, temperatures exceeding 350 K are required. The use of homogeneous

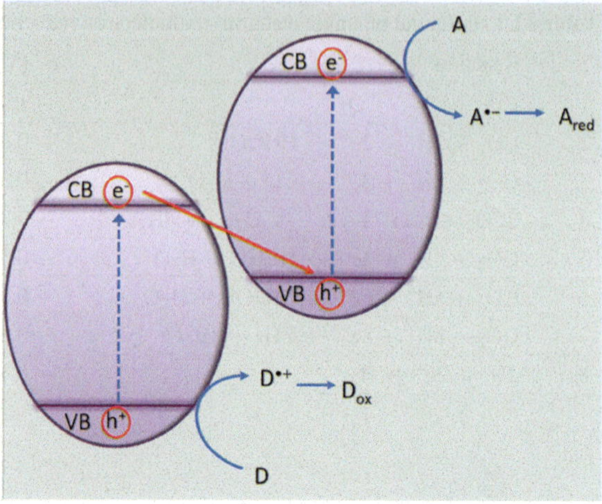

Fig. 12.7 Z-scheme coupling of *p,n*-type semiconductors for limiting charge recombination

photocatalysts based on transition metal complexes helps to drive the selectivity issue as playing on ligands it is possible to tune their redox potential. But, such systems may suffer of short lifetime.

The use of *solar power concentrators-SPC* is another route to the direct use of solar energy. In this case, high temperature dissociative processes that imply both water and CO_2 are promoted by metal oxides. In a typical case, reactions 12.9–12.11 occur with the end production of CO and H_2 that can be used as Syngas in FT processes or in the synthesis of methanol and DME (Eq. 12.12).

$$MO_x + \text{heat} \rightarrow MO_{x-1} + 1/2O_2 \tag{12.9}$$

$$MO_{x-1} + H_2O \rightarrow MO_x + H_2 \tag{12.10}$$

$$MO_{x-1} + CO_2 \rightarrow MO_x + CO \tag{12.11}$$

$$x\text{CO} + y\text{H}_2 \rightarrow \text{CH}_3\text{OH, DME, Gasoline, Diesel, Olefins} \tag{12.12}$$

Dissociative processes require high temperature (> 1100 K) that is produced using concentrators of solar energy of various types [38, 39]: parabolic, hyperboloid, Fresnel, CPC, dielectric totally internally reflecting, flat high concentration, quantum dot. Most of them demand an efficient sun-tracking system. The use of an engineered metal oxide (often modified ceria is used) that can easily release an oxygen atom undergoing reduction of the original oxidation state of the metal (Eq. 12.9), can help to reduce the working temperature. Such technology has been applied to the production of Syngas to be used in FT processes (Eq. 12.12) and because both CO_2 and H_2O can be captured from the atmosphere, the fuel produced was labeled "*Diesel from air*" [40]. Efforts are currently made to capture atmospheric CO_2 (direct air capture-DAC) and use it for the production of liquid fuels, a process that still requires important cost reduction for being economically viable. DAC per se is quite costly, due to the strong antientropic work necessary for the capture of CO_2 from such a diluted gas mixture (415 ppm of CO_2, as for 2022). As a matter of fact, the DAC cost is some six times higher (roughly around 600 US$/$t_{CO2}$) than the recovery from more concentrated sources (flue gases from power plants have an average 14% CO_2 and fermentation plants can reach > 90%). Nevertheless, the DAC process is under deep investigation and a target of 100 US$/$t_{CO2}$ has been set by some companies (Carbon Engineering Ltd. and Climeworks).

Coupling solar energy capture and electrochemistry in a single device such as a *photo-electrochemical cell (PECO2R)* is attracting much interest. PECO2R would combine the advantage of existing expertise in electrochemistry with the direct CO_2–H_2O processing. Electrode processes can be better driven (engineering active sites under controlled potential) than processes on semiconductors. Several attempts have been made and some interesting pioneering results are available in the literature. Nanowires made of "SnO_2 on CuO" coupled with a GaInP/GaInAs/Ge photovoltaic cell, show an interesting solar-to-CO conversion efficiency of 13.4%, higher than any natural process [41]. This is quite encouraging as this technology is simple and works at ambient temperature.

An emerging technology is that based on *photo-bio-electrochemical systems (PBECO2R)* in which solar energy is converted into electrons used by biosystems (whole organisms or enzymes) for the conversion of CO_2 and water into complex molecules. Several different applications can be found that bear to the production of methanol [42], methane [43], acetate [44, 45], methane and acetate [46], or more complex molecules [47].

Even the application of plasma technology for converting CO_2-water into fuels [48] or coupling low temperature plasma with photocatalysis have been suggested as possible ways to low energy CO_2 conversion into chemicals and fuels [49].

12.5 The Efficiency Issue in the Use of Energy

The *efficiency* issue is a key factor for driving the *climate change* problem. If we shall not be able to improve the efficiency of production and use of energy, the overall situation will most probably not significantly change. In this direction, the implementation of innovative technologies that are more environmentally friendly is of paramount importance. In previous paragraphs, a variety of options have been discussed for recycling carbon, within a circular economy strategic frame. Some of them use non-fossil primary energy sources (SWHG) coupled to technologies on stream for making fuels that will be used in existing infrastructures. Several questions are still open and demand a response.

1. To what extent defossilization of the energy sector will be possible by 2050?
2. Is the zero fossil-C a realistic target by 2050?

3. Will recycling of carbon help to mitigate climate change?
4. Can a man-made carbon cycle effectively complement the natural C-cycle and assure energy to our society?
5. Substitution of carbon with hydrogen will mitigate climate change?
6. Which energy-intensive sectors will be decarbonized?
7. To what extent bio-products will help to solve the present climate crisis?
8. The extended use of H_2 as energy vector and the increase of production of high-temperature water vapor will really be climate-neutral?
9. Will be necessary to reinvent the use of energy for reducing the environmental impact of our society?
10. Which role will play efficiency in future of our society?

Figure 12.8 shows the actual trend in production and use of energy, while Fig. 12.9 depicts a possible new frame, with change of the primary energy source from fossil-C to SWHG and continuing to use C-based fuels, based on cycling atmospheric carbon.

Let us make a comparison of substitution of C-based fuels with hydrogen and check the changes and putative benefits, comparing the direct use of H_2 to the conversion of non-fossil H_2 and CO_2 into fuels and use of fuels.

About 3 kg of a $C_{10}H_{22}$ saturated hydrocarbon (M = 142; 21.1 mol of $C_{10}H_{22}$) are necessary for producing the same energy released by burning 1 kg of H_2 (500 mol of H_2, 120 MJ). In the former case, 211 mol of CO_2 and 232.1 mol of H_2O are released, in the latter 500 mol of H_2O. Therefore, using H_2 instead of C_{10}, the

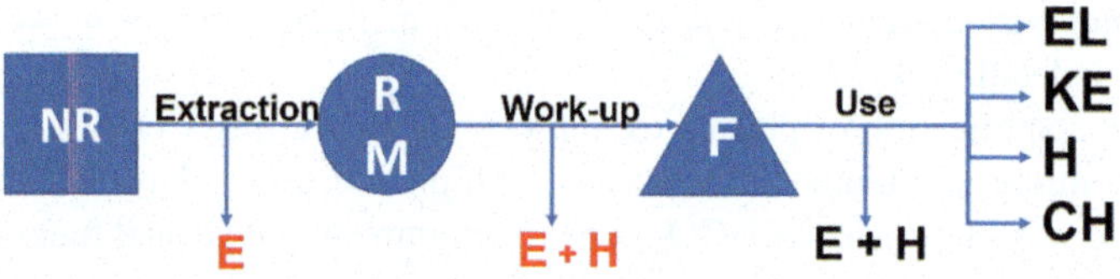

Fig. 12.8 The linear economy model adopted today based on fossil-C that provides 82% of the energy used by mankind (NR = natural resources, E = emissions; RM = raw materials; H = heat; F = fuels; EL = electricity; KE = kinetic energy; CH = chemicals)

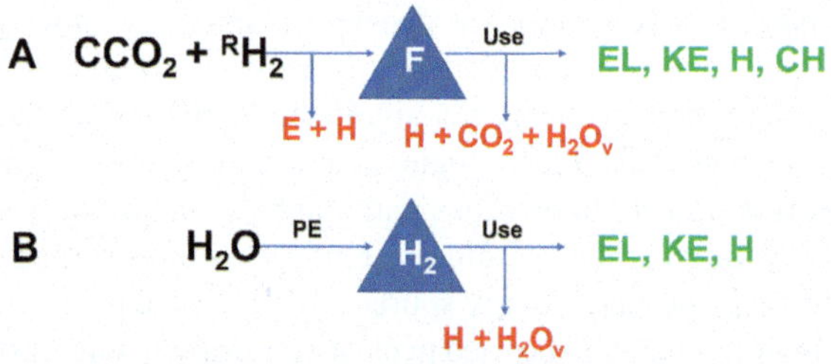

Fig. 12.9 Defossilization of the primary sources of energy (CCO_2 = captured CO_2; $^RH_2$ = renewable H_2 or green-H_2; E = emissions, H = heat; F = fuels; EL = electricity; KE = kinetic energy; CH = chemicals; PE = perennial energies, SWHG)

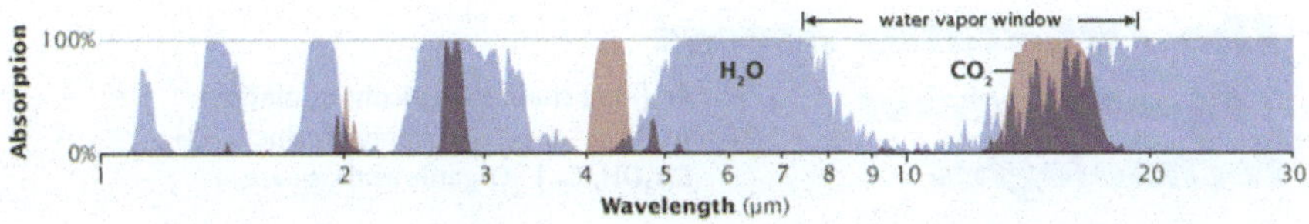

Fig. 12.10 The IR adsorption spectra of water vapor and CO_2. It is evident that water has a larger absorption that CO_2 and is a stronger GHG [50]

number of mol of only water will be 12.8% higher than the sum of $mol_{CO2} + mol_{H2O}$. Water is a stronger GHG than CO_2, as demonstrated by the IR spectra of the two gases (Fig. 12.10). Depending on the application, the temperature of the combustion gases will vary within a large range (500–1200 K), well above the boiling point of water and the gases released at different height over ground. If the released gas stays in the atmosphere, it will cause a worst impact because water is a stronger GHG than CO_2 (Fig. 12.10).

The examination of global surface data [13] from over 15,000 weather stations (terrestrial and marine) during 1975–2004 has shown an increase of the atmospheric relative humidity of 0.5–2% per decade over the central and eastern United States, India, and western China, associated with an increase in temperature and low clouds. Is, thus, the atmospheric water vapor level increasing even at a higher rate than that of CO_2? The higher concentration of water vapor causes heavy rains that we see these days, even in geographical areas which were not used to such phenomena, causing serious disasters, and the reinforcing of such phenomena in equatorial areas. Such trend is not apparently valid for the entire planet [51]. Will such extreme events be reinforced if the amount of water released will be increased, even if the amount of CO_2 will be reduced? How the continued emission of heat and water vapor will impact the climate? Will be enough to recycle atmospheric carbon or even to cut the use of carbon-based fuels for reducing the impact on climate?

All these questions demand a great amount of work in modeling and building scenarios for future energy frame.

12.6 Conclusions

Several options for C-recycling are available at different TRL. Figure 12.11 lists some examples.

As shown in Fig. 12.11, several technologies are available for the synthesis of the same species.

For example, methanol can be synthesized via thermal catalysis from CO_2 and H_2 (a technology on stream) or via a most advanced technology based on a cascade of enzymes that very selectively convert CO_2 and water into methanol at room temperature [53]. The latter technology is at quite low TRL. One can envisage that the former technology could be used in the short time, while investing in R&I may

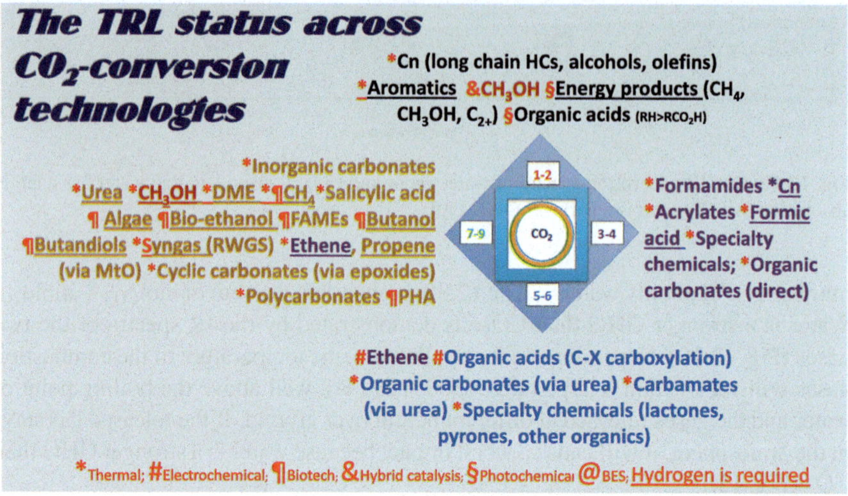

Fig. 12.11 The technology readiness level-TRL [52] of conversion of CO_2 into a variety of chemicals and energy products. The same species can appear at different TRLs, depending on the technology considered

allow the latter to be ready for use in the medium–long term. This would be a major change with respect to the former. This is true also for ethene that can be produced in the short- term from bio-ethanol (skipping fossil-C) and in the medium term by electroreduction of CO_2 in water. A key point is what portion of the chemicals and energy market can be covered by bio-sourced products. Maybe 10%, at the largest. This leaves a great space to fuels and products made directly from CO_2.

The technology innovation will mark the real changes and produce major benefits in terms of reducing the impact on climate.

References

1. Dibenedetto, Zhang J, Trochowski M, Angelini A, Macyk W, Aresta M (2017) Photocatalytic carboxylation of C-H bonds promoted by Popped Graphene Oxide (PGO) either bare or loaded with CuO. JCOU 20:97–104
2. Aresta M, Dibenedetto A (2021) The CO_2 revolution. Springer
3. https://www.reanin.com. Accessed 12 Aug 2023
4. https://www.iea.org. Accessed 13 Aug 2023
5. Gregor B (1985) The mass-age distribution of Phanerozoic sediments. Geol Soc Lond Mem 10:284–289
6. MacKenzie FT, Morse JW (1992) Sedimentary carbonates during the Phanerozoic time. Geochim Cosmochim Acta 56:3282–3295. https://doi.org/10.1016/0016-7037(92)90305-3
7. Bar-On YM, Phillips R, Milo R (2018) The biomass distribution on Earth. Proc Natl Acad Sci USA 115:6506–6511. https://doi.org/10.1073/pnas.171184211
8. https://www.iea.org/reports/hydrogen. Accessed 12 Aug 2023

9. Shiva Kumar S, Himabindu V (2019) Hydrogen production by PEM water electrolysis—a review. Mater Sci Energy Technol 2(3):442–454. https://doi.org/10.1016/j.mset.2019.03.002
10. Grigoriev SA, Fateev VN, Bessarabov DG, Millet P (2020) Current status, research trends, and challenges in water electrolysis science and technology. Int J Hydrogen Energy 45(49):26036–26058. https://doi.org/10.1016/j.ijhydene.2020.03.109
11. Smolinka T, Ojong ET, Garche J (2015) Hydrogen production from renewable energies—electrolyzer technologies. Electrochemical energy storage for renewable sources and grid balancing, pp 103–128. https://doi.org/10.1016/B978-0-444-62616-5.00008-5
12. Keçebaş A, Kayfeci M, Bayat M (2019) In: Calise F, D'Accadia MD, Santarelli M, Lanzini A, Ferrero D (eds) Solar hydrogen production: processes, systems and technologies. Elsevier, pp 299–317
13. Dai (2006) Recent climatology, variability, and trends in global surface humidity. J Clim 19(15):3589–3606. https://doi.org/10.1175/JCLI3816.1
14. Dessler AE, Zhang Z, Yang P (2008) Water-vapor climate feedback inferred from climate fluctuations, 2003–2008. Geophys Res Lett 35:L20704. https://doi.org/10.1029/2008GL035333
15. Ercoli R, Guainazzi M, Silvestri G, Gambino S, Filardo G, Giannici B, Galluzzo M (1973) Chim Ind (Milan) 55:156
16. Gambino S, Filardo G, Silvestri G (1982) Electrochemical carboxylation of organic substrates. Synthesis of carboxylic derivatives of acenaphthylene. J Appl Electrochem 12:549–555. https://doi.org/10.1007/BF00614982
17. Filardo G, Gambino S, Silvestri G, Gennaro A, Vianello E (1984) Electrocarboxylation of styrene through homogeneous redox catalysis. J Electroanal Chem 117:303–309. https://doi.org/10.1016/0022-0728(84)80232-6
18. Roberts FS, Kuhl KP, Nilsson A (2015) High selectivity for ethylene from carbon dioxide reduction over copper nanocube electrocatalysts. Angew Chem Int Ed 127(17):5268–5271. https://doi.org/10.1002/ange.201412214
19. Gambino S, Filardo G, Silvestri G (1987) Electrochemical carboxylation of styrene. J Electrochem Soc 134:2172–2175. https://doi.org/10.1149/1.2100846/meta
20. Jung H, Lee SY, Lee CW, Cho MK, Won DH, Kim C, Oh HS, Min BK, Hwang YJ (2019) Electrochemical fragmentation of Cu_2O nanoparticles enhancing selective C-C coupling from CO_2 reduction reaction. J Am Chem Soc 141(11):4624–4633. https://doi.org/10.1021/jacs.8b11237
21. Lee SY, Jung H, Kim NK, Oh HS, Min BK, Hwang YJ (2018) Mixed copper states in anodized Cu electrocatalyst for stable and selective ethylene production from CO_2 reduction. J Am Chem Soc 140(28):8681–8689. https://doi.org/10.1021/jacs.8b02173
22. Lv JJ, Jouny M, Luc W, Zhu W, Zhu JJ, Jiao F (2018) A highly porous copper electrocatalyst for carbon dioxide reduction. Adv Mater 30(49):1–8. https://doi.org/10.1002/adma.201803111
23. De Luna P, Quintero-Bermudez R, Dinh CT, Ross MB, Bushuyev OS, Todorović P, Regier T, Kelley SO, Yang P, Sargent EH (2018) Catalyst electro-redeposition controls morphology and oxidation state for selective carbon dioxide reduction. Nat Catal 1(2):103–110. https://doi.org/10.1038/s41929-017-0018-9
24. Wan Q, Zhang J, Zhang B, Tan D, Yao L, Zheng L, Zhang F, Liu L, Cheng X, Han B (2020) Boron-doped CuO nanobundles for electroreduction of carbon dioxide to ethylene. Green Chem 22(9):2750–2754. https://doi.org/10.1039/D0GC00730G
25. Ebaid M, Jiang K, Zhang Z, Drisdell WS, Bell AT, Cooper JK (2020) Production of C_2/C_3 oxygenates from planar copper nitride-derived mesoporous copper via electrochemical reduction of CO_2. Chem Mater 32(7):3304–3311. https://doi.org/10.1021/acs.chemmater.0c00761
26. Huang J, Mensi M, Oveisi E, Mantella V, Buonsanti R (2019) Structural sensitivities in bimetallic catalysts for electrochemical CO_2 reduction revealed by Ag–Cu nanodimers. J Am Chem Soc 141(6):2490–2499. https://doi.org/10.1021/jacs.8b12381
27. Dinh C-T, Burdyny T, Kibria G, Seifitokaldani A, Gabardo CM, García de Arquer FP, Kiani A, Edwards JP, De Luna P, Bushuyev OS, Zou C, Quintero-Bermudez R, Pang Y, Sinton D, Sargent EH (2018) CO_2 electroreduction to ethylene via hydroxide-mediated copper catalysis at an abrupt interface. Science 360(6390):783–787. https://doi.org/10.1126/science.aas9100

28. Calle-Vallejo F, Koper MTM (2013) Theoretical considerations on the electroreduction of CO to C2 species on Cu(100) electrodes. Angew Chem Int Ed 125(28):7423–7426. https://doi.org/10.1002/anie.201301470
29. Sun Z, Ma T, Tao H, Fan Q, Han B (2017) Fundamentals and challenges of electrochemical CO_2 reduction using two-dimensional materials. Chem 3(4):560–587. https://doi.org/10.1016/j.chempr.2017.09.009
30. Kortlever R, Shen J, Schouten KJP, Calle-Vallejo F, Koper MTM (2015) Catalysts and reaction pathways for the electrochemical reduction of carbon dioxide. J Phys Chem Lett 6(20):4073. https://doi.org/10.1021/acs.jpclett.5b01559
31. Pérez-Gallent E, Figueiredo MC, Calle-Vallejo F, Koper MTM (2017) Spectroscopic observation of a hydrogenated CO dimer intermediate during CO reduction on Cu(100) electrodes. Angew Chem 129(13):3675. https://doi.org/10.1002/ange.201700580
32. Montoya JH, Shi C, Chan K, Nørskov JK (2015) Theoretical insights into a CO dimerization mechanism in CO_2 electroreduction. J Phys Chem Lett 6(11):2032. https://doi.org/10.1021/acs.jpclett.5b00722
33. Shi C, Chan K, Yoo JS, Nørskov JK (2016) Barriers of electrochemical CO_2 reduction on transition metals. Org Proc Res Dev 20(8):1424. https://doi.org/10.1021/acs.oprd.6b00103
34. Popović S, Smiljanić M, Jovanovič P, Vavra J, Buonsanti R, Hodnik N (2020) Stability and degradation mechanisms of copper-based catalysts for electrochemical CO_2 reduction. Angew Chem Int Ed 59(35):14736. https://doi.org/10.1002/ange.202000617
35. Li X, Wang S, Li L, Sun Y, Xie Y (2020) Progress and perspective for in situ studies of CO_2 reduction. J Am Chem Soc 142(21):9567–9581. https://doi.org/10.1021/jacs.0c02973
36. Neatu S, Maciá-Agullò JA, Concepciòn P, Garcia H (2014) Gold-copper nanoalloys supported on TiO_2 as photocatalysts for CO_2 reduction by water. J Am Chem Soc 136(45):15969–15976. https://doi.org/10.1021/ja506433k
37. Baran T, Wojtyła S, Dibenedetto A, Aresta M, Macyk W (2016) Photocatalytic carbon dioxide reduction at p-type copper (i) iodide. Chemsuschem 9(20):2933–2938. https://doi.org/10.1002/cssc.201600289
38. Jebasingh VK, Herbert GMJ (2016) A review of solar parabolic trough collector. Renew Sustain Energy Rev 54:1085–1091. https://doi.org/10.1016/j.rser.2015.10.043
39. Zhuang Z, Yu F (2014) Optimization design of hybrid Fresnel-based concentrator for generating uniformity irradiance with the broad solar spectrum. Opt Lasers Technol 60:27–33. https://doi.org/10.1016/j.optlastec.2013.12.021
40. https://www.sciencealert.com/audi-have-successfully-made-diesel-fuel-from-air-and-water. Accessed 12 Aug 2023
41. Schreier M, Héroguel F, Steier L, Ahmad S, Luterbacher JS, Mayer MT, Luo J, Graetzel M (2017) Solar conversion of CO_2 to CO using Earth-abundant electrocatalysts prepared by atomic layer modification of CuO. Nat Energy 2(7):17087–17096. https://doi.org/10.1038/nenergy.2017.87
42. Schlager S, Dibenedetto A, Aresta M, Apaydin DH, Dumitru LM, Neugebauer H, Sariciftci NS (2017) Biocatalytic and bioelectrocatalytic approaches for the reduction of carbon dioxide using enzymes. Energy Technol 5(6):812–821. https://doi.org/10.1002/ente.201600610
43. Batlle-Vilanova P, Rovira-Alsina L, Puig S, Dolors Balaguer M, Icaran P, Monsalvo VM, Rogalla F, Colprim J (2019) Biogas upgrading, CO_2 valorisation and economic revaluation of bioelectrochemical systems through anodic chlorine production in the framework of wastewater treatment plants. Sci Total Environ 690:352–360. https://doi.org/10.1016/j.scitotenv.2019.06.361
44. Bajracharya S, van den Burg B, Vanbroekhoven K, De Wever H, Buisman CJ, Pant D, Strik DP (2017) In situ acetate separation in microbial electrosynthesis from CO_2 using ion-exchange resin. Electrochim Acta 237:267–275. https://doi.org/10.1016/j.electacta.2017.03.209
45. Aryal N, Wan L, Overgaard MH, Stoot AC, Chen Y, Tremblay PL, Zhang T (2019) Increased carbon dioxide reduction to acetate in a microbial electrosynthesis reactor with a reduced graphene oxide-coated copper foam composite cathode. Bioelectrochemistry 128:83–93. https://doi.org/10.1016/j.bioelechem.2019.03.011

46. Zeppilli M, Chouchane H, Scardigno L, Mahjoubi M, Gacitua M, Askri R, Majone M (2020) Bioelectrochemical vs hydrogenophilic approach for CO_2 reduction into methane and acetate. Chem Eng J 125243. https://doi.org/10.1016/j.cej.2020.125243
47. López-Garzón CS, Straathof AJJ (2014) Recovery of carboxylic acids produced by fermentation. Biotechnol Adv 32:873–904. https://doi.org/10.1016/j.biotechadv.2014.04.002
48. Bogaerts A, Centi G (2020) Plasma technology for CO_2 conversion: a personal perspective on prospects and gaps. Front Energy Res 8:111. https://doi.org/10.3389/fenrg.2020.00111
49. Ray D, Ye P, Yu JC, Song C (2023) Recent progress in plasma catalytic conversion of CO_2 to chemicals and fuels. Catal Today 1:113973
50. http://climateilluminated.com/climate/climate_science/26_physics_IR_absorption.html. Accessed 12 Aug 2023
51. You Q, Min J, Lin H, Pepin N, Sillanpää M, Kang S (2015) Observed climatology and trend in relative humidity in the central and eastern Tibetan Plateau. J Geophys Res Atmos 120:3610–3621. https://doi.org/10.1002/2014JD023031
52. https://www.nasa.gov/directorates/heo/scan/engineering/technology/technology_readiness_level. Accessed 12 Aug 2023
53. Aresta M, Dibenedetto A, Baran T, Angelini A, Łabuz P, Macyk W (2014) An integrated photocatalytic/enzymatic system for the reduction of CO_2 to methanol in bioglycerol-water. Beilstein J Org Chem 10:2556–2565

The manufacturer's authorised representative in the EU is Springer Nature Customer Service Centre GmbH, Europaplatz 3, 69115 Heidelberg, Germany. If you have any concerns regarding our products, please contact ProductSafety@springernature.com

Printed and bound by CPI Group (UK) Ltd, Croydon, CR0 4YY
01/10/2026
02225872-0008